Studies in Computational Intelligence

Volume 1045

Series Editor

Janusz Kacprzyk, Polish Academy of Sciences, Warsaw, Poland

The series "Studies in Computational Intelligence" (SCI) publishes new developments and advances in the various areas of computational intelligence—quickly and with a high quality. The intent is to cover the theory, applications, and design methods of computational intelligence, as embedded in the fields of engineering, computer science, physics and life sciences, as well as the methodologies behind them. The series contains monographs, lecture notes and edited volumes in computational intelligence spanning the areas of neural networks, connectionist systems, genetic algorithms, evolutionary computation, artificial intelligence, cellular automata, self-organizing systems, soft computing, fuzzy systems, and hybrid intelligent systems. Of particular value to both the contributors and the readership are the short publication timeframe and the world-wide distribution, which enable both wide and rapid dissemination of research output.

This series also publishes Open Access books. A recent example is the book Swan, Nivel, Kant, Hedges, Atkinson, Steunebrink: The Road to General Intelligence https://link.springer.com/book/10.1007/978-3-031-08020-3

Indexed by SCOPUS, DBLP, WTI Frankfurt eG, zbMATH, SCImago.

All books published in the series are submitted for consideration in Web of Science.

Nguyen Hoang Phuong · Vladik Kreinovich
Editors

Biomedical and Other Applications of Soft Computing

Springer

Editors
Nguyen Hoang Phuong
Informatics Division
Thang Long University
Hoang Mai, Hanoi, Vietnam

Vladik Kreinovich
Computer Science Department
University of Texas at El Paso
El Paso, TX, USA

ISSN 1860-949X ISSN 1860-9503 (electronic)
Studies in Computational Intelligence
ISBN 978-3-031-08582-6 ISBN 978-3-031-08580-2 (eBook)
https://doi.org/10.1007/978-3-031-08580-2

This Springer imprint is published by the registered company Springer Nature Switzerland AG
The registered company address is: Gewerbestrasse 11, 6330 Cham, Switzerland

Preface

In medical decision-making, it is very important to take into account the experience of medical doctors—and thus, to supplement traditional statistics-based data processing techniques with methods of computational intelligence, methods that allow us to take this experience into account.

In some cases, the existing computational intelligence techniques—often, after creative modifications—can be efficiently used in biomedical applications. However, biomedical problems are difficult. As a result, in many situations, the existing computational intelligence techniques are not sufficient to solve the corresponding problems. In such situations, we need to develop new techniques—and, ideally, first show their efficiency on other applications, to make sure that these techniques are indeed efficient. The fact that these techniques are efficient in so many different areas makes us hope that they will be useful in biomedical applications as well.

We hope that this volume will help practitioners and researchers to learn more about computational intelligence techniques and their biomedical applications—and to further develop this important research direction.

We want to thank all the authors for their contributions and all anonymous referees for their thorough analysis and helpful comments.

The publication of this volume was partly supported by Thang Long University, Hanoi, Vietnam. Our thanks to the leadership and staff of this institution for providing crucial support. Our special thanks to Prof. Hung T. Nguyen for his valuable advice and constant support.

We would also like to thank Prof. Janusz Kacprzyk (Series Editor) and Dr. Thomas Ditzinger (Senior Editor, Engineering/Applied Sciences) for their support and cooperation with this publication.

Hanoi, Vietnam
El Paso, USA
December 2021

Nguyen Hoang Phuong
Vladik Kreinovich

Contents

Question-Answering System over Knowledge Graphs Using Analogical-Problem-Solving Approach ... 1
Nhuan D. To, Marek Z. Reformat, and Ronald R. Yager

Fuzzy Transform on 1-D Manifolds ... 13
Thi Minh Tam Pham, Jiří Janeček, and Irina Perfilieva

A Systematic Review of Privacy-Preserving Blockchain in e-Medicine ... 25
Usman Ahmad Usmani, Junzo Watada, Jafreezal Jaafar, and Izzatdin Abdul Aziz

Why Rectified Linear Neurons: Two Convexity-Related Explanations ... 41
Jonatan Contreras, Martine Ceberio, Olga Kosheleva, Vladik Kreinovich, and Nguyen Hoang Phuong

How to Work? How to Study? Shall We Cram for the Exams? and How Is This Related to Life on Earth? ... 49
Olga Kosheleva, Vladik Kreinovich, and Nguyen Hoang Phuong

Why Quantum Techniques Are a Good First Approximation to Social and Economic Phenomena, and What Next ... 57
Olga Kosheleva and Vladik Kreinovich

How the Pavement's Lifetime Depends on the Stress Level and on the Dry Density: An Explanation of Empirical Formulas ... 67
Edgar Daniel Rodriguez Velasquez, Vladik Kreinovich, Olga Kosheleva, and Nguyen Hoang Phuong

Freedom of Will, Physics, and Human Intelligence: An Idea ... 73
Miroslav Svítek, Vladik Kreinovich, and Nguyen Hoang Phuong

Why Normalized Difference Vegetation Index (NDVI)? 83
Francisco Zapata, Eric Smith, Vladik Kreinovich,
and Nguyen Hoang Phuong

Binary Image Classification Using Convolutional Neural Network for V2V Communication Systems 93
Hoang-Thong Vo, Ngan-Linh Nguyen, Dao N. Ngoc, Trong-Hop Do,
and Quang-Dung Pham

Topic Model—Machine Learning Classifier Integrations on Geocoded Twitter Data 105
Gillian Kant, Christoph Weisser, Thomas Kneib, and Benjamin Säfken

Shop Product Tracking and Early Fire Detection Using Edge Devices 121
Tuan Linh Dang and Viet Tien Ha

SDNs Delay Prediction Using Machine Learning Algorithms 133
Tuan Linh Dang and Nhat Minh Ngo

A Linear Neural Network Approach for Solving Partial Differential Equations on Porous Domains 143
T. T. V. Le, C.-D. Le, and K. Le-Cao

Accuracy Measures and the Convexity of ROC Curves for Binary Classification Problems 155
Le Bich Phuong and Nguyen Tien Zung

Stochastic Simulations of Airborne Particles in a Fibre Matrix 165
T. T. V. Le, K. Le-Cao, T. Ba-Quoc, and Y. Nguyen-Quoc

Disease Diagnosis Based on Symptoms Description 179
Huong Hoang Luong, Phong Cao Nguyen, and Hai Thanh Nguyen

Chest X-Ray Image Analysis with ResNet50, SMOTE and SafeSMOTE 191
Nam Anh Dao and Xuan Tho Dang

Weakly Supervised Localization of the Abnormal Regions in Breast Cancer X-Ray Images Using Patches Classification 203
Nguyen Hoang Phuong, Ha Manh Toan, Le Tuan Linh,
Nguyen Ngoc Cuong, and Bui My Hanh

Effects Evaluation of Data Augmentation Techniques on Common Seafood Types Classification Tasks 213
Hai Thanh Nguyen, Ngan Kim Thi Nguyen, Chi Le Hoang Tran,
and Huong Hoang Luong

Image Caption Generator with a Combination Between Convolutional Neural Network and Long Short-Term Memory 225
Duy Thuy Thi Nguyen and Hai Thanh Nguyen

Clothing Classification Using Shallow Convolutional Neural Networks 239
Mai Truc Lam Nguyen and Hai Thanh Nguyen

Similar Vietnamese Document Detection in Online Assignment Submission System 251
Hai Thanh Nguyen, Trinh Kieu Nguyen, Minh Tri Pham, Chi Le Hoang Tran, Tran Thanh Dien, and Nguyen Thai-Nghe

A Study of Causal Modeling with Time Delay for Frost Forecast Using Machine Learning from Data 265
Shugo Yoshida, Yosuke Tamura, Kenta Owada, Liya Ding, Kosuke Noborio, and Kazuki Shibuya

Question-Answering System over Knowledge Graphs Using Analogical-Problem-Solving Approach

Nhuan D. To, Marek Z. Reformat, and Ronald R. Yager

Abstract We introduce an analogical-problem-solving based question-answering system, *LingTeQA*. It generates templates from known pairs ⟨*question-SPARQL query*⟩ and uses generated templates to answer newly asked questions. These questions can be of regular/usual form and can contain imprecise concepts represented by linguistic terms. The system works over open-domain Knowledge Graphs on the Web. In addition, *LingTeQA* can generate linguistic summaries to answer questions whose answers contain large amounts of numerical values. This system is accessible at https://www.lingteqa.site.

1 Introduction

Large amounts of structured data is being published on theWeb in the representation format of Resource Description Framework (RDF). The data constitute multitude of RDF datasets in various domains such as publication, life sciences, social networking. The datasets are regarded as Knowledge Graphs (KGs) that provide useful information for a variety of applications, including Question-Answering (QA) systems, i.e., computer programs that answer end-users questions posed in a natural language.

N. D. To
University of Alberta, Canada & Nam Dinh University of Technology Education, Nam Dinh, Vietnam
e-mail: nhuan@ualberta.ca

M. Z. Reformat (✉)
University of Alberta, Canada & University of Social Sciences, Lodz, Poland
e-mail: marek@ualberta.ca

R. R. Yager
Iona College, New Rochelle, NY 10801 & King Abdelaziz University, Jeddah, Saudi Arabia
e-mail: yager@panix.com

N. H. Phuong and V. Kreinovich (eds.), *Biomedical and Other Applications of Soft Computing*, Studies in Computational Intelligence 1045,
https://doi.org/10.1007/978-3-031-08580-2_1

A standard means of accessing/retrieving needed information from KGs is performed by applying queries in the RDF query language called SPARQL.[1] The queries can be regarded as solutions/answers of users' questions. They have to be syntactically and semantically correct, and use a proper vocabulary that is utilized by a given KG. SPARQL allows to construct a wide variety of queries, from simple to complex ones. However, a process of building correct queries is a challenging task even for expert users. It is due to the complexity and expressiveness of the query language (SPARQL) and the heterogeneity of KGs' schema (vocabularies).

Analysis of syntactic structure of KG-based questions used to evaluate QA systems on such datasets as QALD[2] and LC-QuAD[3] has showed that questions with similar structures very often have isomorphic SPARQL queries. This observation together with Analogical-Problem-Solving approach allow us to develop a question-answering system that let users ask questions in English to obtain needed information. The proposed methodology is applied as follows: given a question (source problem) and a corresponding SPARQL query (source solution), our system creates a pair of templates ⟨*question template-SPARQL query template*⟩; then a newly asked question is matched to a single *question template* and following the Analogical-Problem-Solving approach, a SPARQL query is generated based on the corresponding *SPARQL query template*; and executed over a specific KG.

The developed system—*LingTeQA*—will allow for answering not only simple questions but also questions requiring additional processing, and providing more human-like answers. To do so, we aim at developing algorithms, methods, and tools necessary for doing the following tasks: (1) template generation from pairs of question and SPARQL query; (2) question to query translation based on generated templates; (3) construction of definitions of user-based linguistic terms and quantifiers (if any) so questions that contain such terms can be answered accordingly to their understanding of the terms; (4) data summarization in forms of linguistic summaries and aggregated values (if suitable).

2 Background

2.1 RDF Knowledge Graphs and RDF Query Language

The W3C has introduced a graph-based data representation form called Resource Description Framework (RDF) for representing information on the Web. A single RDF triple in the format of subject-predicate-object (can be visualized as a *node-arc-node* link) is used to present a relationship/property that holds between a subject and an object of that triple. An RDF graph is a set of such *triples*. We depict some triples describing *Lionel Messi* in KGs, namely *Wikidata*, *DBpedia*, and *YAGO* in Fig. 1.

[1] https://www.w3.org/TR/sparql11-query/.

[2] http://qald.aksw.org/.

[3] http://lc-quad.sda.tech/.

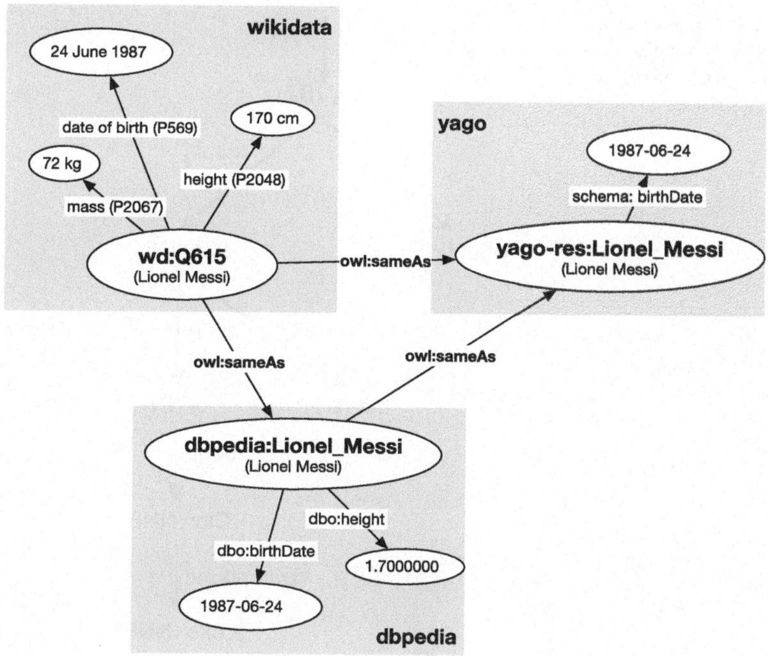

Fig. 1 Multi-RDF graphs answer to query about Lionel Messi (snippet)

This simple set of triples provides a good illustration of what to expect from the RDF data. In the brackets, we included names/IDs/labels of properties and nodes—*P's* and *Q's*, respectively—as they are used in *Wikidata*.

SPARQL is a SQL-like query language for accessing RDF data repositories. It provides a powerful means for constructing queries containing functions, triple patterns, and query modifiers to retrieve information from RDF data stores.

2.2 Dependency Parser

A process of answering questions that utilizes RDF data involves segmentation of the questions into their constituents and mapping them to the KG's vocabularies. This leads to construction of SPARQL queries. Dependency parsers are tools used for preforming analysis of questions. A dependency parser produces grammatical relations between words in a sentence. These relations can be visualized as a tree whose nodes are words in the question and directed, labeled arcs are relations among the words. For example, a relation labeled 'nsubj' indicates a nominal subject, 'dobj' means a direct object, and 'iobj' is an indirect object relation. Besides relations, a dependency parser also labels words in the processed sentence with Part-of-speech

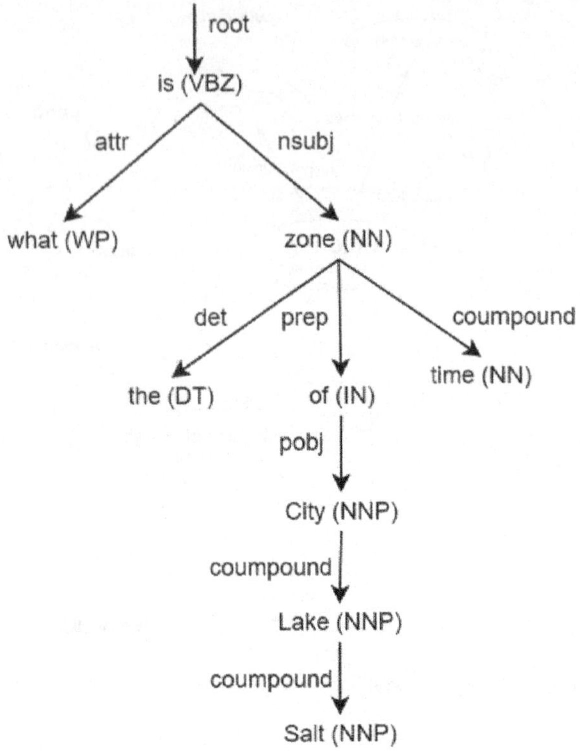

Fig. 2 The dependency tree of "what is the time zone of Salt Lake City?" produced by SpaCy's parser

(POS) tags using Penn Treebank tagsets. Tagsets with 'VB' are used for verbs, ones that begin with 'NN' are used for nouns, and ones that start with 'JJ' are used for adjectives.

We portray a dependency tree (with part-of-speech tags presented in parenthesis) of the sentence "what is the time zone of Salt Lake City?" produced by SpaCy's dependency parser, Fig. 2.

3 *LingTeQA*: An Analogical-Problem-Solving QA System

3.1 LingTeQA: Representing Questions

The proposed *LingTeQA* uses a phrasal dependency tree for representing an English question. The tree is a set of connected nodes, each reachable via a unique path from a distinguished root node. The system performs a two-step procedure to generate such a tree.

Step 1. It applies the *spaCy* dependency parser [7] to obtain typed dependencies that form a word-level dependency tree (see an example in Fig. 2).

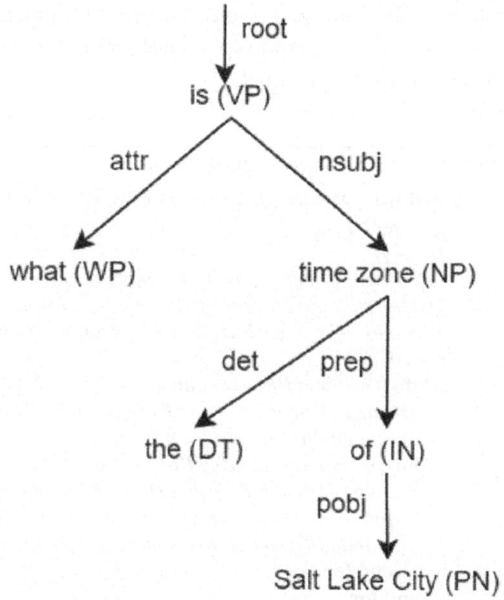

Fig. 3 The phrase-level dependency tree of "what is the time zone of Salt Lake City?" generated based on tree in Fig. 2

Step 2. It refines the word-level dependency tree to produce a phrase-level counterpart using specialized heuristics. In particular, it combines words that involve a multiword-expression relation such as 'compound', 'mwe'. We depict a phrasal dependency tree in Fig. 3 as the result of Step 2 obtained from the tree shown in Fig. 2.

3.2 *LingTeQA*: Generating Templates

Given a pair of an English question (Qs) and a corresponding SPARQL query (Qr) w.r.t a Knowledge Graph (K), *LingTeQA* generates a template consists of a question template (Qst) paired with a query template (Qrt). First, it represents the question in a phrase-level dependency tree. Second, *LingTeQA* explores the tree using a pre-order algorithm that produces, at each visited node, an accessing path (a sequence of arc labels) together with a POS tag of the phrase stored in the node. Based on that a question template is created. Third, *LingTeQA* extracts the phrase stored in the visited node and uses WordNet to find its extensions such as synonyms and other word surfaces. Next, *LingTeQA* maps all obtained phrases to items from the K and other query items such as constants and comparative signs using heuristics. For each item, it creates a function-like placeholder to encode the mapping process. *LingTeQA* adds the obtained item and the placeholder as a new entry to a mapping dictionary (D) if the entry does not belong to it. Finally, the system extracts specific items in the given query using regular expressions and searches their existences in the mapping dictionary. If found, *LingTeQA* replaces the extracted item in the query with the corresponding value (placeholder) in the dictionary, D. The process is formally described in Algorithm 1.

Algorithm 1 Template generation

```
procedure TEMPLATE_GENERATION(Qs, Qr, K )
    Qrt ← Qr
    D ← {}
    dependencyTree ← systemGenerated(Qs)
    Qst ← systemGenerated(dependencyTree)
    Phrases ← systemExtrated(dependencyTree)
    i ← 0
    while i <= len(Phrases) do
        extendedPhrases ← getExtensions(WordNet, Phrases[i])
        extendedPhrases.adds(Phrases[i])
        for phrase ∈ extendedPhrases do
            sourceElements, placeHolders ← mapping(phrase, K)
            for sourceElement ∈ sourceElements do
                D.add({sourceElement, placeHolder})
            end for
        end for
        i ← i + 1
    end while
    targetElements ← extractedFrom(Qr)
    for targetE ∈ targetElements do
        if targetE ∈ D then
            placeHolder ← D[targetE]
            Qrt ← Qrt.replace(targetE, placeHolder)
        end if
    end for
    return Qst, Qrt
end procedure
```

For example, given the question "what is the time zone of Salt Lake City?" whose phrasal dependency is shown in Fig. 3 and its correspond query w.r.t DBpedia is:

SELECT ?answer WHERE{dbr:Salt_Lake_City dbo:timeZone ?answer.}

the proposed *LingTeQA* generates a template pair ⟨Qst, Qrt⟩:

Qst: root(VP)root.attr(WP)root.nsubj(NP)root.nsubj.det(DT)_root.nsubj.prep(IN)root.nsubj.prep.pobj(PN)
Qrt: *SELECT ?answer WHERE{Res(root.nsubj.prep.pobj) Pro(root.nsubj) ?answer.}*

4 *LingTeQA*: Answering Questions

To answer a question, the system—*LingTeQA*—processes the user's question. At first, it builds a dependency tree of the question and creates a question template. Next, it uses this template as a key to retrieve a corresponding SPARQL query template in a template repository. After that, the system extracts placeholders in the retrieved template. Based on each of extracted placeholders, *LingTeQA* extracts a phrase stored in the node whose location in the tree is specified by the placeholder. The system maps it into a target item also indicated by the placeholder. Then it replaces the placeholder in the query template by the obtained target item, and creates an executable SPARQL query. *LingTeQA* executes the SPARQL query and collects data from the K (knowledge graph). Finally, it detects linguistic term(s). If there is no linguistic term in the question, the system provides the obtained data to the user as an answer. Otherwise, it invokes a process for answering questions with linguistic terms (see Sect. 4.1).

4.1 *LingTeQA: Answering Question Containing Linguistic Terms*

LingTeQA uses generated templates and user's assistance in defining the meaning, via fuzzy subsets, of linguistic terms to answer questions that contains such terms. We briefly describe the process of answering this type of questions in Table 1.

Table 1 A process of answering a question containing a linguistic term

No.	Agent	Action(s)	Result(s)
1	User	Selects a KG and submits a question containing a linguistic term	A question (qs), a linguistic term (t), a KG (K)
2	LingTeQA	(1) collects data from K and stores it in an RDF graph; (2) prepares an interface (GUI) for the user to enter a fuzzy set describing the term t	Collected data stored in a graph G
3	User	enters and submits a fuzzy set (A) by drawing its membership function using the GUI	A set of ordered pairs of the form $\{x, A(x)\}$
4	LingTeQA	(1) fits A to a trapezoid; (2) calculates membership grades for objects stored in G; (3) updates the graph G; (4) selects and provides five objects from G with the highest-membership-grade value to the user	Top five objects as an answer

4.2 LingTeQA: Answering Questions by Generating Linguistic Summaries

LingTeQA can also answer questions by generating linguistic summaries using a template 'Q obj are/have S', where obj represents summarized entities, Q is a linguistic quantifier such as FEW, MOST, and S is a summarizer such as SMALL, LARGE. We briefly present the process of answering such questions in Table 2.

Table 2 A process of answering a question by generating a linguistic summary

No.	Agent	Action(s)	Result(s)
1	User	Selects a KG and submits a question containing a linguistic term	A question (qs), a linguistic term (t), a KG (K)
2	LingTeQA	(1) collects data from K and stores it; (2) prepares an interface (GUI) for the user entering summarizers and fuzzy sets describing them	Collected data
3	User	Enters and submits a list of summarizers, each summarizer S_i plus a fuzzy set μ_{Si} drawn on the GUI at a time	A set of summarizers $S_1, S_2, ..., S_N$ together a list of fuzzy sets $\mu_{S1}, \mu_{S2}, ..., \mu_{SN}$ each of them is a set of ordered pairs of the form $\{x, \mu_S(x)\}$
4	LingTeQA	(1) fits each μ_{Sj} to a trapezoid; (2) calculates membership grades for stored objects (o); (3) calculates $\alpha = min(hgt(\mu_{S1}), hgt(\mu_{S2}), ..., hgt(\mu_{SN}))$; (4) selects a summarizer S_i whose index $i^* = argmax_i\ card(\{o \vert \mu_{Si}(o) \geq \alpha\})$; (5) computes $c = card(\{o \vert \mu_{S_{i*}}(o) > 0\})$	Selected summarizer S_{i*}, number of supporting object c
5	User	Enters and submits a list of quantifiers, each quantifier Q_j plus a fuzzy set μ_{Qj} drawn on the GUI at a time	A set of quantifiers $Q_1, Q_2, ..., Q_M$ presented by fuzzy sets $\mu_{Q1}, \mu_{Q2}, ..., \mu_{QM}$
6	LingTeQA	(1) fits each fuzzy set μ_{Qk} to a trapezoid; (2) calculates membership grades for c w.r.t μ_{Qk}; (3) select a quantifier Q_j whose index $j_* = argmax_j\ \mu_{Qj}(c)$	Selected quantifier Q_{j_*}, membership grade $T = \mu_{Qj_*}(c)$
7	LingTeQA	(1) forms a linguistic summary: 'Q_{j_*} obj are S_{i^*}' as an answer; (2) provides the answer and the validity degree T to the user	

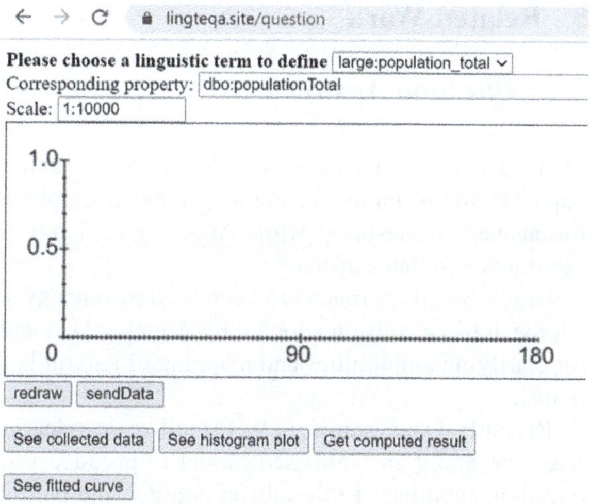

Fig. 4 TiFS-based Web Interface for Defining Linguistic Terms

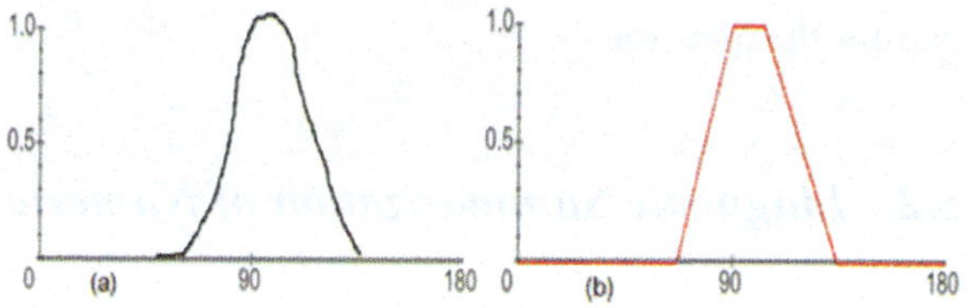

Fig. 5 User-drawn membership function (**a**); and system-fitting one (**b**)

4.3 *LingTeQA: Defining Linguistic Terms with a User-friendly Web Interface*

Linguistic terms present in the user's question require user's assistance in defining the terms' meaning. For that purpose, *LingTeQA* uses a user-interface *TiFS* [20] to allow the user to enter membership functions defining the terms. After collecting data from a KG, the system prepares a coordinate plane for users to draw shapes representing their understanding of the terms. To make the x-axis relevant, the system determines a range of possible values: it finds the minimum and the maximum of the obtained results, and uses them to scale the x-axis properly. A screenshot of the interface when answering the question "give me large cities in Poland by population" with data collected from DBpedia is shown in Fig. 4.

The user can select **See collected data** to see actual data or **See histogram plot** to see the distribution of data. The user draws/redraws a shape of membership function representing the defined term on the provided coordinate plane. Then the users submits it (via **sendData**) when he/she is satisfied with the shape.

The user-submitted 'shape' is then fitted to a trapezoid. We show an example of the user-drawn membership function and the corresponding function obtained after the fitting process in Fig. 5.

5 Related Work

5.1 Question Answering

Many QA systems have been developed. In a template-based approach, OQA [5] and Aqqu [1], for example, use manually created templates to construct SPARQL queries for answering questions, while AquaLog [13] and Platypus [15] use automatically generated templates instead.

Graph-based approach has been used by other systems [8, 21, 22]. An interactive approach has also been adopted by Aqualog [13] and FREyA [3] to solicit the user for clarifying ambiguities and mapping of natural language expressions to semantic items.

Recently, QA systems [6, 9, 14] utilize deep learning networks. A common practice is to apply an NN-based model to produce vector representations of a given question, candidate KG's subject entities, and predicates associated with the question. Then a similarity evaluation is used to find the most matched subject-predicate pair for the question.

5.2 Linguistic Summarization of Numeric Data

Dr. Yager has introduced an approach to data summarization that summarizes the data in terms of three values: a summarizer, a quantity in agreement, and a truth value [19]. Other authors have further developed tools and algorithms inspired by the idea. For example, Kacprzyk and his collaborators presented FQuery [11] that allows for the use of linguistic terms in fuzzy queries in Microsoft Access. Rasmussen and Yager introduced Summary SQL [16] whereas Bosc and Pivert proposed SQLf [2] as extensions of SQL.

Dubois and Prade [4] presented a gradual linguistic summarization capturing progressive change between entities while Wilbik and Kaymak [17] proposed protoform-based gradual summaries. Wu et al. [18] developed IF-THEN rule-like linguistic summary. Kacprzyk et al. [10], and Kobayashi et al. [12] have studied linguistic summaries aiming at discovering relations in time series. In linguistic summarization methods have been introduced, fuzzy sets presenting linguistic terms are predefined and maintained by systems. Inversely, in our system, they are user-provided on the fly.

6 Result and Conclusion

We validate *LingTeQA*'s performance on the QALD dataset. *LingTeQA* generated 107 pairs *<question template – SPARQL query template>* from the 408 pairs of *<question-SPARQL query>* in the QALD-9th training set. Using generated tem-

Table 3 LingTeQA results: QALD-9 DBpedia & QALD-7 Wikidata

Dataset	microP	microR	microF1
DBpedia	0.526	0.642	0.535
Wikidata	0.634	0.735	0.642

plates, it answered 69/150 questions over DBpedia, and 55/100 questions over Wikidata. Table 3 shows precision (P), recall (R), and F1 values that are calculated over answerable questions.

In addition to answering regular questions as given in QALD datasets, *LingTeQA* can answer questions whose syntactic structure is identical to that of the question "list LARGE cities in Canada by population" and those of its paraphrases. It can also generate linguistic summaries to answer questions, for example, "how BIG are Canadian cities by population?". Dear readers, please ask *LingTeQA* those questions at www.lingteqa.site.

In this paper, we have introduced *LingTeQA* that allows users to ask questions of higher complexity and to obtain results as they would interact with another human. In particular, *LingTeQA* is capable of answering questions that contain imprecise linguistic terms, and provide answers that are summarization of numerical responses. The user-based and known-solution-based answering questions approach is a distinct property of the system.

References

1. H. Bast, E. Haussmann, More accurate question answering on freebase, in *Proceedings of the International Conference on Information and Knowledge Management*, vol. 19–23, pp. 1431–1440 (2015)
2. P. Bosc, O. Pivert, SQLf: a relational database language for fuzzy querying, in *IEEE Transactions on Fuzzy Systems*, vol. 3, no. 1, pp. 1–17 (1995). ISSN: 19410034
3. D. Damljanovic, M. Agatonovic, H. Cunningham, FREyA: an interactive way of querying linked data using natural language, in *Lecture Notes in Computer Science* (including subseries Lecture Notes in Artificial Intelligence and Lecture Notes in Bioinformatics) LNCS, vol. 7117, pp. 125–138 (2012). ISSN: 03029743
4. D. Dubois, H. Prade, Gradual inference rules in approximate reasoning. Inf. Sci. **61**(1–2), 103–122 (1992). ISSN: 00200255
5. A. Fader, L. Zettlemoyer, O. Etzioni, Open question answering over curated and extracted knowledge bases, in *Proceedings of the ACM SIGKDD International Conference on Knowledge Discovery and Data Mining*, pp. 1156–1165 (2014)
6. Y. Hao, H. Liu, S. He, K. Liu, J. Zhao, Pattern-revising enhanced simple question answering over knowledge bases, in *International Conference on Computational Linguistics*, pp. 3272–3282 (2018)
7. M. Honnibal, M. Johnson, An improved non-monotonic transition system for dependency parsing, in *Conference Proceedings—EMNLP 2015: Conference on Empirical Methods in Natural Language Processing*, no. September, pp. 1373–1378 (2015)
8. S. Hu, L. Zou, J. X. Yu, H. Wang, D. Zhao, Answering natural language questions by subgraph matching over knowledge graphs (extended abstract), in *Proceedings—IEEE 34th Interna-*

tional Conference on Data Engineering, ICDE 2018, vol. 30, no. 5, pp. 1815–1816 (2018). https://doi.org/10.1109/ICDE.2018.00265
9. S. Iyer, I. Konstas, A. Cheung, J. Krishnamurthy, L. Zettlemoyer, Learning a neural semantic parser from user feedback, in *ACL 2017—55th Annual Meeting of the Association for Computational Linguistics, Proceedings of the Conference (Long Papers)*, vol. 1, pp. 963–973 (2017). arXiv: 1704.08760
10. J. Kacprzyk, A. Wilbik, Linguistic summarization of time series using linguistic quantifiers: augmenting the analysis by a degree of fuzziness, in *IEEE International Conference on Fuzzy Systems*, pp. 1146–1153 (2008). ISSN: 10987584
11. J. Kacprzyk, A. Ziolkowski, Database queries with fuzzy linguistic quantifiers. IEEE Trans. Syst. Man Cybern. **1**(3), 474–479 (1986)
12. M. Kobayashi, I. Kobayashi, An approach to linguistic summarization based on comparison among multiple time-series data, in *The 6th International Conference on Soft Computing and Intelligent Systems, and The 13th International Symposium on Advanced Intelligence Systems* (IEEE, 2012), pp. 1100–1103
13. V. Lopez, E. Motta, Ontology-driven question answering in AquaLog, in *Lecture Notes in Computer Science* (including subseries Lecture Notes in Artificial Intelligence and Lecture Notes in Bioinformatics), vol. 3136, pp. 89–102 (2004). ISSN: 16113349
14. D. Lukovnikov, A. Fischer, J. Lehmann, S. Auer, "Neural network-based question answering over knowledge graphs on word and character level, in *26th International World Wide Web Conference, WWW 2017*, pp. 1211–1220 (2017)
15. T. Pellissier Tanon, M.D. de Assunção, E. Caron, F.M. Suchanek, Demoing platypus—a multilingual question answering platform for wikidata, in *Lecture Notes in Computer Science* (including subseries Lecture Notes in Artificial Intelligence and Lecture Notes in Bioinformatics) LNCS, vol. 11155, pp. 111–116 (2018). ISSN: 16113349
16. D. Rasmussen, Summary SQL—a fuzzy tool for data mining. Intell. Data Anal. **1**(98), 49–58 (1997)
17. A. Wilbik, U. Kaymak, Gradual linguistic summaries. Commun. Comput. Inf. Sci. **443** CCIS, no. PART 2, 405–413 (2014). ISSN: 18650929
18. D. Wu, J.M. Mendel, J. Joo, Linguistic summarization using IF-THEN rules, in *IEEE World Congress on Computational Intelligence* WCCI, vol. 2010 (2010)
19. R.R. Yager, A new approach to the summarization of data. Inf. Sci. **28**(1), 69–86 (1982). ISSN: 00200255. https://doi.org/10.1016/0020-0255(82)90033-0
20. R.R. Yager, M.Z. Reformat, N.D. To, Drawing on the iPad to input fuzzy sets with an application to linguistic data science. Inf. Sci. **479**, 277–291 (2019). ISSN: 00200255. https://doi.org/10.1016/j.ins.2018.11.048
21. C. Zhu, K. Ren, X. Liu, H. Wang, Y. Tian, Y. Yu, A graph traversal based approach to answer non-aggregation questions over DBpedia, in *Lecture Notes in Computer Science* (including subseries Lecture Notes in Artificial Intelligence and Lecture Notes in Bioinformatics), vol. 9544, pp. 219–234 (2016). ISSN: 16113349. arXiv: 1510.04780
22. L. Zou, R. Huang, H. Wang, J. X. Yu, W. He, D. Zhao, Natural language question answering over RDF—a graph data driven approach, *Proceedings of the ACM SIGMOD International Conference on Management of Data*, no. Feb. 2015, pp. 313–324 (2014). ISSN: 07308078. https://doi.org/10.1145/2588555.2610525

Fuzzy Transform on 1-D Manifolds

Thi Minh Tam Pham, Jiří Janeček, and Irina Perfilieva

Abstract In this contribution, we establish a connection between a well-known mathematical structure called a manifold and a fuzzy partitioned space. We extend the apparatus of the theory of fuzzy (F-) transforms to 1-dimensional manifolds and thus make the first step in constructing data-driven F-transforms.

1 Introduction

With the overwhelming success of neural networks, we have witnessed a certain paradigm shift associated with the concept of data-driven versus model-driven. Generally speaking, this means that the data determines how it is processed. This manifests itself, for example, in a change in the focus of modeling from functions to features and from approximation to the statistical learning. It appears in the use of new proximity spaces, where a relaxed concept of closeness comes into play instead of traditional metric or normed vector spaces.

One of the reasons for this paradigm shift is the well-known mathematical phenomenon of the existence of continuous operators that map elements of the space F to elements of the space G one-to-one, but do not have corresponding inverse continuous operators. Simply saying, in relation to the above text, this means that many continuous data models (provided they are developed globally for the entire dataset) cannot continuously capture data differences.

T. M. T. Pham · J. Janeček
Department of Mathematics, Faculty of Science, University of Ostrava, 30. dubna 22, 701 03 Ostrava, Czech Republic
e-mail: p19171@student.osu.cz

J. Janeček
e-mail: Jiri.Janecek@osu.cz

I. Perfilieva (✉)
Institute for Research and Applications of Fuzzy Modeling, University of Ostrava, 30. dubna 22, 701 03 Ostrava, Czech Republic
e-mail: Irina.Perfilieva@osu.cz

N. H. Phuong and V. Kreinovich (eds.), *Biomedical and Other Applications of Soft Computing*, Studies in Computational Intelligence 1045,
https://doi.org/10.1007/978-3-031-08580-2_2

This fact shows that the corresponding inverse problem is unstable[1] and if it happens, then the direct problem (modeling) is "ill-posed". And this is not only a mathematical phenomenon. It turned out that it appears in real-life problems, when for example, one tries to estimate unknown reasons from observed consequences.

An important step in understanding the structure of ill-posed problems was made by A. Tikhonov who proved the so-called inverse operator lemma [7]: if A is a continuous one-to-one operator from E_1 to E_2, then the inverse operator A^{-1}, defined on the images of any compact sets in E_2, is stable. Without many technical details, we think that this fact justifies the interest to manifolds, i.e. spaces that are locally "modelled on" Euclidean spaces.[2] Below, we give a semantic description of this notion.

A smooth n-dimensional manifold in the Euclidean space $\mathbb{R}^\ell$ is such a subset that in a neighborhood of each of its points coincides with the graph of a smooth map from $\mathbb{R}^n$ to $\mathbb{R}^{\ell-n}$, where $n < \ell$. Thus, a manifold denotes a set (collection) of points that constitute a smooth curve/surface lying on/in a surface/three-dimensional space.

In this description, it is important to point out that manifolds establish new space structures where locality plays a leading role. In our contribution, we aim to show that this is similar to spaces with fuzzy partitions. The latter structures are actively discussed in scientific publications connected with fuzzy rule bases and fuzzy transforms (F-transforms). In particular, the F-transform methodology proposes approximation models with the best approximation quality on local compact subsets of Euclidean subspaces. The latter property fully agrees with the idea of manifoldness. The discrete version of the F-transform maps any collection of data points to its compressed representation as a set of vector-features. This has been successfully used in a new design of deep neural networks with F-transform kernels [4].

To summarize: to get rid of ill-posedness of inverse problems, the universe of discourse should be partitioned into a set of locally compact subspaces that "coincide" on non-empty open intersections. The known mathematical structure for this trick is named "manifold". The less known is a space with a fuzzy partition, which we aim to connect with a manifold by establishing a set of homeomorphic transformations.

The contribution is structured as follows: after the introduction and preliminary information, we establish a fuzzy partition on manifolds (Sect. 3). Then, in Sect. 5, we define a higher degree F-transform on the manifold, where the main approximation theorem is proved.

2 Preliminaries

The simplest kind of manifold to define is a topological manifold, which locally looks like some "ordinary" Euclidean space $\mathbb{R}^n$. Likewise, a fuzzy partitioned universe is

[1] The problem expressed by the equation $g = Kf$ is unstable if any solution g is determined up to an additive function in the null-space of the operator K.

[2] https://en.wikipedia.org/wiki/Manifold.

a collection of fuzzy sets with supports covering the universe. Both concepts were introduced independently in a related language. In this section, we give a short introduction to both concepts using the original terminology.

2.1 Fuzzy Partition

The notion of fuzzy partition has been evolved in the theory of fuzzy sets being adjusted to various requests to a space structure. The closest form to that which we use in this paper, has been introduced in [5].

Definition 1 *(Fuzzy partition)* Let $[a, b]$ be an interval on $\mathbb{R}$, $n \geq 2$, and let $x_0, x_1, \ldots, x_n, x_{n+1}$ be nodes such that $a = x_0 \leq x_1 < \ldots < x_n \leq x_{n+1} = b$. We say that fuzzy sets $A_1, \ldots, A_n : [a, b] \to [0, 1]$, which are identified with their membership functions, constitute a *fuzzy partition* of $[a, b]$ if for $k = 1, \ldots, n$, they fulfill the following conditions:

1. locality: $A_k(x) = 0$ if $x \in [a, x_{k-1}] \cup [x_{k+1}, b]$,
2. continuity: $A_k(x)$ is continuous,
3. positivity: $A_k(x) > 0$ if $x \in (x_{k-1}, x_{k+1})$,
4. Ruspini condition: $\sum\limits_{k=1}^{n} A_k(x) = 1$ if $x \in [x_1, x_{n-1}]$.

The membership functions $A_1, \ldots, A_n$ are called *basic functions*.
We say that the fuzzy partition $A_1, \ldots, A_n, n \geq 2$, is h-uniform if nodes $x_0, \ldots, x_{n+1}$ are h-equidistant, i.e. for all $k = 1, \ldots, n+1$, $x_k = x_{k-1} + h$, where

$$h = \frac{b-a}{n+1},$$

and the following three additional properties are fulfilled:

5. for all $k = 1, \ldots, n$, $A_k(x)$ strictly increases on $[x_{k-1}, x_k]$ and strictly decreases on $[x_k, x_{k+1}]$,
6. for all $k = 1, \ldots, n$, and for all $x \in [0, h]$, $A_k(x_k - x) = A_k(x_k + x)$,
7. for all $k = 2, \ldots, n$, and for all $x \in [x_{k-1}, x_{k+1}]$, $A_k(x) = A_{k-1}(x - h)$.

It is easy to see that if a fuzzy partition $A_1, \ldots, A_n$, of $[a, b]$ is h-uniform, then there exists an even function $A_0 : [-1, 1] \to [0, 1]$ such that for all $k = 1, \ldots, n$,

$$A_k(x) = A_0\left(\frac{x - x_k}{h}\right), \quad x \in [x_{k-1}, x_{k+1}].$$

We call A_0 a *generating function* of the uniform fuzzy partition.

2.2 Topological Manifold

By definition, all manifolds are topological, so the phrase "topological manifold" is usually used to emphasize that a manifold lacks additional structure, or that only its topological properties are being considered.

Definition 2 (*Topological manifold* [3]) An n-dimensional *topological manifold* is a topological space M that can be covered by a collection of open subsets $\{U_i\}$ (which are called *local coordinate neighborhoods*) with bi-continuous, one-to-one mappings (homeomorphisms) $\phi_i : U_i \to \mathbb{R}^n$, which are called *coordinate maps (or charts)*. A collection of charts which covers manifold M is called an *atlas* of M. Since all subsets U_i's cover M, we write $M = \bigcup U_i$. Also, since ϕ_i is invertible, the ϕ_i^{-1} exists and it is continuous as well.

In this research, we consider a *connected, 1-dimensional manifold M with boundary points* $p_0, p_1, \ldots, p_n$ that is located on a smooth, connected curve embedded in $\mathbb{R}^2$. The manifold M is described by open charts (U_i, ϕ_i), $i = 1, \ldots, n$. We assume that $U_i \cap U_j = \emptyset$ for all $i \neq j$. Every chart has the form $U_i = \{(h_i(t), g_i(t)) | t \in (a_i, b_i)\}$, where $h_i, g_i : \mathbb{R} \to \mathbb{R}$ are smooth maps, and the coordinate map $\phi_i : U_i \to (a_i, b_i)$, so that for all $t \in (a_i, b_i)$, $\phi_i(h_i(t), g_i(t)) = t$. Boundary points are points of the manifold that are not covered by any chart.

Additionally, we assume that the considered manifold M has a finite atlas, and that the index set of the collection $\{U_i\}$ of charts is $N_n = \{1, 2, \ldots, n\}$, $n \geq 2$. Moreover, let the boundary points be p_0, $p_1 \in \partial U_1$, $p_0 \in \partial U_1$ is the other boundary point of U_1 than p_1, $p_1 \in \partial U_1 \cap \partial U_2$, $p_2 \in \partial U_2 \cap \partial U_3$, ..., $p_{n-1} \in \partial U_{n-1} \cap \partial U_n$, i.e. $\partial U_i \cap \partial U_{i+1} = \{p_i\}$ is singleton for all $i = 1, \ldots, n-1$, and $p_n \in \partial U_n$ is the other boundary point of U_n than p_{n-1}.

We assume that the indexation of charts enables the above described relations among all coordinate neighbourhoods and all corresponding boundary points.

Recalling that for all $i = 1, \ldots, n$, $\phi_i : U_i \to (a_i, b_i)$ is a coordinate map, we have

$$\lim_{t \to a_1^+} (h_1(t), g_1(t)) = p_0, \quad \lim_{t \to b_n^-} (h_n(t), g_n(t)) = p_n \quad \text{and}$$

$$\lim_{t \to a_i^+} (h_i(t), g_i(t)) = \lim_{t \to b_{i-1}^-} (h_{i-1}(t), g_{i-1}(t)) = p_{i-1} \quad \text{for all } i = 2, \ldots, n.$$

Throughout the manuscript, we assume that the 1-dimensional manifold M with $n + 1$ boundary points $p_0, p_1, \ldots, p_n$,

$$M = \{(U_1, \phi_1), (U_2, \phi_2), \ldots, (U_n, \phi_n), p_0, p_1, \ldots, p_n\},$$

is fixed and satisfies the above assumptions.

2.3 Properties of Manifold M

Let $M = \{(U_1, \phi_1), (U_2, \phi_2), \ldots, (U_n, \phi_n), p_0, p_1, \ldots, p_n\}$ be fixed. Below are the topological properties of particular charts. We will use them later when the measure on M is defined.

Lemma 1 *Let $\phi_i(U_i) = (a_i, b_i)$, and $c, d \in (a_i, b_i)$ be such that $c < d$. Then, $\phi_i^{-1}[c, d]$ is path-connected.*

Lemma 2 *If $X,\ Y \subset U_i$ are such that $X \cap Y = \emptyset$, then $\phi_i(X) \cap \phi_i(Y) = \emptyset$.*

Lemma 3 *If $X,\ Y \subset U_i$, then $\phi_i(X) \cup \phi_i(Y) = \phi_i(X \cup Y)$.*

3 Fuzzy Partition of a Manifold

In this section, we connect the two introduced structures: manifolds and universes with fuzzy partitions by introducing a manifold with a fuzzy partition.

We assume that $M = \{(U_1, \phi_1), (U_2, \phi_2), \ldots, (U_n, \phi_n), p_0, p_1, \ldots, p_n\}$ is as above. Let

$$\lim_{p \to p_0} \phi_1(p) = a, \quad \lim_{p \to p_n} \phi_n(p) = b\,,$$

and $a < b$. Let in addition, on the interval $[a, b]$, there is a fuzzy partition $A_1, \ldots, A_n$ with nodes $x_0, \ldots, x_n$ as defined in Definition 1. Without loss of generality, we assume that intervals (a_i, b_i) and $(x_{i-1}, x_{i+1}), i = 1, \ldots, n$, coincide (otherwise, we impose additional transformations).

We set:

$$\overline{A}_i(p) = \begin{cases} A_i \circ \phi_i(p) & p \in U_i \\ 0 & p \subset M \setminus U_i\,, \end{cases} \tag{1}$$

as a map on M such that $\overline{A}_i : M \to [0, 1], i = 1, \ldots, n$, and we say that fuzzy sets $\overline{A}_1, \ldots, \overline{A}_n$ establish a *fuzzy partition on manifold M*.

Lemma 4 *Let $\overline{A}_1, \ldots, \overline{A}_n$ be a fuzzy partition of M. Then for every $i = 1, \ldots, n$, we have that $\overline{A}_i$ is continuous on M.*

Remark 1 A fuzzy partition of the topological space M determines a new manifold, say M^f, given by $M^f = \{(U_1, \overline{A}_1), \ldots (U_n, \overline{A}_n), p_0, p_1, \ldots, p_n\}$. We assume ([2]) that fuzzy partition $\overline{A}_1, \ldots, \overline{A}_n$ is a "partition of unity" that subordinate to the open cover of M, which means that for every $x \in (x_i, x_{i+1}), i = 1, \ldots, n$

$$\overline{A}_i \circ \phi_i^{-1}(x) + \overline{A}_{i+1} \circ \phi_{i+1}^{-1}(x) = 1.$$

4 Hilbert Space on a Manifold

A manifold is a structured universe serving as a domain of functions. Our goal is to develop a functional analysis on a manifold with a fuzzy partition. In particular, our goal is to establish a Hilbert space in this universe.

Let $M = \{(U_1, \phi_1), (U_2, \phi_2), \ldots, (U_n, \phi_n), p_0, p_1, \ldots, p_n\}$, and $\overline{A}_1, \ldots, \overline{A}_n$, $n \geq 2$, its fuzzy partition. We fix i from N_n and denote Σ_i a power set of U_i. Then, (U_i, Σ_i) is a measurable space.

Lemma 5 *Function $\mu_i : \Sigma_i \to \mathbb{R}$ such that for every $E \subseteq U_i$,*

$$\mu_i(E) = \frac{\int_{\phi_i(E)} A_i(x)\,\mathrm{d}x}{\int_{x_{i-1}}^{x_{i+1}} A_i(x)\,\mathrm{d}x}, \tag{2}$$

is a measure of (U_i, Σ_i).

Let $L_2(\overline{A}_i)$ be a set of functions $f : U_i \to \mathbb{R}$ s.t. $f \circ \phi_i^{-1} : (x_{i-1}, x_{i+1}) \to \mathbb{R}$ is square-integrable. Denote $L_2(\overline{A}_1, \ldots, \overline{A}_n)$, the set of functions $f : M \to \mathbb{R}$ such that for all $i = 1, \ldots, n$, $f|_{U_i} \in L_2(\overline{A}_i)$.

Lemma 6 *For all $f, g \in L_2(\overline{A}_i)$, the following integral*

$$\int_{x_{i-1}}^{x_{i+1}} f \circ \phi_i^{-1}(x) g \circ \phi_i^{-1}(x) A_i(x)\,\mathrm{d}x$$

is well defined.

We denote

$$\overline{s}_i = \int_{x_{i-1}}^{x_{i+1}} A_i(x) dx, \tag{3}$$

and consider

$$\frac{\int_{x_{i-1}}^{x_{i+1}} f \circ \phi_i^{-1}(x) g \circ \phi_i^{-1}(x) A_i(x)\,\mathrm{d}x}{\overline{s}_i},$$

as a Lebesgue integral $\int_{U_i} f(p)g(p)\,\mathrm{d}\mu_i$, where μ_i is defined by (2).

The inner product of f and g in $L_2(\overline{A}_i)$ is defined as

$$\langle f, g \rangle_i = \int_{U_i} f(p)g(p)\,\mathrm{d}\mu_i = \frac{\int_{x_{i-1}}^{x_{i+1}} f \circ \phi_i^{-1}(x) g \circ \phi_i^{-1}(x) A_i(x)\,\mathrm{d}x}{\overline{s}_i}.$$

Then, norm $\|.\|_i$ and distance ρ_i are defined by

$$\|f\|_i = \sqrt{\langle f, f \rangle_i} \text{ and } \rho_i(f, g) = \|f - g\|_i.$$

The functions $f, g \in L_2(\overline{A}_i)$ are orthogonal in $L_2(\overline{A}_i)$ if $\langle f, g\rangle_i = 0$. The function $f \in L_2(\overline{A}_i)$ is orthogonal to a subspace B of $L_2(\overline{A}_i)$ if $\langle f, g\rangle_i = 0$ for all $g \in B$. The latter is denoted by $f \in B^{\perp}$. It is not difficult to show that $L_2(\overline{A}_i)$ is a Hilbert space.

Theorem 1 *Let H be a general Hilbert space with the norm $\|\cdot\|$, and let B be its subspace. Then, for every element $f \in H$, there exists a unique best approximation $h_0 \in B$ in the sense that h_0 fulfills*

$$\|f - h_0\| = \inf\{\|f - h\| \mid h \in B\}.$$

Moreover, $f - h_0 \in B^{\perp}$ and h_0 is called an orthogonal projection of f on B.

4.1 Subspace $L_2^m(\overline{A}_i)$

For all $l \in \mathbb{N}$, let us denote a function

$$\begin{aligned} \Gamma_l : U_i &\to \mathbb{R} \\ p &\mapsto \Gamma_l(p) = (\phi_i(p))^l . \end{aligned} \tag{4}$$

Lemma 7 *For all $p \in U_i$, $\{\Gamma_0(p), \Gamma_1(p), \Gamma_2(p), \ldots, \Gamma_m(p)\}$ is a linearly independent system restricted to U_i.*

To obtain an orthogonal system in $L_2(\overline{A}_i)$, we apply the Gram–Schmidt orthogonalization to the system $\{\Gamma_0(p), \Gamma_1(p), \Gamma_2(p), \ldots, \Gamma_m(p)\}$. The resulting orthogonal polynomials are: $\overline{P}_i^0, \overline{P}_i^1, \overline{P}_i^2, \ldots, \overline{P}_i^m$.

Let us denote as $L_2^m(\overline{A}_i)$ a linear subspace of $L_2(\overline{A}_i)$ with the basis $\overline{P}_i^0, \overline{P}_i^1, \ldots, \overline{P}_i^m$. Clearly, we have

$$L_2^0(\overline{A}_i) \subset L_2^1(\overline{A}_i) \subset L_2^2(\overline{A}_i) \subset \ldots \subset L_2^m(\overline{A}_i) \subset \ldots \subset L_2(\overline{A}_i). \tag{5}$$

5 F^m-transform

In this section, we introduce the F^m-transform on M, following [6]. Now, we construct some basic concepts. The following lemma characterizes an orthogonal projection of a function $f \in L_2(\overline{A}_i)$ on the subspace $L_2^m(\overline{A}_i)$.

Lemma 8 *Let function F_i^m be the orthogonal projection of $f \in L_2(\overline{A}_i)$ on $L_2^m(\overline{A}_i)$. Then,*

$$F_i^m = \overline{c}_{i,0}\overline{P}_i^0 + \overline{c}_{i,1}\overline{P}_i^1 + \ldots + \overline{c}_{i,m}\overline{P}_i^m, \tag{6}$$

where for all $j = 0, 1, \ldots, m$,

$$\overline{c}_{i,j} = \frac{\langle f, \overline{P}_i^j \rangle_i}{\langle \overline{P}_i^j, \overline{P}_i^j \rangle_i}. \tag{7}$$

Proof Since F_i^m is the orthogonal projection of f on $L_2^m(\overline{A}_i)$, by Theorem 1, we have $(f - F_i^m) \in L_2^m(\overline{A}_i)^\perp$. Therefore, $\langle f - F_i^m, \overline{P}_i^j \rangle_i = 0$ for every $j = 0, 1, \ldots, m$. And because $F_i^m \in L_2^m(\overline{A}_i)$, F_i^m has the form

$$F_i^m = \overline{c}_{i,0}\overline{P}_i^0 + \overline{c}_{i,1}\overline{P}_i^1 + \ldots + \overline{c}_{i,m}\overline{P}_i^m,$$

where $\overline{c}_{i,0}, \overline{c}_{i,1}, \ldots, \overline{c}_{i,m}$ are unknown components that we need to find. Therefore, we have

$$\langle f - F_i^m, \overline{P}_i^j \rangle_i = \langle f - \overline{c}_{i,0}\overline{P}_i^0 - \overline{c}_{i,1}\overline{P}_i^1 - \ldots - \overline{c}_{i,m}\overline{P}_i^m, \overline{P}_i^j \rangle_i = 0.$$

Since $(\overline{P}_i^j)_{j=0,1,\ldots,m}$ is orthogonal system, we have

$$\langle f - F_i^m, \overline{P}_i^j \rangle_i = \langle f, \overline{P}_i^j \rangle_i - \overline{c}_{i,j}\langle \overline{P}_i^j, \overline{P}_i^j \rangle_i = 0.$$

Hence,

$$\overline{c}_{i,j} = \frac{\langle f, \overline{P}_i^j \rangle_i}{\langle \overline{P}_i^j, \overline{P}_i^j \rangle_i},$$

$j = 0, 1, \ldots, m$.

Definition 3 (F^m*-transform*) Let $f : M \to \mathbb{R}$ be a function from $L_2(\overline{A}_1, \ldots, \overline{A}_n)$, and let $m \geq 0$ be a fixed integer. By F_i^m, we denote the i-th orthogonal projection of f on $L_2^m(\overline{A}_i)$, $i = 1, \ldots, n$. We say that the n-tuple $(F_1^m, \ldots, F_n^m)$ is an F^m*-transform* of f with respect to $\overline{A}_1, \ldots, \overline{A}_n$, or formally,

$$F^m[f] = (F_1^m, \ldots, F_n^m).$$

The component F_i^m is given by the Eq. (6) where its coefficients are defined by (7).

Below, we indicate three main properties of F^m-transform, they are valid for all $m \geq 0$.

(A) By Theorem 1, the F^m-transform component F_i^m, $i = 1, \ldots, n$, minimizes the function

$$\Phi_i : L_2^m(\overline{A}_i) \to \mathbb{R}$$
$$h \mapsto \Phi_i(h) = \int_{x_{i-1}}^{x_{i+1}} (f \circ \phi_i^{-1}(x) - h \circ \phi_i^{-1}(x))^2 A_i(x)\, dx\,.$$

Therefore, F_i^m is the best approximation of f in $L_2^m(\overline{A}_k)$.

(B) By (6), every F^m-transform component F_i^m, $i = 1, \dots, n$, fulfills the following recurrent equation:

$$F_i^m = F_i^{m-1} + \overline{c}_{i,m} \overline{P}_i^m\,,$$

for $m = 1, 2, \dots$

(C) If we consider the F^m-transform of f as an image of a mapping $\mathcal{F}^m : L_2(\overline{A}_1, \dots, \overline{A}_n) \to L_2^m(\overline{A}_1) \times \dots \times L_2^m(\overline{A}_n)$ such that $\mathcal{F}^m(f) = (F_1^m, \dots, F_n^m)$, then, by (7), $\mathcal{F}^m$ is linear, i.e., for all $f, h \in L_2(\overline{A}_1, \dots, \overline{A}_n)$, it holds:

$$\forall \alpha, \beta \in \mathbb{R} : \mathcal{F}^m(\alpha f + \beta h) = \alpha \mathcal{F}^m(f) + \beta \mathcal{F}^m(h)\,.$$

By Theorem 1, the restriction $f|_{U_i}$ of $f \in L_2(\overline{A}_1, \dots, \overline{A}_n)$ can be approximated by any of the F^m-transform components $F_i^0, F_i^1, \dots, F_i^m$. In the next step, we consider the quality of approximation.

Lemma 9 *Let $m \geq 0$, and let F_i^m, F_i^{m+1} be orthogonal projections of $f \in L_2(\overline{A}_i)$ on $L_2^m(\overline{A}_i)$ and on $L_2^{m+1}(\overline{A}_i)$. Then*

$$\| f|_{U_i} - F_i^{m+1} \|_i \leq \| f|_{U_i} - F_i^m \|_i\,.$$

Proof By Theorem 1,

$$\| f|_{U_i} - F_i^m \|_i = \inf\{\| f|_{U_i} - h\|_i \mid h \in L_2^m(\overline{A}_i)\}\,.$$

Since, due to (5), $L_2^m(\overline{A}_i) \subset L_2^{m+1}(\overline{A}_i)$, then F_i^m also belongs to $L_2^{m+1}(\overline{A}_i)$. Therefore, $\inf\{\| f|_{U_i} - h\|_i \mid h \in L_2^{m+1}(\overline{A}_i)\} \leq \inf\{\| f|_{U_i} - h\|_i \mid h \in L_2^m(\overline{A}_i)\}$, and then we have

$$\begin{aligned}\| f|_{U_i} - F_i^{m+1} \|_i &= \inf\{\| f|_{U_i} - h\|_i \mid h \in L_2^{m+1}(\overline{A}_i)\} \\ &\leq \inf\{\| f|_{U_i} - h\|_i \mid h \in L_2^m(\overline{A}_i)\} = \| f|_{U_i} - F_i^m \|_i\,.\end{aligned}$$

This completes the proof.

In the next subsection, we estimate the error of f and F_i^0. Since the error of F_i^l is smaller then the error of F_i^d if $l > d$, then, for $m > 0$, the error of F_i^m is smaller than the error of F_i^0.

5.1 F^0-transform and its Inverse

Based on (6), (7) and the fact that $\overline{P}_i^0 = 1$, the component F_i^0 is equal to $\overline{c}_{i,0}$ and

$$\overline{c}_{i,0} = \frac{\int_{x_{i-1}}^{x_{i+1}} f \circ \phi_i^{-1}(x) \cdot \overline{P}_i^0 \circ \phi_i^{-1}(x) A_i(x)\,\mathrm{d}x}{\overline{s}_i} = \frac{\int_{x_{i-1}}^{x_{i+1}} f \circ \phi_i^{-1}(x) A_i(x)\,\mathrm{d}x}{\overline{s}_i}, \tag{8}$$

where $\overline{s}_i$ is given by (3).

Let us compose the inverse F-transform on M. We first define the function $\gamma : [a, b] \to \mathbb{R}$,

$$\gamma(x) = \sum_{i=1}^{n} \overline{c}_{i,0} A_i(x). \tag{9}$$

Obviously, γ is continuous on $[a, b]$.

Then for any $i = 1, \ldots, n$, the *inverse F-transform* $\hat{f}$ of f on M is defined for each $p \in M$ as follows:

1. If $p \in U_i$, then $p = \phi_i^{-1}(x)$ and

$$\hat{f}(p) = \gamma \circ \phi_i(p). \tag{10}$$

2. If $p = p_i$, $1 \le i \le n-1$, then

$$\hat{f}(p_i) = \frac{\gamma(x_i) + \gamma(x_{i+1})}{2}.$$

Following Eq. (9), this value is equal to $\frac{\overline{c}_{i,0} + \overline{c}_{i+1,0}}{2}$.

3. $\hat{f}(p_0) = \hat{f}(p_n) = 0$.

5.1.1 Approximation Quality

In this section, we will estimate the quality of approximation of a function f on M by its inverse F-transform $\hat{f}$ on M.

The following theorem and its proof are based on [1]. Let d denote the Euclidean distance in $\mathbb{R}^2$ in which the manifold M is embedded.

Theorem 2 *Let $M = \{(U_1, \phi_1), \ldots, (U_n, \phi_n), p_0, \ldots, p_n\}$ be a manifold and its fuzzy partition formed by $\overline{A}_1, \ldots, \overline{A}_n$ and let $\delta = \max_{i \in 1,\ldots,n} \sup_{p,q \in U_i} d(p, q)$. Then the following error estimate holds: for any $p \in U_i$, $i = 2, \ldots, n-1$ and for any p_i, $i = 1, \ldots, n-1$, we have*

$$|f(p) - \hat{f}(p)| \leq \omega(f, \delta),$$

where $\hat{f}$ is the inverse F-transform of a smooth function f given by Eq. (10) *and $\omega(f, \delta)$ is modulus of continuity of the function $f : M \to \mathbb{R}$.*

Proof We give the proof for the case $p \in U_i$, where $i = 2, \ldots, n-1$, only. There exists a unique $t_p \in (x_{i-1}, x_{i+1})$, such that $t_p = \phi_i(p)$, since ϕ_i is one-to-one. Then,

$$\begin{aligned}
&|f(p) - \hat{f}(p)| \\
&= \left| \sum_{i=1}^{n} f(p) A_i(t_p) - \sum_{i=1}^{n} \frac{\int_{x_{i-1}}^{x_{i+1}} f \circ \phi_i^{-1}(x) A_i(x)\, \mathrm{d}x}{\overline{s}_i} A_i(t_p) \right| \\
&= \left| \sum_{i=1}^{n} \frac{\int_{x_{i-1}}^{x_{i+1}} \left(f(p) - f \circ \phi_i^{-1}(x) \right) A_i(t_p) A_i(x)\, \mathrm{d}x}{\overline{s}_i} \right| \\
&\leq \sum_{i=1}^{n} \frac{\int_{x_{i-1}}^{x_{i+1}} |f(p) - f \circ \phi_i^{-1}(x)| A_i(t_p) A_i(x)\, \mathrm{d}x}{\overline{s}_i} \\
&\leq \sum_{i=1}^{n} \frac{\int_{x_{i-1}}^{x_{i+1}} A_i(t_p) A_i(x) \omega\big(f, d(p, \phi_i^{-1}(x))\big)\, \mathrm{d}x}{\overline{s}_i}.
\end{aligned}$$

Therefore,

$$|f(p) - \hat{f}(p)| \leq \sum_{i=1}^{n} A_i(t_p) \omega(f, \delta) = \omega(f, \delta).$$

6 Conclusion

We have shown that a smooth 1-D manifold whose open cover admits a partition of unity subordinate to this cover establishes a fuzzy partition on the corresponding topological space. Based on this fact, we have extended the apparatus of higher degree F-transforms to manifolds. We have shown that any smooth function on a 1-D manifold can be approximated by the inverse F-transform. We also received an estimate of the quality of the approximation.

Acknowledgements The work was supported from ERDF/ESF by the project "Centre for the development of Artificial Inteligence Methods for the Automotive Industry of the region" No. CZ.02.1.01/0.0/0.0/17_049/0008414.

References

1. B. Bede, *Mathematics of Fuzzy Sets and Fuzzy Logic*, 1st edn. (Springer, Berlin, Heidelberg, 2013), pp. 226–227
2. L. Conlon, *Differentiable Manifolds* (Springer, Birkhauser Advanced Texts, Birkhauser Boston, 2008)
3. J.P. Fortney, *A Visual Introduction to Differential Forms and Calculus on Manifolds* (Springer, Birkhauser, Cham, 2018)
4. V. Molek, I. Perfilieva, Deep learning and higher degree f-transforms: interpretable kernels before and after learning. Int. J. Comput. Intell. Syst. **13**(1), 1404–1414 (2020)
5. I. Perfilieva, Fuzzy transforms: theory and applications. Fuzzy Sets Syst. **157**(8), 993–1023 (2006)
6. I. Perfilieva, M. Daňková, B. Bede, Towards a higher degree F-transform. Fuzzy Sets Syst. **180**, 3–19 (2011)
7. A.N. Tikhonov, V.Y. Arsenin, *Solutions of Ill-Posed Problems* (Winston, New York, 1977)

A Systematic Review of Privacy-Preserving Blockchain in e-Medicine

Usman Ahmad Usmani, Junzo Watada, Jafreezal Jaafar, and Izzatdin Abdul Aziz

Abstract Blockchains provide a decentralized, permanent, and verifiable ledger that can record transactions having digital properties, leading to a fundamental shift in various revolutionary scenarios, such as smart cities, eHealth, or eGovernment. Blockchain has a wide variety of applications in healthcare that can enhance mobile health applications, tracking devices, exchanging, and storing electronic medical records, clinical trial data, and insurance information storage. The survey covers privacy strategies in public and unauthorized blockchains, e.g., Bitcoin and Ethereum, and privacy-preserving research ideas and solutions in both public and private blockchains. We also take into account various blockchain scenarios such as privacy-preserving identity management systems and platforms.

1 Introduction

The disintermediation given by Blockchain is changing the democratization, verifiability, and universal access to tokenized digital assets of any sort, leading to a revolution in various types of scenarios [1] beyond cryptocurrencies, such as healthcare [2], smart cities [3], decentralized Internet of Things (IoT) [4], intelligent transport systems [5] or e Administration [6], to name a few. Blockchain enables the transfer of digital assets in a decentralized manner using the ledger, without the intermediary

U. A. Usmani (✉) · J. Jaafar · I. A. Aziz
Universiti Teknologi Petronas, UTP, Seri Iskandar, 32610 Perak, Malaysia
e-mail: usman_19001067@utp.edu.my

J. Jaafar
e-mail: jafreez@utp.edu.my

I. A. Aziz
e-mail: izzatdin@utp.edu.my

J. Watada
Waseda University, Tokyo 82610, Japan

N. H. Phuong and V. Kreinovich (eds.), *Biomedical and Other Applications of Soft Computing*, Studies in Computational Intelligence 1045,
https://doi.org/10.1007/978-3-031-08580-2_3

of central third parties, while at the same time allowing public authentication as well as the origin of digital transactions and records.

However, Blockchain implementations are subject to a range of different concerns, such as compliance with legal regulations (e.g., General Data Protection Regulation (GDPR) [7]), scalability and response times [8], security threats [9], or privacy issues that threaten user anonymity, confidentiality and privacy control in their Ledger transactions. These privacy issues raise concerns among people and companies who are still wary of implementing Blockchain in their processes and businesses. It could mean sharing (even if encrypted or anonymized) their data and transactions in a publicly accessible database. Although the use of pseudonyms prevents link-ing transactions with real identity, users are not fully anonymous in their movements because all uses behind these pseudonyms can be traceable and linkable, particularly when managing multiple-entry transactions with multiple addresses from different accounts belonging to the same person.

In this context, emerging privacy-preserving proposals for Blockchain and platforms such as MediLedger Network or Coral Health propose enhanced decentralized ledgers that empower users with privacy-preserving mechanisms in their digital transactions. Its privacy-preserving existence distinguishes the management of relevant user information in approved Blockchains. With the emergence of Blockchain, Identity Management (IDM) frameworks move from conventional web-centric or identity federation approaches to the model of self-sovereign identity (SSI). Self-governing identities empower people to control their data at any time in any online situation. In this approach, personal user data is no longer kept raw in third-party services, neither in Service Providers nor Identity Providers. Details on purchases and user experiences in services can be anonymized.

It prevents third parties from leaking personal data and, in the worst case, from being a possible source of other, more significant, threats, such as identity-related cybercrimes (e.g., identity theft). However, considering the outstanding features and benefits offered by SSI, as analyzed in this paper, blockchain scenarios still need to resolve a range of privacy challenges, such as transaction linkability, blockchain P2P network privacy, private key management and recovery, quantum computing cryptographic algorithms, malicious trust third parties (TTP), malicious smart contracts.

To overcome these problems, specific plans for blockchain services, such as mixers, aim to include a third party responsible for concealing a transaction among many unrelated transactions. Sensitive details, such as the payer, the payer, or the sum paid [10], may also be entirely anonymized [11], although often at the risk of delays in the transaction and higher costs. Some other privacy-preserving crypto solutions incorporate SSI together with protected multi-party computing [12] or Zero Information Proofs (ZKPs), e.g. [13], and anonymous credential systems such as [14] on the blockchain network. Some other blockchain implementations use ring signatures [15] to mask user transactions. This paper comprehensively analyzes blockchain privacy concerns and privacy-preserving research ideas, strategies, and solutions in healthcare-based blockchain networks that tend to resolve privacy issues in this promising technology.

Blockchain already has some other security surveys [9], but they are not explicitly privacy-based. Also, some additional security and privacy analysis papers in distributed ledgers, such as [16], primarily concentrate on bitcoin [17] or cryptocurrencies. There is another recent survey on privacy in blockchains [18]; however, privacy concerns have not been established, unlike this paper. The risks they cover are exclusively linked to confidentiality issues in transactions. Also, neither the blockchain platforms nor the application scenarios evaluated the privacy solutions.

Our survey paper focuses on a broader scale, including a study of privacy-preserving crypto solutions in bitcoin and a review of privacy-preserving research proposals and platforms for different types of blockchain scenarios, namely eGovernment eHealth, cryptocurrency, Smart Cities, and Cooperative ITS [19–21]. This paper's contributions are manifold. First, we suggest different scenarios for healthcare applications and evaluate current implementations of technology that could be used to bring the scenarios into effect.

The rest of the paper is organized as follows. Section 2 describes the Blockchain and the Electronic Health Records Model. Section 3 gives the Privacy-Preserving Identity Management Systems and platforms in the Blockchain. It also discusses the limitations of the current techniques. Then, Sect. 4 discusses the future directions in the Blockchain EHRs. Finally, the conclusions are being drawn in Sect. 5.

2 Blockchain and the Electronic Health Records Model

Although cloud-assisted eHealth systems make these advantages more attractive than ever before, crucial concerns regarding privacy and protection in EHR outsourcing have been raised seriously, see [22]. From the point of view of EHR owners, including patients and medical institutions, the content of EHRs should not be leaked for privacy purposes because EHRs are one of the most sensitive and personal data for EHRs [23]. However, existing cloud service providers would not be responsible for protecting the privacy of EHRs against adversaries through their Service Level Agreements (SLAs). They would only pledge to protect privacy as much as possible [24].

2.1 Conventional Electronic Health Records Model

Moreover, unlike the conventional EHR management model, where medical institutions or patients store their EHRs locally, both medical institutions and patients will not physically own their EHRs once EHRs have been outsourced to cloud servers. As such, the accuracy and credibility of outsourced EHRs are put at risk in practice [25]. We emphasize that correctness and honesty mean that the contents of the EHRs are not changed and that the period when the EHRs were produced and outsourced is not altered. Traditional cryptographic primitives that have been commonly used in cloud storage systems for data confidentiality and integrity security, such as public-key encryption [26], digital signature [27], and message authentication code [28], cannot be explicitly used in cloud-assisted eHealth systems for the following reasons.

For example, an encrypted cloud system was proposed in [29], a stable and equal payment system was proposed in [30], a hierarchical multi-authority and attribute-based encryption for protection and privacy of social networks was proposed in [31], and a prototype of a secure EHR could be proposed in [32]. First, unlike conventional cloud storage systems, the owner of EHRs is not always the maker of EHRs for the generation of EHRs. Specifically, the patient's EHRs are created and outsourced by a delegated physician. The patient will not sign the EHRs before outsourcing to minimize to be in with a chance of selecting, verifying & validating transactions. The patient's contact and computation costs. Second, since the EHRs are outsourced to the doctor without the patient's presence, conventional encryption algorithms cannot be used straightforwardly.

In particular, the size of the EHRs can be large and cannot be encrypted by public-key encryption schemes for efficiency reasons. It is also difficult to agree between the doctor and the patient on the key to the symmetric key encryption algorithm. Also, maintaining the credibility and correctness of outsourced EHRs is more complex than ever when the doctor outsources the EHRs on behalf of the patient. The doctor has faith in the patient only during the treatment period and could be malicious after the treatment period [33]. Usually, a doctor can forge, alter, or delete outsourced EHRs to cover up their errors in medical malpractice. To ensure the confidentiality of outsourced EHRs, the current scheme [34] uses a smartphone-based key agreement to create a secure channel between the patient and the doctor (Fig. 1).

However, it allows the patient to equip a robust diagnostic smartphone, which is not always practical. Current schemes [35, 36] use an authentication method to ensure the correctness and credibility of outsourced EHRs. However, there is an apparent belief in existing schemes that the cloud server will not collide with the doctor to tamper with the outsourced EHRs. If the doctor allows the cloud server to change external EHRs, it isn't easy to detect such misbehavior. A malicious doctor can compromise a cloud server because it is a reasonable entity [36–38]. Therefore, it deviates from the pre-scribed schemes if such a strategy increases its system income. A trivial solution is to set up a trusted server to authenticate the doctor to resist collaboration between a misbehavior doctor and an irresponsible cloud server. If the doctor authenticates the trusted server, they are allowed to outsource the EHRs. However, such a mechanism's security relies on the security and reliability of the trusted server and is faced with a single-point failure problem. It is challenging to resist collaboration between a doctor and a cloud server without adding any trusted individual. In this paper, the authors propose a safe cloud-assisted eHealth framework, called TP-EHR, which safeguards the confidentiality of outsourced EHRs and protects outsourced EHRs from unauthorized modification without the implementation of any trusted agency.

The TP-EHR framework is described both in its implementation and conceptual model. The confidentiality of TP-EHR is ensured even if the doctor is in touch with the cloud server. The central concept is to use the blockchain methodology (i.e., Blockchain-based currencies) [39], which allows for the tamper-proofing and distribution of transactions without a central authority, see [40]. In TP-EHR, the EHR

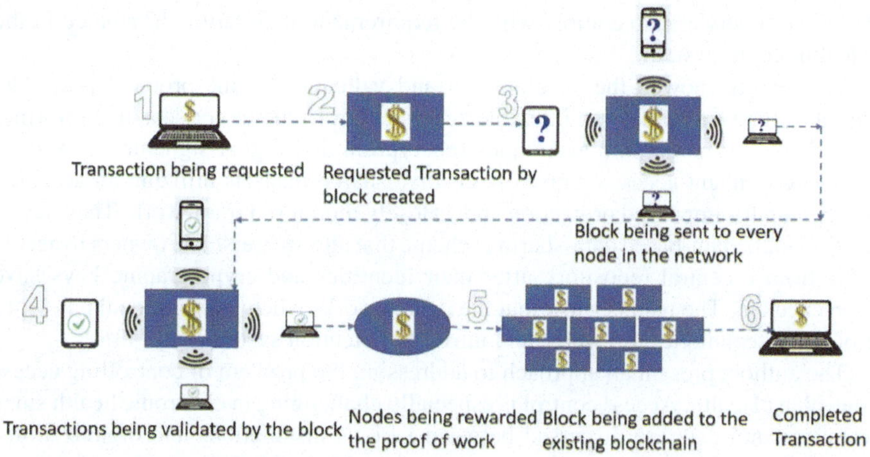

Fig. 1 Execution of a transaction on a blockchain network. (1) Authentication-This is done using cryptographic keys, a string of data (like a password) that identifies a user and gives access to their "account" or "wallet" of value on the system. (2) Authorization-Once the users agree upon the transaction, it needs to be approved or authorized before adding it to a block in the chain. (3) Proof of Work requires the people who own the computers in the network to solve a complex mathematical problem to be able to add a block to the chain. (4) Proof of Stake Later blockchain networks have adopted "Proof of Stake" validation consensus protocols, where participants must have a stake in the Blockchain—usually by owning some of the cryptocurrency

created by a doctor is integrated into a blockchain transaction. The cloud server will accept the EHR created by the doctor if and only if the corresponding transaction is registered in the Blockchain. TP-EHR uses a password-based key agreement system to create secure channels between patients and physicians who are patient-friendly without requiring additional patient devices. TP-EHR can withstand attacks by guessing passwords and thus have a better security guarantee compared to the current system.

2.2 *Blockchain-Based Privacy-Preserving Models in Electric Health Records*

E-health has been one of the major research subjects since the advent of the Internet of Things (IoT). Thanks to the accessibility of medical data, it seems complicated to protect patient privacy. Patient data are typically stored in the cloud in healthcare systems, making it impossible for patients to control their data properly. Nevertheless, according to the GDPR, the data subject's right to know where, how, who can access, and to what degree the data is stored. In [41], it proposes a new protocol for the security of patient privacy, called Pseudonym Based Encryption with Different Authorities (PBE-DA), applying the Blockchain principle to health communication operators on the eHealth network, to comply with the requirement of structure distributed in the eHealth record system.

The results showed the development and validation of the protocol [42]. The authors suggest a reliable and flexible access control scheme for sensitive information. Stable cryptographic techniques (encryption and digital signatures) are used to ensure efficient access control of sensitive shared datasets utilizing an accepted blockchain for improved protection and a closely monitored framework. They developed a blockchain-based data-sharing scheme that allows users/data owners to access EHR from a central repository after their identities and cryptographic keys have been checked. The results show that the device works where conventional password protection techniques, firewalls, and intrusion detection systems fail [43].

The authors present an approach to addressing the problem of controlling access control in eHealth. Access control is especially challenging in electronic health since resources and data are dispersed between various installations and organizations. They suggest an approach that takes advantage of Blockchain to store transactional information about eHealth records and access control policies to address this difficulty. Overall, the results demonstrate that the solution is feasible, which provides a range of advantages over current structures.

In [44], the authors suggest a DABS scheme for Blockchain in healthcare, which allows for efficient verification of the validity and identity of the signatory EHR data. They also define a comprehensive, collaborative on-chain and off-chain storage framework for efficient storage and verification of EHR data. This blockchain-based storage mechanism ensures that stored and exchanged EHR data is not limited. The combination of on-chain and off-chain storage effectively provides the safe sharing of large-scale distributed EHR data. The experimental results indicate that the proposed protocol is both efficient and practical. Reference [45] proposed a stable EHR framework based on attributes and blockchain technology.

The framework used Attribute-based Encryption (ABE) and Identity-based Encryption (IBE) to encrypt medical data using Identity-based Signature (IBS) to enforce digital signatures. This significantly improves the system's management and does not entail the implementation of various cryptographic schemes for different security requirements. Also, blockchain strategies ensure the confidentiality and traceability of medical data. In [46], the authors suggest a blockchain-based, safe, and privacy-preserving personal health information (PHI) sharing (BSPP) scheme enhance diagnostics in e-health systems. In this scenario, the Blockchain and the blockchain alliance are included. The private Blockchain is responsible for storing the PHI, while the blockchain consortium keeps the PHI-protected indexes. Block generators must have a conformity test to connect new blocks to the Blockchain, ensuring the framework's availability. The results show that the proposed protocol will fulfill the safety objectives.

Some cloud-based access control schemes have been introduced in [47, 48] to achieve data security during the EHR sharing process. A new fine-grained access control method called ciphertext-policy-based signaling and secure sharing of personal health records in cloud computing has been proposed in [49]; an efficient and secure fine-grained access control scheme has been introduced in [50] that allows authorized users to access EHRs in cloud storage. Supports specific physicians for writing EHRs; [51] proposes a hierarchical comparison-based encryption scheme and develops a dynamic policy update scheme using proxy re-encryption technology to achieve dynamic access control in cloud-based EHR systems.

To improve the searchability and interoperability of EHR sharing, [52] proposed a new cloud-based EHR support for a fuzzy keyword search for secure data sharing and effective use of EHRs; [53] used conjunctive keyword search with proxy re-encryption to build a secure EHR search scheme for data sharing between different medical institutions. Also, [54] proposed a general framework for securely sharing EHRs that allow patients to store and share their EHR on a cloud server and enable physicians to access EHRs in the cloud.

With the development of blockchain technology, its decentralized, traceability, and anonymous characteristics have been widely affected by the medical industry's applications. Many scholars are currently focusing on the privacy and security of EHR sharing based on blockchain technology. To help patients use and share their health data conveniently and securely, Amofa et al. [55] presented a blockchain architecture designed to ensure the security control of personal data in the exchange of health information by matching smart contracts with user-generated acceptable policies. The architecture minimized data security risks by designing a shared data control mechanism.

Zheng et al. [56] have proposed a conceptual design for personal continuous dynamic health data sharing based on blockchain technology. Cloud storage is complemented to share information related to personal health securely and transparently. An identity and access management system using blockchain technology to support digital system entities' authentication and authorization has been proposed in [57]. This system described Blockchain in the Hyperledger Fabric Identity Authentication and Access Management Framework. Also, Guo et al. [58] proposed a multi-agency signature scheme to ensure the effectiveness of encapsulated EHRs in the Blockchain. In this scheme, the patient accepted the message based on the qualities and only presented proof that he had attested it (Fig. 2).

Some schemes merge cloud technology with blockchain technology to boost EHR sharing protection. Cao et al. [59] suggested a cloud-assisted safe eHealth framework that would use blockchain technology to protect outsourced EHRs in the cloud from unauthorized alteration. This method's core concept was that EHRs should only be outsourced to authenticated participants, and every activity on the outsourced EHRs was incorporated as a transaction into the public Blockchain. Liu et al. [60] suggested a blockchain-based privacy protection scheme for data sharing, namely BPDS. EMR stands for Electronic Medical Records, a digital version of a paper record or a chart in a clinician's office. EMRs usually contain general information, such as treatment and patient history, obtained by specific medical practices.

In BPDS, the cloud was used to store the original EMRs safely, and the tamper-proof blockchain consortium was designed to share the EMR indexes. The scheme used this way to reduce the chance of leakage of medical data. The use of the consortium blockchain means that the EMRs cannot be changed on a discretionary basis. A storage regime and service system for storing, exchanging, and using medical data

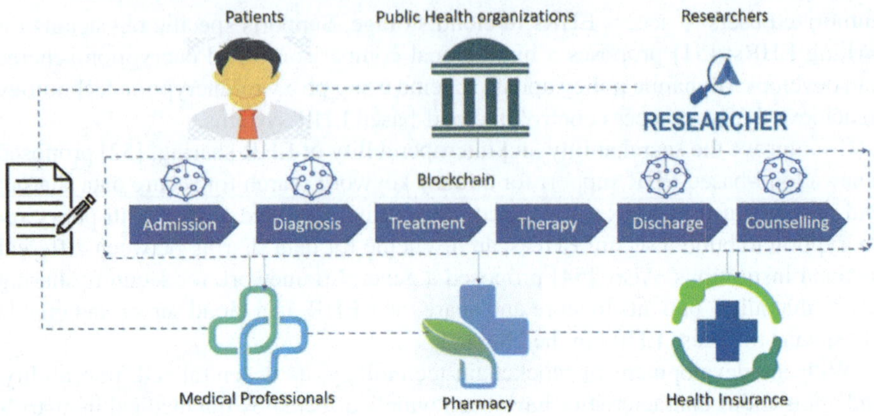

Fig. 2 A blockchain-based electronic health record system. Blockchain technology allows patients to assign access rules for their medical data, for example, permitting specific researchers to access parts of their data for a fixed period. With blockchain technology, patients can connect to other hospitals and collect their medical data automatically

based on Blockchain and cloud were proposed [61]. In this scheme, blockchain-based personal medical data applications can provide a patient medical information service without violating privacy concerns. Another work line focused on managing the privacy and access control of EHR sharing on Blockchain. Reference [62] introduced a sensitive data sharing model to support a personal health record system based on blockchain technology and proxy re-encryption.

The model discussed three critical issues: online data protection, limited comprehensive medical data storage, and consent revocation. Reference [63] proposed Blockchain-based system architecture to achieve auditable medical data exchange and healthcare data access handling. In other ways, Chen et al. [64] suggested a blockchain-based searchable encryption system for electronic medical record sharing to enhance data searchability. In this case, creating the EHR indexes stored in the Blockchain was complex logical expressions so that data users could use certain logical expressions to search for indexes. Taking advantage of Blockchain's decentralized properties, data owners had full power over who could see their EHRs. Blockchain technology maintains data confidentiality, anti-interference, and traceability.

Unlike the above works, Zhang and Lin et al. [65] suggested a multi-type safe and privacy-preserving PHI sharing (BSPP) blockchain boost diagnostics. In BSPP, the private Blockchain was used to store PHI for the hospital, and the blockchain consortium was responsible for documenting protected PHI indices. The scheme used public-key encryption with keyword search to ensure the blockchain consortium's data protection and sharing safety. The above works suggested various methods for sharing EHRs from different aspects. Generally, they introduced an idea or principle without comprehensive solutions for particular application scenarios. In our work, the authors combine keyword searchable encryption and proxy re-encryption technology to achieve privacy-preserving and secure EHR data sharing based on consortium blockchain technology and cloud storage. Also, they develop the protocol in detail.

3 Privacy-Preserving Identity Management Systems and Platforms in Blockchain

The MediLedger Network allows you to maintain an immutable record of and enforce cross-industry business rules without exposing your critical private data. This makes it easy to certify the authenticity of raw materials and medications, stop counterfeit products from entering your supply chain, and handle payment contract terms easily, secure your business intelligence so that your data remains behind your firewall and under your power. Use permission-based private messaging to share only the data you want to share with the partners you want to share, communicate with trading partners and trusted service providers to the forefront of new solutions for the pharmaceutical industry today. The MediLedger Project was launched in 2017.

It brought together pharmaceutical manufacturers and wholesalers in a working group to explore the potential of Blockchain to fulfill the requirements of the Drug Supply Chain Protection Act for a track and trace framework for U.S. drugs by 2023. With industry guidance, the MediLedger Project has become the MediLedger Network, a completely decentralized peer-to-peer and Blockchain network that could allow real value between companies. Network nodes are to be distributed and managed by in- industry members and technology providers representing the industry. For the first time, business rules for transactions and share data between companies can be applied via Blockchain without revealing private data. The MediLedger Network is being set up as a forum to potentially enable the open creation of participants and third parties to unlock disruptive solutions beyond today's technology.

Coral Health believes that for patient-centered designs, distributed leather technology is critical. The persistence, accessibility, and immutability of medical records are essential. All nodes for data sharing must reach a consensus about which events occurred first and ensure that no missing or duplicate records are available.

These properties are especially strong at distributed ledgers and offer dramatic changes to almost all healthcare systems currently in place. Coral Health has already announced that it will make its platform publicly accessible and start selling tokens on 28 September 2018. Coral Health that the sale of the token would be compatible with the availability of all mobile app users, not just those receiving the invitations for the test. Our healthcare partners have validated our HIPAA-compliant approach to EHR integration. At the same time, our approved blockchain infrastructure ensures the ongoing preservation of patient data, and our pre-cleared utility token offers the required incentives to make that information available.

Curisium is a fully managed software-as-a-service network that has minimal integration requirements. Without any I.T. pressure, one may benefit from the continuous growth of Blockchain-based EHR software and technology. At a fraction of the cost of conventional solutions, the data federation technology can handle vast quantities of data. The methods of cryptography ensure that knowledge is used exclusively for its intended and permitted purposes. HIPAA and GDPR are fully compliant with the platform, and our Blockchain maintains an automated audit trail of all activities. It helps to enjoy the opportunity to create and manage patient-level, complex contracts.

ADLTTM is an Enterprise Blockchain technology that complements current structures and processes, breaks down data silos, ensures data source, and provides an in- interoperable and high-performance ecosystem focused on enhancing patient outcomes. Our state-based health data model enables complex contracts to be applied concisely, readable, and auditable. With the first step, all journeys begin. It is a blockchain solution for manufacturers of biopharmaceuticals, healthcare, medical devices, and life sciences. ADLTTM connects manufacturers to patients and patients to the rest of the healthcare system and can combine IoMT, machine learning, and big data to improve patient outcomes and help guide R&D efforts. MedicalChain: The aim is to put the patient in charge of their medical data, allowing them the ability to share a single, most accurate version of their record with each agency within their organization. Fragmented, siloed medical records across the complexity of the healthcare system cause inefficiencies and inaccuracies.

Medicalchain uses blockchain technology to securely manage clinical information in a shared, innovative approach to healthcare. It helps to get in contact with clinics or hospitals for details. A database of patients who have agreed to be contacted by researchers via Medicalchain can be accessed. For your particular inquiries, the data you receive is up-to-date, accurate and configured. With a dynamic health record that remains with the medical, patient follow-up is streamlined. Patients have the option of allowing other users access to their electronic health records (EHRs) and of withdrawing access by creating a time-limited portal, thus improving their experience and ensuring data security.

This cuts out the middle parties and helps you access their reliable and up-to-date data in a more cost-effective and time-efficient way, accessing timely, verified health information directly from patients. It allows them to find out more about the monetary value of health data to contact us. Tech is at the cutting edge of healthcare technology, and they are eager to collaborate with other leaders and early adopters to help shape the digital healthcare future. To optimize clinical outcomes, patients, physicians, and healthcare providers are encouraged to safely access and transfer protected health information while offering actionable insights.

The patient uses blockchain technology from PTOYMatrix to ensure end-to-end encryption while adhering to regulatory guidelines and specifications for compliance. It empowers you with unique suggestions based on your health records. It helps to share the full medical history with doctors, all in one location, by aggregated data from multiple providers and medical facilities. Physicians can easily monitor their schedule, electronic position orders and submit or receive online referrals. Tools and reminders for automation will help your doctors save you time. Through our secure messaging platform, users can securely exchange information with their physicians, family members, caregivers, specialists and nurses, or chat. Encrypted middleware to fulfill state-of-the-art health I.T. high-volume standards. HIPAA-compliant secure data storage that complies with regional regulatory guidelines.

The PokitDok platform-as-a-service makes bringing new applications and services to the market quicker and simpler for healthcare organizations. To leverage real-time, member-specific health insurance data on a scale, it links directly through more than 700 payers. From a single source, it can access 93% of U.S.-covered lives. To innovate, there's no need to rip and replace legacy systems. The healthcare API suite allows you to migrate existing systems into modern times quickly.

The new features are easily incorporated into your existing workflows, ranging from cross-scheduling to patient portal eligibility and identity management. The Eligibility Solution from PokitDok helps the consumer directly integrate medical insurance eligibility and benefit verification into your workflow. There is no minimum spending cap per month. The Pharmacy Solutions of PokitDok provides real-time access to verify pharmacy benefits through Medicare and Commercial Plan. It is the first solution before the submission of a prescription to report the form and benefits information.

4 Future Research Directions

Based on the previous analysis, this section addresses some of the key research directions for Blockchain on privacy aspects. New regulatory frameworks should be compliant with the existing and future blockchain technologies. The inherent distribution of blockchain technology poses a range of obstacles to creating compliant solutions to current data protection regulations, such as the GDPR, as defined in the E.U. Blockchain Observatory and Forum Report 2018.

Despite attempts to create more privacy-respectful solutions, it is still essential to preserve privacy as denied by certain legal instruments. The applicability of privacy-preserving techniques due to the cost of cryptographic operations is one of the key problems described in the previous study. Nevertheless, in some large-scale scenarios, the research community is working on new crypto-privacy protocols to resolve the cost-effective computational operations demanded by emerging blockchain privacy techniques that undermine the scalability and broad-based adoption of Blockchain.

In situations in which resource-constraint devices or systems are commonly regarded, such as the IoT, these issues are compounded. Therefore, the development of new privacy-preserving methods would resolve these concerns to facilitate the adoption of blockchain technologies. A further line of research is the denial of novel crypto-algorithms resistant to quantity. Furthermore, future research will need to build strategies that maintain privacy to protect privacy in the execution of smart contracts while ensuring that they are formally checked.

Recent work in this regard, for example, explores potential solutions to make current implementations of blockchains resistant to quantum computing technologies. The usability aspects, as already stated, are critical to ensuring that end-users can control their privacy effectively. We conclude that there is a lack of systematic methods for this purpose, based on our research. It is particularly applicable in contexts such as eHealth, where it is possible to share confidential data for customized healthcare services.

Using user-friendly software, some operations should be automated for this purpose. The presence of many technologies and applications designed to provide privacy features of blockchain systems is another significant thing emerging from our study. Therefore, to ensure large-scale blockchain scenarios, it is necessary to ensure interoperability between these implementations. For this purpose, through a shared forum where different blockchains can protect their privacy preferences, intermediate approaches can help mitigate potential interoperability problems.

In this respect, it will be essential to research the use of inter-leading techniques in the coming years to enhance interoperability and mitigate performance problems. The use of blockchain technology has been widely regarded every day in many different contexts in recent years. In these cases, most of these emerging ideas are not compatible with existing requirements. C-ITS, where there is an explicit agreement between government agencies and industry to use PKI as the basis for security asset provision, is one of the prominent examples. Most recent research recommendations, however, consider Blockchain to be the only method for this purpose. Therefore, to ensure universal adoption of new techniques, the implementation of Blockchain and the incorporation of privacy-preserving techniques should be in accordance (not only) with established regulations but with current requirements in specific scenarios.

5 Conclusions

This study conducted a comprehensive review of the literature on EHRs within the Blockchain. To recognize and tackle key concerns, problems, and future benefits. In the field of healthcare, through the introduction of Blockchain. The usage of Blockchain has exceeded the reach of different fields, and the importance of Blockchain for healthcare has now been highlighted. After reviewing the literature review findings, we believe that Blockchain technology can be an effective solution in the future to growing healthcare industry problems. As an EHR interoperability, building confidence between healthcare providers, auditability, safety, and pro- viding access to health data control for patients to enable them to choose who they want to trust. However, additional experiments, trials, and evaluations must be conducted to ensure a robust and well-established programme introduced as a customer before the wide-scale use of Blockchain technology in healthcare. Confidential, extremely confidential, and critical information is health data. For future work and review, this research may act as a base or inspiration. Our research questions have been answered, and taxonomy may contribute to the design of architecture or model. In addition, research into the combination of Blockchain and the Internet of Things (IoT) in healthcare is a potential guide for future research. The aim was to do a good literature review of the algorithms in the current EHR systems and then make a theoretical and architectural comparison with the algorithms' weaknesses taken into account.

References

1. D.D.F. Maesa, P. Mori, L. Ricci, A Blockchain-based approach for the definition of auditable access control systems. Comput. Secur. **84**, 93–119 (2019)
2. N.S. Artzi, S. Shilo, E. Hadar, H. Rossman, S. Barbash-Hazan, A. Ben-Haroush, R.D. Balicer, B. Feldman, A. Wiznitzer, E. Segal, Prediction of gestational diabetes based on nationwide electronic health records. Nat. Med. **26**(1), 71–76 (2020)
3. S. Tanwar, K. Parekh, R. Evans, Blockchain-based electronic healthcare record system for healthcare 4.0 applications. J. Inf. Secur. Appl. **50**, 102407–102407 (2020)
4. C. Sohrabi, Z. Alsafi, N. O'Neill, M. Khan, A. Kerwan, A. Al-Jabir, C. Iosifidis, R. Agha, World Health Organization declares global emergency: a review of the 2019 novel coronavirus (COVID-19). Int. J. Surg. (2020)
5. S. Henry, K. Buchan, M. Filannino, A. Stubbs, O. Uzuner, 2018 n2c2 shared task on adverse drug events and medication extraction in electronic health records. J. Am. Med. Inform. Assoc. **27**(1), 3–12 (2020)
6. M. Muzny, A. Henriksen, A. Giordano, J. Muzik, A. Grøttland, H. Blixgård, G. Hartvigsen, E. Årsand, Wearable sensors with possibilities for data exchange: analyzing status and needs of different actors in mobile health monitoring systems. Int. J. Med. Inform. **133**, 104017–104017 (2020)
7. J.J. Hathaliya, S. Tanwar, An exhaustive survey on security and privacy issues in Healthcare 4.0 (2020)
8. W.M. Kersemaekers, K. Vreeling, H. Verweij, M. van der Drift, L. Cillessen, D. van Dierendonck, A.E.M. Speckens, Effectiveness and feasibility of a mindful leadership course for medical specialists: a pilot study. BMC Med. Educ. **20**(1), 34–34 (2020)
9. M. Segarra-Oña, A. Peiró-Signes, R. Verma, Fostering innovation through stake- holders' engagement at the healthcare industry: tapping the right key. Health Policy **124**(8), 895–901 (2020)
10. B. Moazzami, N. Razavi-Khorasani, A.D. Moghadam, E. Farokhi, N. Rezaei, COVID- 19 and telemedicine: Immediate action required for maintaining health- care providers well-being. J. Clin. Virol. **126**, 104345–104345 (2020)
11. G.T. Kitching, M. Firestone, B. Schei, S. Wolfe, C. Bourgeois, P. O'Campo, M. Rotondi, R. Nisenbaum, R. Maddox, J. Smylie, Unmet health needs and discrimination by healthcare providers among an Indigenous population in Toronto Canada. Can. J. Public Health **111**(1), 40–49 (2020)
12. A. Balapour, H.R. Nikkhah, R. Sabherwal, Mobile application security: role of perceived privacy as the predictor of security perceptions. Int. J. Inf. Manage. **52**, 102063–102063 (2020)
13. L. Lenert, B.Y. McSwain, Balancing health privacy, health information exchange, and research in the context of the COVID-19 pandemic. J. Am. Med. Inform. Assoc. **27**(6), 963–966 (2020)
14. N.C. Ghasi, D.C. Ogbuabor, V.A. Onodugo, Perceptions and predictors of organizational justice among healthcare professionals in academic hospitals in South-Eastern Nigeria. BMC Health Serv. Res. **20**(1), 1–12 (2020)
15. M. Wolderslund, P.E. Kofoed, R. Holst, K. Waidtløw, J. Ammentorp, Out-patients' recall of information when provided with an audio recording: a mixed-methods study. Patient Educ. Couns. **103**(1), 63–70 (2020)
16. A. Alorwu, N.V. Berkel, J. Goncalves, J. Oppenlaender, M.B. López, M.Seethara- man, S. Hosio, (2020)
17. S.M. Ahmed, A. Rajput, Threats to patients' privacy in smart healthcare environment. Innov. Health Inform. 375–393 (2020)

18. A. Gauhar, N. Ahmad, Y. Cao, S. Khan, H. Cruickshank, E.A. Qazi, A. Ali, xDBAuth: Blockchain-based cross-domain authentication and authorization framework for internet of things. IEEE Access **8**, 58800–58816 (2020)
19. Z.B. Miled, K. Haas, C.M. Black, R.K. Khandker, V. Chandrasekaran, R. Lipton, M.A. Boustani, Predicting dementia with routine care EMR data. Artif. Intell. Med. **102**, 101771–101771 (2020)
20. F. Shahid, A. Khan, Smart digital signatures (SDS): a post-quantum digital signature scheme for distributed ledgers. Futur. Gener. Comput. Syst. **111**, 241–253 (2020)
21. N.S. Key, A.A. Khorana, N.M. Kuderer, K. Bohlke, A.Y. Lee, J.I. Arcelus, S.L. Wong, E.P. Balaban, C.R. Flowers, C.W. Francis, L.E. Gates, A.K. Kakkar, M.N. Levine, H.A. Liebman, M.A. Tempero, G.H. Lyman, A. Falanga, Venous thromboembolism prophylaxis and treatment in patients with cancer: ASCO clinical practice guideline update. J. Clin. Oncol. **38**(5), 496–520 (2020)
22. M. Sorbello, K. El-Boghdadly, I.D. Giacinto, R. Cataldo, C. Esposito, S. Falcetta, G. Merli, G. Cortese, R.M. Corso, F. Bressan, S. Pintaudi, R. Greif, A. Do-nati, The Italian coronavirus disease 2019 outbreak: recommendations from clinical practice. Anaesthesia **75**(6), 724–732 (2020)
23. I. Lin, L. Wiles, R. Waller, R. Gouache, Y. Nagree, M. Gibberd, L. Straker, C.G. Maher, P.P.B. O'Sullivan, What does best practice care for musculoskeletal pain look like? eleven consistent recommendations from high-quality clinical practice guidelines: systematic review. Br. J. Sports Med. **54**(2), 79–86 (2020)
24. W.B. Issa, I.A. Akour, A. Ibrahim, A. Almarzouqi, S. Abbas, F. Hisham, J. Griffiths, Privacy, confidentiality, security and patient safety concerns about electronic health records. Int. Nurs. Rev. **67**(2), 218–230 (2020)
25. W. Liang, Y. Fan, K.C. Li, D. Zhang, J.L. Gaudiot, Secure data storage and recovery in industrial Blockchain network environments. IEEE Trans. Industr. Inf. **16**(10), 6543–6552 (2020)
26. M. Ahmed, O. Kazar, S. Benharzallah, L. Kahloul, A. Merizig, An intelligent and secure health monitoring system based on agent, in *2020 IEEE International Conference on Informatics, IoT, and Enabling Technologies (ICIoT)*, pp. 291–296 (2020)
27. G. Srivastava, R.M. Parizi, A. Dehghantanha, The future of blockchain technology in healthcare internet of things security, in *Blockchain Cybersecurity, Trust and Privacy*. (Springer, 2020), pp. 161–184
28. H. Yoon, Y. Jang, P.W. Vaughan, M. Garcia, Older adults' internet use for health information: digital divide by race/ethnicity and socioeconomic status. J. Appl. Gerontol. **39**(1), 105–110 (2020)
29. B. Ali, N.P. Dimoska, N. Marina, (2020) 495–514 of 3027—Fiscal decentralization and the role of local government inlocal...
30. H.C. Nunes, R.C. Lunardi, A.F. Zorzi, R.A. Michelin, S.S. Kanhere, Context-based smart contracts for appendable-block blockchains, in *2020 IEEE International Conference on Blockchain and Cryptocurrency (ICBC)*, pp. 1–9 (2020)
31. Y. Du, H. Tu, H. Yu, S. Lukic, Accurate consensus-based distributed averaging with variable time delay in support of distributed secondary control algorithms. IEEE Trans. Smart Grid **11**(4), 2918–2928 (2020)
32. J.D. Vyas, M. Han, L. Li, S. Pouriyeh, J.S. He, Integrating Blockchain technology into healthcare, in *Proceedings of the 2020 ACM Southeast Conference*, pp. 197–203 (2020)
33. P. Pandey, R. Litoriya, Ensuring elderly well-being during COVID-19 by using IoT (2020)
34. Y. Zheng,Q. Li, C. Wang, X. Li, B. Yang, A magnetic based indoor positioning method on fingerprint and confidence evaluation (2020)
35. A. Builders, R. Laanoja, A. Truu, S.A. Guardtime, Blockchain-assisted hash-based data signature system and method. U.S. (2020)
36. V. Gramoli, From blockchain consensus back to Byzantine consensus. Futur. Gener. Comput. Syst. **107**, 760–769 (2020)
37. D.V. Le, L.T. Hurtado, A. Ahmad, M. Minaei, B. Lee, A. Kate, A tale of two trees: one writes, and other reads: optimized oblivious accesses to bitcoin and other UTXO-based Blockchains. Proc. Priv. Enhancing Technol. **2020**, 519–536 (2020)

38. M. Alharby, A.V. Moorsel, BlockSim: an extensible simulation tool for Blockchain systems, in *Frontiers Blockchain*, vol. 3 (2020)
39. S.T. Yalla, P.N.S. Nikhilendra, An overview on Blockchain technology and its applications, in *ICDSMLA 2019*. (Springer, 2020), pp. 1030–1035
40. H. Wang, Y. Jin, X. Tan, Study on sustainable development of the transnational power grid interconnection projects under diversified risks based on variable weight theory and bayesian network (2020)
41. D. Bumblauskas, A. Mann, B. Dugan, J. Ritter, A blockchain use case in food distribution: do you know where your food has been? Int. J. Inf. Manage. **52**, 102008–102008 (2020)
42. D. Lee, N. Park, Blockchain-based privacy-preserving multimedia intelligent video surveillance using secure Merkle tree. Multimed. Tools Appl. (2020)
43. R. Hu, W.Q., Yan, Design and implementation of visual Blockchain with merkle tree, in *Handbook of Research on Multimedia Cyber Security*, pp. 282–295 (2020)
44. O. Ersoy, Z. Erkin, R.L. Lagendijk, Decentralized incentive-compatible and sybil-proof transaction advertisement, in *Mathematical Research for Blockchain Economy* (Springer, 2020), pp. 151–165
45. D.J. Moroz, D.J. Aronoff, N. Narula, D.C. Parkes, Selfish behavior in the Tezos proof-of-stake protocol 10736, 11 (2020). arXiv:2002
46. Z. Cui, X.U.E. Fei, S. Zhang, X. Cai, Y. Cao, W. Zhang, J. Chen, A hybrid BlockChain-based identity authentication scheme for multi-WSN. IEEE Trans. Serv. Comput. **13**(2), 241–251 (2020)
47. T.T. Thwin, S. Vasupongayya, Blockchain based secret-data sharingmodel for personal health record system, in *2018 5th International Conference on Advanced Informatics: Concept Theory and Applications (ICAICTA)* (IEEE, 2018), pp. 196–201
48. F. Aljaafari, L.C. Cordeiro, M.A. Mustafa, EBF: a hybrid verification tool for finding software vulnerabilities in IoT cryptographicprotocols (2020)
49. F. Aljaafari, L.C. Cordeiro, M.A. Mustafa, R. Menezes (2020) arXiv:2103.11363
50. D.R. High, B.W. Wilkinson, T. Mattingly, B.G. Mchale, V.J.J. O'Brien, R. Cantrell, J. Jurich, Walmart Apollo LLC, 2020. Verifying authenticity of computer-readable information using the Blockchain. U.S. Patent 10,495–495 (2020)
51. F.A. Khan, M. Asif, A. Ahmad, M. Alharbi, H. Aljuaid, Blockchain technology, improvement suggestions, security challenges on smart grid and its application in healthcare for sustainable development. Sustain. Cities Soc. **55**, 102018–102018 (2020)
52. V. Dedeoglu, R. Jurdak, A. Dorri, R.C. Lunardi, R.A. Michelin, A.F. Zorzo, S.S. Kanhere, Blockchain technologies for IoT, in *Advanced Applications of Blockchain Technology* (Springer, 2020), pp. 55–89
53. D. Lizcano, J.A. Lara, B. White, S. Aljawarneh, Blockchain-based approach to create a model of trust in open and ubiquitous higher education. J. Comput. High. Educ. **32**(1), 109–134 (2020)
54. A. Dolgui, D. Ivanov, S. Potryasaev, B. Sokolov, M. Ivanova, F. Werner, Blockchain-orienteddynamic modelling of smart contract design and execution in the supply chain. Int. J. Prod. Res. **58**(7), 2184–2199 (2020)
55. H. Amofa, H. Qin, M. Zhao, X. Wei, H. Shen, W. Susilo, Blockchain-based fair payment smart contract for public cloud storage auditing. Inf. Sci. **519**, 348–362 (2020)
56. L. Zheng, Z. Li, S. Hou, B. Xiao, S. Guo, Y. Yang, A survey of IoT applications in blockchain systems: architecture, consensus, and traffic modeling. ACM Comput. Surv. (CSUR) **53**(1), 1–32 (2020)
57. T. Mitani, A. Otsuka, Traceability in permissioned Blockchain. IEEE Access **8**, 21573–21588 (2020)
58. A. Guo, R.M. Parizi, M. Han, A. Dehghantanha, H. Karimipour, K.K.R. Choo, Public blockchains scalability: an examination of sharding and segregated witness, in *Blockchain Cybersecurity, Trust and Privacy* (Springer, 2020), pp. 203–232
59. J.B. Cao, R.E. Sibai, K. Kambhampaty, J. Demerjian, Permissionless reputation-based consensus algorithm for Blockchain. Internet Technol. Lett. **3**(3), 151–151 (2020)

60. S. Liu, K. Parekh, R. Evans, Blockchain-based electronic healthcare record system for healthcare 4.0 applications. J. Inf. Secur. Appl. **50**, 102407–102407 (2020)
61. S.F. Wamba, J.R.K. Kamdjoug, R.E. Barack, J.G. Keogh, Bitcoin, Blockchain and Fintech: a systematic review and case studies in the supply chain. Prod. Plan. Control **31**(2–3), 115–142 (2020)
62. L. Ewen, S. Reddy, P. Patel, A. Kundal, P. Patel, S. Mohammed, Utilizing Blockchaintechnology in social media bot identification. Technology **1**(5), 6–6
63. A. Rejeb, J.G. Keogh, H. Treiblmaier, Leveraging the internet of things and Blockchain technology in supply chain management. Futur. Internet **11**(7), 161–161 (2019)
64. I. Chen, S. Tanwar, S. Tyagi, N. Kumar, Blockchain for 5G-enabled IoT for in-dustrial automation: a systematic review, solutions, and challenges(2020)
65. H. Lin, Y. Song, Secure cloud-based EHR system using attribute-based cryptosystem and Blockchain. J. Med. Syst. **42**(8), 152–152 (2018)

Why Rectified Linear Neurons: Two Convexity-Related Explanations

Jonatan Contreras, Martine Ceberio, Olga Kosheleva, Vladik Kreinovich, and Nguyen Hoang Phuong

Abstract At present, the most efficient machine learning technique is deep learning, in which non-linearity is attained by using rectified linear functions $s_0(x) = \max(0, x)$. Empirically, these functions work better than any other nonlinear functions that have been tried. In this paper, we provide a possible theoretical explanation for this empirical fact. This explanation is based on the fact that one of the main applications of neural networks is decision making, when we want to find an optimal solution. We show that the need to adequately deal with situations when the corresponding optimization problem is feasible—i.e., for which the objective function is convex—uniquely selects rectified linear activation functions.

1 Formulation of the Problem

Rectified linear neurons are very successful. At present, the most successful machine learning technique is deep neural networks; see, e.g., [4]. In general, in neural networks, signals go through two types of transformations: linear transformations and non-linear transformation described by the so-called *activation function* $x \mapsto s(x)$.

Deep neural networks mostly used *rectified linear* (ReLU) activation functions

---

J. Contreras · M. Ceberio · O. Kosheleva · V. Kreinovich (✉)
University of Texas at El Paso, 500 W. University El Paso, El Paso, TX 79968, USA
e-mail: vladik@utep.edu

J. Contreras
e-mail: jmcontreras2@utep.edu

M. Ceberio
e-mail: mceberio@utep.edu

O. Kosheleva
e-mail: olgak@utep.edu

N. H. Phuong
Division Informatics, Math-Informatics Faculty, Thang Long University, Nghiem Xuan Yem Road, Hoang Mai District, Hanoi, Vietnam

N. H. Phuong and V. Kreinovich (eds.), *Biomedical and Other Applications of Soft Computing*, Studies in Computational Intelligence 1045,
https://doi.org/10.1007/978-3-031-08580-2_4

$$s_0(x) = \max(0, x). \tag{1}$$

The main reason for this choice is that empirically, these activation function have been most successful.

But why are they successful? From the theoretical viewpoint, this empirical success is a challenge: why are these activations functions more successful than others? Are there activation functions that we have not tried yet—which will be even more successful?

Important comment. Before we start analyzing this question, we should mention that the fact that we have linear transformations before and after each application of an activation function implies that the same results that we obtain by using rectified linear activation function $s_0(x)$ can also obtained by neurons that use shifted and scale versions of this function:

$$s_1(x) = b_0 + b_1 \cdot x + b_2 \cdot s_0(a_0 + a_1 \cdot x), \tag{2}$$

i.e., if we use functions of the type:

$$s_1(x) = b_0 + a_- \cdot (x - x_0) \text{ for } x \le x_0; \tag{3}$$

$$s_1(x) = b_0 + a_+ \cdot (x - x_0) \text{ for } x \ge x_0, \tag{4}$$

corresponding to some values b_0, a_-, a_+, and x_0.

What is known and what we do in this paper. There are some theoretical explanations of why rectified linear neurons are so successful: e.g., in [2, 7, 8], it was proven that the rectified linear activation functions are, in some reasonable sense, optimal. This explanation is based on the idea that the relative quality of different data processing techniques—in particular, the relative quality of neural networks using different activation functions—should not change if we change all the numerical values by changing the measuring units and/or the starting points for measuring the corresponding quantities.

In this paper, we provide yet another theoretical explanation for this empirical success—this time, an explanation based on computational efficiency and convexity.

2 Why Convexity

Need for optimization. In practice, we always want to find the best possible solution. In precise terms, which solution is better and which is worse is usually described in numerical terms, by assigning a number to each possible solution, so that a solution with the largest (or smallest) value of this numerical characteristic is the best. The mapping that assigns such a number to each alternative x is known as the *objective function* $f(x)$. For example, a company tries to maximize its profit, an environmental agency tries to minimize the overall pollution, etc.

In general, as the above examples show, we can have both maximization and minimization problems. However, the problem of maximizing an objective function $f(x)$ is equivalent to minimizing the function $g(x) \stackrel{\text{def}}{=} -f(x)$. Thus, all optimization problems can be easily reduced to minimizations. So, without losing generality, mathematicians usually only talk about minimization problems.

Need for convex optimization. In general, optimization is NP-hard (see, e.g., [9, 14]), meaning that unless P = NP (which most computer scientists believe to be impossible), no feasible algorithm can solve all optimization problems. There is an important class of optimization problems for which optimization is feasible: the class of all *convex* optimization problems (see, e.g., [10, 11, 14]), in which the minimized functions $f(x)$ is convex, i.e., satisfies the condition

$$f(\alpha \cdot x + (1 - \alpha) \cdot x') \le \alpha \cdot f(x) + (1 - \alpha) \cdot f(x') \tag{5}$$

for all x, x', and for all $\alpha \in [0, 1]$.

Moreover, it has been proven that convex functions are, in some reasonable sense, the largest class of functions for which optimization is feasible: once we add some non-convex functions to this problem, the optimization problem becomes NP-hard; see [6].

This result will underlie our two explanations.

3 First Convexity-Related Explanation

How is all this related to neural networks. One of the main applications of neural networks is to make decisions. For this purpose, we need to train the neural network to predict, for each possible action, the consequences of this action. In other words, we want, given the parameters x that characterize the possible decision, to compute the value $f(x)$ of the objective function that characterizes this decision. For the simplest neural networks, this means that we approximate the original function $f(x_1, \ldots, x_n)$ by a linear combination of the outputs of non-linear neurons:

$$f(x_1, \ldots, x_n) = \sum_{k=1}^{K} W_k \cdot s\left(\sum_{i=1}^{n} w_{ki} \cdot x_i - w_{k0}\right) - W_0. \tag{6}$$

For multi-layer neural networks, the corresponding expression is more complicated.

Towards resulting natural requirements on the activation function. Once we train the neural network to compute the value of the objective function, a natural next step is to find the alternative x that minimizes this objective function. Since, as we have mentioned, optimization is only feasible for convex objective functions, it makes sense to make sure that the expression (6)—and a similar expression for multi-layer neural networks—preserve convexity as much as possible.

In other words, if the actual activation function is convex, we want to make sure that this convexity is, in some reasonable sense, preserved in an approximating expressions like (6).

First requirement. The above idea means, in particular, that for the simplest case when one neuron is sufficient, the activation function $s(x)$ itself must be convex.

Comment. The rectified linear activation function (1) itself is convex, so it satisfies this requirement.

On the other hand, there are many other convex functions, so this requirement does not uniquely determine the rectified linear function. For this unique determination, we need to come up with additional requirement(s).

Second requirement. It is known that if functions $f_1(x)$, ..., $f_n(x)$ are convex, then their convex combination

$$f(x) = w_1 \cdot f_1(x) + \ldots + w_K \cdot f_K(x), \tag{7}$$

where $w_k \geq 0$ and $\sum\limits_{k=1}^{K} w_k = 1$, is also convex. Moreover, any linear combination with non-negative coefficients is convex, even when the sum of these coefficients is different from 1. On the other hand, if we allow even one of the coefficients to be negative, then we already get non-convex functions. So, the only way to make sure that a linear combination of convex functions is convex is to make sure that all the coefficients w_k are non-negative.

It is therefore reasonable to require that every convex function $f(x)$—at least every convex function of one variable—be representable as a linear combination of activation functions with non-negative coefficients. This is our second requirement.

Let us analyze what are the activation functions that satisfy this requirement.

Let us recall the usual calculus-based characteristics of convexity. It is known that a differentiable function $f(x)$ is convex if and only if its second derivative $f''(x)$ is everywhere non-negative $f''(x) \geq 0$. Not all convex functions are everywhere differentiable—e.g., the rectified linear activation function $s_0(x)$ is not differentiable at the point $x = 0$. However, for such function, we can consider, as derivatives, *generalized functions* (also known as *Schwartz distributions*), which are limits of usual functions; see, e.g., [3, 5]. The most well-known generalized function is a *delta-function* $\delta(x)$ which is equal to 0 for all $x \neq 0$ and which tends to ∞ at $x = 0$; such functions are used in physics to describe, e.g., point-wise particles and objects; see, e.g., [1, 12]. In particular, the derivative $s_0'(x)$ of the rectified linear function is equal to 0 for $x \leq 0$ an to 1 for $x > 0$, and the second derivative is exactly the delta-function.

For a linear combination of functions (7), its second derivative is equal to the linear combination of its second derivatives, with exactly the same coefficients w_k:

$$f''(x) = w_1 \cdot f_1''(x) + \ldots + w_K \cdot f_K''(x). \tag{8}$$

So, in terms of the second derivatives, the above second requirement means that every non-negative (generalized) function can be represented as a linear combination of the functions corresponding to second derivative of the activation function $s(x)$—and of its shifted and scaled versions $s(a_0 + a_1 \cdot x)$.

Now we can prove that only rectified linear activation function satisfies both our requirements. If the second derivative $s''(x)$ of an activation function $s(x)$ differs from 0 for at least two different values $x \neq x'$, then this property remains true for any convex combination of shifted and scaled versions of this activation function. Thus, this way, we will never get a convex function for which the second derivative is non-zero only for one value x—e.g., the rectified linear function $s_0(x)$.

On the other hand, if we select the rectified linear function $s_0(x)$ as an activation function, then we have $s_0''(x) = \delta(x)$. In this case, any non-negative function $f''(x)$ can be represented as a linear combination of shifted versions of $s_0''(x)$: indeed,

$$f''(x) = \int f''(y) \cdot \delta(x - y)\, dy = \int f''(y) \cdot s_0''(x - y)\, dy, \tag{9}$$

and thus, the function $f(x)$ can be represented as a similar linear combination of the shifted versions of $s_0(x)$—plus possibly some linear terms:

$$f(x) = b_0 + b_1 \cdot x + \int f''(y) \cdot s_0(x - y)\, dy. \tag{10}$$

In general, our second requirement is satisfied by any convex function for which the second derivative is different from 0 only for one value $x = x_0$. This second derivative can therefore be described as

$$s''(x) = c \cdot \delta(x - x_0), \tag{11}$$

for some $c > 0$. Integrating twice the equality (11), we conclude that

$$s(x) = b_0 + b_1 \cdot x + c \cdot s_0(x - x_0), \tag{12}$$

for some values s_0 and s_1. One can check that this is exactly the expression (2–4), i.e., that indeed, the above two natural convexity-related requirements naturally lead to the rectified linear activation functions.

4 Second Convexity-Related Explanation

Let us consider a more general setting. Out of the above two requirements, the first one looks more convincing, the second one is somewhat less convincing. Let us therefore consider a more general setting, when we still postulate the first requirement—i.e., we still consider only convex activation functions—but instead of postulating the second requirement, we want to find the activation function which is the best in some sense, i.e., for which the corresponding objective functional $F(s)$ describing the relative qualities of different convex activation functions $s(x)$—attains its smallest possible value.

What calculus tells us. In general, a maximum or minimum of a function on a multi-D domain is attained either inside this domain—in which case it is a stationary point of this function—or on its boundary. When the domain is relatively small, the probability that a global stationary point is inside this domain is very small, so it is reasonable to assume that the minimum is attained on the boundary.

This general conclusion can be applied to our case when we optimize a functional $F(s)$ on the domain of all convex functions s. Indeed, most functions are not convex. So, in the space of all possible functions, the domain of all convex functions is indeed small.

Similarly, if the domain's boundary contains a flat face-type part—as when the domain is a polytope—then it is reasonable to assume that the minimum is attained not in the interior of this face, but on its boundary. If this boundary also contains a flat part—as in the case of a polytope where the boundary of a face consists of edges—we can similarly conclude that the minimum is most probably attained at the boundary of this part—e.g., for a 3-D polytope, at one of the vertices.

In general, we can conclude that the minimum is most probably attained at one of the *extreme points* of the original domain—i.e., a point that cannot be represented as a convex combination of other points from this domain.

Comment. For a precise mathematical description of this idea, see [13].

What this implies for optimal activation functions. We want to select an activation function. In this case, the domain is the set of all convex functions. What are the extreme elements of this domain?

We have already shown that any convex function $s(x)$ whose second derivative differs from 0 at least 2 different points can be represented as a convex combination of other convex functions—namely, shifted rectified linear functions. Hence, such functions $s(x)$ are not extreme elements of our domain. Thus, the only extreme elements of this domain are convex functions whose second derivative differs from 0 only at one point—which are, as we have shown, exactly rectified linear functions.

Since, with high probability, only extreme elements can be optimal, we conclude that with high probability, only rectified linear functions can be optimal—no matter what optimality criterion we used. Thus, we have indeed provided a second theoretical justification for the success of rectified linear activation functions.

Acknowledgements This work was supported in part by the National Science Foundation grants 1623190 (A Model of Change for Preparing a New Generation for Professional Practice in Computer Science), and HRD-1834620 and HRD-2034030 (CAHSI Includes), and by the AT&T Fellowship in Information Technology. It was also supported by the program of the development of the Scientific-Educational Mathematical Center of Volga Federal District No. 075-02-2020-1478.

References

1. R. Feynman, R. Leighton, M. Sands, *The Feynman Lectures on Physics* (Addison Wesley, Boston, Massachusetts, 2005)
2. O. Fuentes, J. Parra, E. Anthony, V. Kreinovich, Why rectified linear neurons are efficient: a possible theoretical explanations, in *Beyond Traditional Probabilistic Data Processing Techniques: Interval, Fuzzy, etc.* ed. by O. Kosheleva, S. Shary, G. Xiang, R. Zapatrin (Methods and Their Applications, Springer, Cham, Switzerland, 2020), pp. 603–613
3. I. M. Gel'fand, G.E. Shilov, *Generalized Functions* (Academic Press, New York, 1966–1969)
4. I. Goodfellow, Y. Bengio, A. Courville, *Deep Learning* (MIT Press, Cambridge, Massachusetts, 2016)
5. G. Grubb, *Distributions and Operators* (Springer, New York, 2009)
6. R.B. Kearfott, V. Kreinovich, Beyond convex? global optimization is feasible only for convex objective functions: a theorem. J. Glob. Optim. **33**(4), 617–624 (2005)
7. V. Kreinovich, O. Kosheleva, Deep learning (partly) demystified, *Proceedings of the 2020 4th International Conference on Intelligent Systems, Metaheuristics & Swarm Intelligence ISMSI'2020*, Thimpu, Bhutan, 18–19 Apr. 2020
8. V. Kreinovich, O. Kosheleva, Optimization under uncertainty explains empirical success of deep learning heuristics, in *Black Box Optimization*. ed. by P. Pardalos, V. Rasskazova, M.N. Vrahatis (Machine Learning and No-Free Lunch Theorems, Springer, Cham, Switzerland, 2021), pp. 195–220
9. V. Kreinovich, A. Lakeyev, J. Rohn, P. Kahl, *Computational Complexity and Feasibility of Data Processing and Interval Computations* (Kluwer, Dordrecht, 1998)
10. G. Nocedal, S.J. Wright, *Numerical Optimization* (Springer, New York, 2006)
11. R.T. Rockafeller, *Convex Analysis* (Princeton University Press, Princeton, New Jersey, 1997)
12. K.S. Thorne, R.D. Blandford, *Modern Classical Physics: Optics, Fluids, Plasmas, Elasticity, Relativity, and Statistical Physics* (Princeton University Press, Princeton, NJ, 2017)
13. B.S. Tsirel'son, A geometrical approach to maximum likelihood estimation for infinite-dimensional Gaussian location. I. Theory Prob. Appl. **27**, 411–418 (1982)
14. S.A. Vavasis, *Nonlinear Optimization: Complexity Issues* (Oxford University Press, New York, 1991)

How to Work? How to Study? Shall We Cram for the Exams? and How Is This Related to Life on Earth?

Olga Kosheleva, Vladik Kreinovich, and Nguyen Hoang Phuong

Abstract If we follow the same activity for a long time, our productivity decreases. To increase productivity, a natural idea is therefore to switch to a different activity, and then to switch back and resume the current task. On the other hand, after each switch, we need some time to get back to the original productivity. As a result, too frequent switches are also counterproductive. Natural questions are: shall we switch? if yes, when? In this paper, we use a simple model to provide approximate answers to these questions.

1 When to Switch Activities: Formulation of the Problem

Need to switch activities. People get tired when doing the same work for a long time, or studying the same material for a long time. As time goes, their productivity decreases. The best way to restore productivity is to switch to a different activity—or to some relaxation—and then get back to the original activity.

Too many switches are counterproductive too. On the other hand, too many switches decrease productivity as well, since a person needs some time to become productive when switching to a new activity.

There are many examples of such a decrease in productivity. For example, it is a common knowledge that constant interruptions—like immediate replies to emails and/or to phone calls—decrease productivity. Historically, this was one of the reasons why switching from a 6-day work week to a 5-day work week increased productivity without increasing the number of work hours: crudely speaking, the first hour of each

O. Kosheleva · V. Kreinovich (✉)
University of Texas at El Paso, El Paso, TX 79968, USA
e-mail: vladik@utep.edu

O. Kosheleva
e-mail: olgak@utep.edu

N. H. Phuong
Division Informatics, Math-Informatics Faculty, Thang Long University, Nghiem Xuan Yem Road, Hoang Mai District, Hanoi, Vietnam

N. H. Phuong and V. Kreinovich (eds.), *Biomedical and Other Applications of Soft Computing*, Studies in Computational Intelligence 1045,
https://doi.org/10.1007/978-3-031-08580-2_5

work day is not very productive, so the fewer such unproductive hours per week, the better.

This effect drastically varies from one person to another. This effect is different for different people.

Some students cram for the exam by studying for many hours in a row—and do well. Other students try cramming and fail. During a 2-hour-long class, some students urge the instructor for a break after the first hour, since their ability to understand decreases, while others urge to continue, since they do not want to lose the track. Some workers prefer to work through the lunch break and go home earlier, while others need the whole lunch break to restore their productivity.

A recent pandemic, during which people worked from home, showed that people switch to different strategies: some work for 8 h every day, others work for a longer time some days, and relax in some other days.

Problem. It takes some time for people to find their best switching schedule. During this time, their productivity is not the best: they may be switching too rarely getting less productive at the end of each work spurt, or, vice versa, switching too frequently wasting too much time on switching.

It is therefore desirable to help people by providing individualized recommendations on how to switch. Coming up with such recommendations is the main objective of this paper.

2 Let Us Formulate This Problem in Precise Terms

How we get tired. As we start performing some activity, after a short period of adjustment, we reach a reasonable productivity level p_0—the day's maximum productivity level. Let us take the moment of time when we reach this productivity level as the starting point $t = 0$ for measuring time. So, the productivity $p(t)$ at moment $t = 0$ is equal to p_0: $p(0) = p_0$. As we continue performing the same activity, our productivity $p(t)$ decreases, so its derivative $\dot{p}(t)$ is negative. How can we describe this decrease?

The rate $\dot{p}(t)$ at which productivity decreases, in general, depends on the original productivity level: $\dot{p}(t) = f(p(t))$ for some function $f(p)$.

We are not considering extreme cases, when a person works at the limit of his/her abilities—these situations are rare, since it is not possible to maintain such an extreme productivity all the time. Usually, our productivity is much smaller that this maximum amount. Since the usual productivity p is reasonably small, we can expand the dependence $f(p)$ in Taylor series and keep only the few first terms in this expansion. In particular, if we only keep linear terms, we conclude that $f(p) = a_0 + a_1 \cdot p$ for some constants a and b.

When the person is so tired that his/her productivity is close to 0, this productivity will stay at close to 0—there is no room for any further decrease. So, we have $f(0) = 0$, which implies that $a_0 = 0$ and thus, $f(p) = a_1 \cdot p$. Since productivity decreases, we have $f(p) < 0$, i.e., $a_1 < 0$. Thus, $f(p) = -q \cdot p$, where we denoted $q \stackrel{\text{def}}{=} |a_1|$. From the equation $\dot{p}(t) = -q \cdot p(t)$, taking into account that $p(0) = p_0$, we conclude that

$$p(t) = p_0 \cdot \exp(-q \cdot t).$$

This formula is similar to the usual decay formulas—e.g., to the formulas describing the radioactive decay; see, e.g., [1, 3]. The rate of radioactive decay is usually described by *half-life*, the time h at which we are left with the half of the original amount. Similarly, let us gauge our rate of becoming tired by the time h at which our productivity decreases to the half $p_0/2$ of the original amount. This time is related to the value q by the formula $p_0 \cdot \exp(-q \cdot h) = p_0/2$, i.e., $\exp(-q \cdot h) = 1/2$ and thus, $q = \dfrac{\ln(2)}{h}$.

How we recover. Once we switch to a new activity, we need some time to gain the optimal productivity. Let us denote the switch-caused lost time by t_0.

Formulation of the problem. Suppose we plan an activity for which we allocated time T. If we perform it without taking a break, then the overall productivity P during this time can be obtained by integrating the productivity $p(t)$:

$$P = \int_0^T p_0 \cdot \exp(-q \cdot t)\, dt = p_0 \cdot \left. \frac{\exp(-q \cdot t)}{-q} \right|_0^T = p_0 \cdot \frac{1 - \exp(-q \cdot T)}{q}. \quad (1)$$

On the other hand, if we take a break after time T_1, then we lose time t_0 on adjustment, and continue working for time $T - t_0 - T_1$. Our overall productivity us then the sum of the productivities during these two periods of time, and is, thus, equal to

$$p_0 \cdot \frac{1 - \exp(-q \cdot T_1)}{q} + p_0 \cdot \frac{1 - \exp(-q \cdot (T - t_0 - T_1))}{q}. \quad (2)$$

Natural questions:

- When is it beneficial to take a break? Clearly, it is not beneficial if the time T is short, and it is beneficial if T is long, but what is the threshold value T_0 starting from which the break will be beneficial?
- If it is beneficial to take a break, when should we take it? What is the value T_1 that leads to the largest overall productivity?

3 Analysis of the Problem

If a break, when? Let us first find the optimal value T_1. Possible values T_1 comes from the interval $[0, T - t_0]$. According to calculus, the optimal value of T_1 is:

- either attained at one of the endpoints, when either the duration T_1 of the first phase is 0, or the duration of the first phase is $T_1 = T - t_0$, and the duration T_2 of the second phase is $T_2 = T - t_0 - T_1$ is equal to 0,
- or attained inside the interval, when the derivative of the expression (2) with respect to T_1 is equal to 0.

Equating the derivative of the expression (2) to 0, we get

$$p_0 \cdot \exp(-q \cdot T_1) - p_0 \cdot \exp(-q \cdot (T - t_0 - T_1)) = 0,$$

which implies that $T_1 = T - t_0 - T_1$ and thus, that

$$T_1 = T_2 = \frac{T - t_0}{2}. \tag{3}$$

The productivity corresponding to $T_1 = 0$ or $T_2 = 0$ is smaller: indeed, for the first half of the interval of length $T - t_0$, it coincides with what we have for $T_1 = T_2$, and after that:

- in the $T_1 = T_2$ case, we start afresh, with productivity p_0,
- while in the $T_i = 0$ cases, we start with a tired state.

So, the optimal value T_1 is inside the interval, when $T_1 = T_2$.

Thus, if we need a break, we need to make it right in the middle of the activity, so that the work time T_1 before the break is equal to the work time T_2 after the break. In this case, the overall productivity is equal to

$$2 \cdot p_0 \cdot \frac{1 - \exp(-q \cdot (T/2 - t_0/2))}{q}. \tag{4}$$

What if we need several breaks? If we schedule B breaks, then similarly, we can show that the maximal productivity is attained when the corresponding work time intervals $T_1, \ldots, T_{B+1}$ are equal:

$$T_1 = \ldots = T_{B+1} = \frac{T - B \cdot t_0}{B + 1}. \tag{5}$$

In this case, the overall productivity is equal to

$$(B + 1) \cdot p_0 \cdot \frac{1 - \exp(-q \cdot (T/(B + 1) - B \cdot t_0/(B + 1)))}{q}. \tag{6}$$

Do we need a break? And if yes, how many breaks do we need? The overall time of breaks $B \cdot t_0$ cannot exceed the allocated time T, so we only need to consider values B for which $B \cdot t_0 < T$, i.e., values $B < T/t_0$.

To achieve the maximal productivity, we need to select the value

$$B = 0, 1, 2, \ldots, \lfloor T/t_0 \rfloor$$

for which the value (6) is the largest.

All these expressions (6) are proportional to p_0 and inverse proportional to q, so to decide which one if larger, it is sufficient to compare coefficients at p_0/q at these expressions, i.e., the values

$$(B+1) \cdot (1 - \exp(-q \cdot (T/(B+1) - B \cdot t_0/(B+1))). \tag{7}$$

In particular, to decide whether we need a break at all, we need to compare the values corresponding to $B = 0$ (no breaks) and $B = 1$ (one break). We need a break if the value corresponding to $B = 1$ is larger, i.e., if

$$2 \cdot (1 - \exp(-q \cdot (T/2 - t_0/2))) > 1 - \exp(-q \cdot T). \tag{8}$$

If we denote $z \stackrel{\text{def}}{=} \exp(-q \cdot (T/2))$, then this inequality takes the form

$$2 - 2\alpha \cdot z > 1 - z^2, \tag{9}$$

where we denoted $\alpha \stackrel{\text{def}}{=} \exp(q \cdot t_0/2)$, i.e., equivalently, the form $z^2 - 2\alpha \cdot z + 1 > 0$.

This inequality is satisfied if z is:

- either smaller that the smaller α_- of the two roots of the corresponding quadratic equation $z^2 - 2\alpha \cdot z + 1 = 0$,
- or larger than the larger toot α_+.

The roots of this quadratic equation are equal to

$$\alpha_\pm = \alpha \pm \sqrt{\alpha^2 - 1}. \tag{10}$$

Here, $\alpha = \exp(q \cdot t_0/2) > 1$, so $\alpha_+ > 1$, but $z = \exp(-q \cdot T/2) < 1$, so we cannot have $z > \alpha_+$. Thus, the break is needed if z is smaller than the smaller of the two roots, i.e., if

$$\exp(-q \cdot T/2) < \alpha_- = \alpha - \sqrt{\alpha^2 - 1}. \tag{11}$$

The decrease in productivity during the break time t_0 is small, so $\exp(-q \cdot t_0) \approx 1$ and thus, the product $q \cdot t_0$ is small. Thus, we can safely consider only the first few terms in the Taylor expansions when analyzing this formula. Hence,

$$\alpha = \exp(q \cdot t_0/2) \approx 1 + q \cdot t_0/2,$$

$$\alpha^2 - 1 = \exp(q \cdot t_0) - 1 \approx 1 + q \cdot t_0 - 1 = q \cdot t_0,$$

and thus,

$$\alpha_- = \alpha - \sqrt{\alpha^2 - 1} \approx 1 + q \cdot t_0/2 - \sqrt{q \cdot t_0}. \quad (12)$$

Since the product $q \cdot t_0$ is small, its square root is much larger than the value itself. So, in comparison with the square root, the term $q \cdot t_0/2$ can be safely ignored, and we get

$$\alpha_- = \alpha - \sqrt{\alpha^2 - 1} \approx 1 - \sqrt{q \cdot t_0}. \quad (13)$$

So, the inequality (11) takes the form

$$\exp(-q \cdot T/2) < 1 - \sqrt{q \cdot t_0}. \quad (14)$$

Taking the logarithm of both sides and taking into account that for small $q \cdot t_0$, we get

$$\ln(1 - \sqrt{q \cdot t_0}) \approx -\sqrt{q \cdot t_0},$$

we conclude that

$$-q \cdot T/2 < -\sqrt{q \cdot t_0},$$

i.e., equivalently, that

$$T > 2 \cdot \sqrt{\frac{t_0}{q}}.$$

Substituting $q = \ln(2)/h$ into this formula, we conclude that

$$T > \frac{2}{\ln(2)} \cdot \sqrt{t_0 \cdot h}. \quad (15)$$

So, we arrive at the following recommendations.

4 Resulting Recommendations

What is given.

- Let h be the time during which a person's productivity drops to half of its original value;
- let t_0 be the time needed to get to speed when switching to a new activity, and
- let T be the time allocated to a certain activity.

Notations. We will denote $q \stackrel{\text{def}}{=} \ln(2)/h$.

What is the optimal number of breaks. In general, the number of breaks B can be between 0 (no breaks) and the largest possible value T/t_0. The optimal number of breaks B_{opt} is attained when the value (7) is the largest:

$$B_{\text{opt}} = \arg\max_B (B+1) \cdot (1 - \exp(-q \cdot (T/(B+1) - B \cdot t_0/(B+1))). \quad (16)$$

When do we need a break in the first place. In particular, we need a break at all if the time T exceeds the following threshold value:

$$T_0 = \frac{2}{\ln(2)} \cdot \sqrt{t_0 \cdot h}. \quad (17)$$

Here, the ratio $2/\ln(2)$ is approximately equal to 3.

Examples. If the recovery time t_0 is 1 h, and the half-life is $h = 4$ h—half or the usual workday, then we need a break when the overall time is larger than $3 \cdot \sqrt{1 \cdot 4} = 6$ h. This explains why most people need a full lunch break during a usual 8-h working day.

In studying, when the recovery time is $t_0 = 10$ min (typical interval between classes), and $h = 50$ min—a typical class time, then we need a break when the class time is larger than $3 \cdot \sqrt{10 \cdot 50} \approx 70$ min. In effect, we need a break during each class which is longer than normal—definitely we need a break for a 2-h class.

If we need breaks, when do we schedule them? Once we selected the optimal number of breaks B_{opt}, and it is positive—which means that we do need at least one break—then, we need to divide the original task into $B + 1$ smaller parts $T_1, \ldots, T_{B+1}$, the optimal productivity is when we divide the time $T - B \cdot t_0$ (that remains after subtracting the breaks time) into $B + 1$ equal durations

$$T_1 = \ldots = T_{B+1} = \frac{T - B \cdot t_0}{B + 1}. \quad (18)$$

5 How Is This Related to Life on Earth?

In the previous sections, we talked about people needing time to get up to speed when switching to a new activity. However, this phenomenon is generic, it is typical to all the living creatures.

In particular, it turned out that bacteria that produce oxygen need some time to switch to the most productive regime. As a result, when many years ago, the Earth was rotating faster and a day lasted only 6 h, a big proportion of that time was spent on adjusting. When the Earth's rotation slowed down to the current 24-h day, this drastically increased the bacteria productivity, and the resulting drastic increase in the amount of oxygen in the Earth's atmosphere led to a boost of other life forms; see, e.g., [2].

Acknowledgements This work was supported in part by the National Science Foundation grants 1623190 (A Model of Change for Preparing a New Generation for Professional Practice in Computer Science), and HRD-1834620 and HRD-2034030 (CAHSI Includes), and by the AT&T Fellowship in Information Technology. It was also supported by the program of the development of the Scientific-Educational Mathematical Center of Volga Federal District No. 075-02-2020-1478, and by a grant from the Hungarian National Research, Development and Innovation Office (NRDI).

References

1. R. Feynman, R. Leighton, M. Sands, *The Feynman Lectures on Physics* (Addison Wesley, Boston, MA, 2005)
2. J.M. Klatt, A. Chennu, B.K. Arbic, B.A. Biddanda, G.J. Dick, Possible link between Earth's rotation rate and oxygenation. Nat. Geosci. **14**, 564–570 (2021). https://doi.org/10.1038/s41561-021-00784-3
3. K.S. Thorne, R.D. Blandford, *Modern Classical Physics: Optics, Fluids, Plasmas, Elasticity, Relativity, and Statistical Physics* (Princeton University Press, Princeton, NJ, 2017)

Why Quantum Techniques Are a Good First Approximation to Social and Economic Phenomena, and What Next

Olga Kosheleva and Vladik Kreinovich

Abstract Somewhat surprisingly, several formulas of quantum physics—the physics of micro-world—provide a good first approximation to many social phenomena, in particular, to many economic phenomena, phenomena which are very far from micro-physics. In this paper, we provide three possible explanations for this surprising fact. First, we show that several formulas from quantum physics actually provide a good first-approximation description for many phenomena in general, not only to the phenomena of micro-physics. Second, we show that some quantum formulas represent the fastest way to compute nonlinear dependencies and thus, naturally appear when we look for easily computable models; in this aspect, there is a very strong similarity between quantum techniques and neural networks. Third, due to numerous practical applications of micro-phenomena, many problems related to quantum equations have been solved; so, when we use quantum techniques to describe social phenomena, we can utilize the numerous existing solutions—which would not have been the case if we use other nonlinear techniques for which not many solutions are known. All this provides an explanation of why quantum techniques work reasonably well in economics. However, of course, economics is different from quantum world, quantum equations only provide a first approximation to economic situations. In this paper, we use the ideas behind our explanations to speculate on what should be the next—not-exactly-quantum—approximation to social and economic phenomena.

1 Formulation of the Problem

In general, different levels are described by different equations. Many processes in our world occur at different scales:

O. Kosheleva · V. Kreinovich (✉)
University of Texas at El Paso, 500 W. University, El Paso, TX 79968, USA
e-mail: vladik@utep.edu

O. Kosheleva
e-mail: olgak@utep.edu

N. H. Phuong and V. Kreinovich (eds.), *Biomedical and Other Applications of Soft Computing*, Studies in Computational Intelligence 1045,
https://doi.org/10.1007/978-3-031-08580-2_6

- cosmological processes describe what is happening on the level of Universe as a whole,
- astrophysics describes what is happening on the level of Galaxies and stars,
- Earth sciences describe what is happening on the level of a planet,
- macrophysics, biology, and social sciences describes what is happening to our-size ("macro") level, and
- finally, microphysics—mostly quantum physics—describes what is happening on the level of molecules, atoms, and elementary particles.

Of course, there are some similarities between different levels—after all, all these processes obey general laws of physics (see, e.g., [12, 31])—but still, the differences between different levels are usually much larger than these similarities. As a result, different level usually use different techniques, different methodologies.

When people naively try to apply equations and ideas from one level to other levels, they rarely succeed. Naive 19 century attempts to describe electrons orbiting nuclei in a similar way as planets orbiting Earth immediately led to paradoxical conclusions—e.g., that, due to tidal forces (enhanced by electric charges), all electrons will fall on their nuclei after a few seconds. So, quantum physics was invented to avoid this paradox.

Similarly, attempts to naively apply Newtonian physics to the world as a whole led to paradoxes—e.g., if we assume that the stars are uniformly distributed in the Universe, then the overall intensity of light from the stars would be very large, and at night, it would be as light as in daytime. So, General Relativity was invented to avoid such paradoxes.

Even for different aspects of the same level, naive transitions rarely work: e.g., while Darwinism is a great way to describe evolution of species, attempts of Social Darwinism to explain social behavior the same way were not very successful.

But there is an important exception. Interestingly, there is an important case when, unexpectedly, equations that describe phenomena on one level are strangely successful in a completely different level: this is the case of *quantum economics*, a successful application of quantum physics to the description of economic (and, more generally, social) phenomena; see, e.g., [1, 3–7, 15–18, 24–26, 28, 29] (see also [20, 22, 30]).

This success is even more puzzling if we take into account that—as we have mentioned earlier—attempts to use even closer-in-level phenomena like biological were much less successful.

We need a quantitative explanation for this important exception. It is easy to come up with *qualitative* explanations for the success of quantum methods in social studies. For example, when we study a social phenomenon, this study often changes the phenomenon itself: e.g., it is known that the very fact that a patient is visited by a doctor and has some tests done already makes many patients feel better. Of all other phenomena, only quantum processes have the same feature—that a measurement changes the state. So, it is reasonable to expect some analogies between social and quantum phenomena.

However, it is not just qualitative quantum ideas like this which are successful in studying social phenomena, it is also quantitative quantum equations which are successful. Why quantum methods are quantitatively successful in describing economic phenomena is a challenge. In this paper, we provide several explanations for this seemingly strange success—explanations that will hopefully make this success less puzzling.

2 Where Can Such an Explanation Come from: General Analysis

In general, we can distinguish between three aspects of a physical theory:

- First, there is a mathematical aspect: equations that the real-world phenomena satisfy. For Newtonian physics, there are Newton's equations. For quantum physics, there are Schrödinger's equations. For General Relativity, there are Einstein's equations.
- However, equations are not all. To be useful, we must have techniques for solving these equations—and coming up with such techniques is usually as difficult (or even more difficult) than coming up with equations themselves. Newton would not have been very famous if all he did was write a system of differential equations describing how the planets move—and then wait several centuries until computers would appear that could solve this system. Einstein would not have been that famous if all he did was write down a system of complex partial differential equations describing space-time geometry, with no clue on how to solve them and how to compare his predictions with observations. This actually almost happened: David Hilbert, the leading mathematician of his time, independently discovered the same equations—and submitted his paper two weeks after Einstein. If he submitted it two weeks before—would we value Einstein's contribution at all? Actually, yes: all Hilbert did was came us with equations, while Einstein also proposed some solutions—and a way to experimentally test these equations, which in a few years led to a spectacular success.
- Finally, for the theory to become widely used, it is not enough to just have techniques for solving the corresponding equations—this would have limited this theory's use to academe where we have enough researchers and graduate students to apply these techniques and defend their theses and dissertations. We cannot hire a PhD student for every single practical problem. To be practically useful, we need to have a large corpus of already solved problems that practitioners can use. For example, all cell phones take relativistic effects into account when dealing with GPS signals—but the cell phone company does not have to hire a physicist every time a new model of a cell phone is designed—they can use known solutions.

This applies to all physical theories. This applies, in particular, to quantum physics—there are equations, there are techniques for solving these equations, and there are numerous solutions of these equations. We will show that each of the three aspects of quantum physics provides some explanation of why quantum equations can be successfully applied to social phenomena—and, taken together, all three explanations form a reasonably convincing case. So let us consider these aspects one by one.

3 First Explanation: Quantum Formulas Provide a Good Description for Many Phenomena in General

First, let us consider the mathematical aspect of quantum physics—the corresponding mathematical equations. We will show that the mathematical formulas of quantum physics provide a good approximate description for many phenomena—not only phenomena from micro-world and from economics.

Towards a general description of real-life phenomena. In most real-life situations, we have many objects of similar type:

- a galaxy consists of many stars,
- a species consists of many individuals,
- a macro-object consists of many molecules,
- a country or a firm is formed by many people, etc.

Each of these objects is characterized by the values of several quantities. The more quantities we study, the more accurate picture of this object we get.

The number of objects is usually very large, so it is not realistic to keep track of all these objects. A more realistic idea is to keep track of the corresponding distributions:

- what is the proportion of stars of given brightness,
- what is the proportion of employees whose salary is within a given range, etc.

Most practical situations are complex, each quantity is determined by many independent factors. For example, in a big multi-national corporation, a person's salary:

- depends on the person's skills,
- depends on the number of years with the company,
- depends on the geographic location—employees located in more expensive-to-live areas usually get higher salary—etc.

It is known that the distribution of a joint effect of a large number independent factors is close to Gaussian. This follows from the Central Limit Theorem, according to which, when the number of relatively small independent random variables increases, the distribution of their sum tends to Gaussian; see, e.g., [27]. So, we can conclude that the joint distribution of quantities $v_1, \ldots, v_n$ characterizing individual objects is (close to) Gaussian.

Need for an approximation. In general, a multi-D Gaussian distribution is uniquely determined by its first two moments, i.e., by its means $m_i \stackrel{\text{def}}{=} E[v_i]$ and by its covariance matrix $C_{ij} \stackrel{\text{def}}{=} E[(v_i - m_i) \cdot (v_j - m_j)]$. So, to describe the distribution, we need to know n values of the means and $\frac{n \cdot (n+1)}{2}$ values of the symmetric matrix C_{ij}, the total of $V = n + \frac{n \cdot (n+1)}{2}$ parameters. These values need to be determined experimentally, and herein lies a problem.

In general, according to statistics, based on N observations, we can estimate the value of a parameter with relative accuracy $\varepsilon \approx 1/\sqrt{N}$. So, to find the value of a parameter with given relative accuracy $\varepsilon > 0$, we need to perform $N(\varepsilon) \approx \varepsilon^{-2}$ observations. To find the values of V parameters, we therefore need to perform $V \cdot N(\varepsilon) \approx V \cdot \varepsilon^{-2}$. For large n, this becomes too large—e.g., if we are interested in comparing countries, and we want to characterize even $n = 3$ quantities with accuracy $\varepsilon \approx 20\%$, then we need a sample of 225 countries—and there are not that many countries in the world.

To be more precise, means m_i are not a problem, we can determine them, the problem is to determine the elements of the covariance matrix.

This simple argument shows that often, we cannot experimentally determine the actual Gaussian distribution—which depends on too many parameters. We therefore need to find a lower-parametric family of distributions that we will use for an approximate description of the phenomena of interest.

How can we find such an approximation? How can we find a natural, intuitively clear approximation? Most of us do not have a good intuition about probability distributions, but we do have a good intuition about geometric descriptions. Good news is that there is a natural geometric description of a multi-D Gaussian distribution. Namely, it is known that we can represent the components $\Delta v_i = v_i - m_i$ of a multi-D Gaussian distribution with 0 means ($E[\Delta v_i] = 0$) as linear combinations of standard independent Gaussian random variables $\xi_1, \ldots, \xi_n$ for which $E[\xi_k] = 0$, $E[\xi_k^2] = 1$, and $E[\xi_k \cdot \xi_\ell] = 0$ for all $k \neq \ell$: $\Delta v_i = \sum_{k=1}^{n} v_{ik} \cdot \xi_k$. In this representation, the covariance $E[\Delta v_i \cdot \Delta v_j]$ takes the form $E[\Delta v_i \cdot \Delta v_j] = \sum_{k=1}^{n} v_{ik} \cdot v_{jk}$.

This is exactly the formula for the dot (scalar) product of the two n-dimensional vectors. So, we conclude that each difference Δv_i is represented by an n-dimensional vector $\mathbf{v}_i = (v_{i1}, \ldots, v_{in})$, and the covariance is equal to the dot products of these vectors: $E[\Delta v_i \cdot \Delta v_j] = \mathbf{v}_i \cdot \mathbf{v}_j$. In particular, the variance $V[\Delta v_i] \stackrel{\text{def}}{=} E[(\Delta v_i)^2]$ has the form $V[\Delta v_i] = (\mathbf{v}_i)^2 = \|\mathbf{v}_i\|^2$, where $\|\mathbf{a}\|$ denotes the length of the vector $\mathbf{a}$. This provides an exact n-dimensional representation of the situation.

As we have mentioned, we often do not have enough experimental data to determine this exact n-dimensional representation. So, a natural idea is to have a lower-dimensional approximation. This is indeed natural: for example, when we do not have enough data to find a full 3D picture of some object, we often have enough data to determine its 2-D projection. In other words, instead of the original (ideal) multi-D vectors $\mathbf{v}_i$, we use lower-dimensional approximate vectors $\mathbf{V}_i = (V_{i1}, \ldots, V_{id})$, for $d \ll n$, and we try to find the vectors $\mathbf{V}_i$ so that the corresponding dot products are close to the ideal ones: $\mathbf{V}_i \cdot \mathbf{V}_j \approx \mathbf{v}_i \cdot \mathbf{v}_j$.

The simplest such approximation leads, in effect, to a quantum description. The most intuitively clear pictures are 2-D ones, with $d = 2$. In this case, each quantity v_j is represented as by a 2-D vector $\mathbf{V}_i = (V_{i1}, V_{i2})$, and its variance is approximately equal to $\|\mathbf{V}_i\|^2 = V_{i1}^2 + V_{i2}^2$.

How is all this related to complex numbers—one of the main techniques of quantum physics? The relation is straightforward: there is a natural geometric representation of complex numbers, where each number $a + b \cdot \mathrm{i}$ (and $\mathrm{i} \stackrel{\text{def}}{=} \sqrt{-1}$) is represented by a point (a, b) on a plane. In this representation, the absolute value $|a + b \cdot \mathrm{i}|$ of the complex number is equal to the length $\|(a, b)\| = \sqrt{a^2 + b^2}$ of the corresponding vector (a, b).

Thus, each vector $\mathbf{V}_i = (V_{i1}, V_{i2})$ characterising a quantity v_i can be naturally represented as a complex number $V_{i1} + V_{i2} \cdot \mathrm{i}$, and the standard deviation of the quantity v_i is equal to the absolute value of this complex number. So, in this approximation, indeed we naturally get a quantum-like description of general objects.

Comment. This idea was first described in [30]; see also [13].

4 Second Explanation: Quantum Formulas Are the Computationally Fastest Way to Describe Nonlinear Phenomena

Now let us concentrate on the computational, algorithmic aspect of quantum physics—and why namely the computational aspects of quantum physics (and not of any other area of physics) turn out to be helpful in describing social phenomena.

To provide this explanation, let us consider the computational aspects of physics from the most general viewpoint. From the purely theoretical viewpoint, once we have an equation, we can find its solution—worse comes to worse, we can try all possible values, all possible combinations of values, etc. Of course, on the abstract level, there are infinitely many possible values of each physical quantity, so it is not possible to try all infinitely many values—but in practice, since measurements are never absolutely accurate, we cannot distinguish between close values. For example, if our measuring instrument only works on the range from 0 to 10, and measures the value of the corresponding quantity with accuracy 0.1, then we can, in effect, only have values 0, 0.1, 0.2, ..., 9.9, 10—any other value will be indistinguishable from one of these.

Of course, trying all possible combinations is not a practically feasible approach—for many unknowns, it can take an astronomical amount of time, up to the time larger than the lifetime of the Universe. So, from the practical viewpoint, the question is not to find *an* algorithm, the problem is to find a *reasonably fast* algorithm. This is a big challenge. For example, we have learned to predict tomorrow's weather reasonably well—it takes several hours on a high-performance computer. Almost the same algorithms can, in principle, predict where a devastating tornado will turn in the next 15 minutes—but, unfortunately, this prediction would also take several hours, which makes it useless. From this viewpoint, an important question is how to perform computations as fast as possible.

Computations consist of several elementary steps, steps on which we compute some elementary functions. For example, current computers use min, max, addition, and multiplication as such steps, everything else—whether it is computing the inverse $1/a$ or computing the value of the sine function—is implemented as a sequence of these operations.

So, to make computations faster, it is important to select elementary computational steps—i.e., corresponding functions—which are the fastest to compute. Which functions are the fastest to compute? The first idea is that functions are either linear or non-linear. Of course, linear functions are faster to compute than nonlinear ones, so we should use computing linear functions as elementary computational operations.

We cannot limit ourselves to linear functions only: if we only perform linear combinations, then, since the composition of linear function is linear, we will only get linear functions—but many real-life processes are nonlinear. So, some nonlinear elementary operations are needed—but, since we want to speed up overall computations, we should have as few non-linear elements as possible.

So, we end up with a computational scheme in which most operations are linear, but sometimes some nonlinear operations are needed. Which non-linear operations should we use?

To predict the results of deterministic processes, a natural idea is to use some nonlinear functions. So, we end up with a computation scheme in which we interchangingly apply linear transformations and some simple nonlinear ones. This is exactly what neural networks are doing (see, e.g., [14]): at each layer, we:

- first form a linear combination $s = w_0 + w_1 \cdot s_1 + \cdots + w_n \cdot s_n$ of signals coming from the previous layer, and
- then, we apply some nonlinear function $F(x)$ (known as *activation function*) to this result s, returning the value $F(s)$.

Many processes, however, are nondeterministic, in the sense that, based on the available information, we can only predict the results with some probabilities. In this case, it makes sense to have non-deterministic nonlinear components, that return different results with different probabilities. This is exactly what is happening according to the quantum physics, where interchanhingly, we have

- linear transformations—which correspond to normal dynamics as described by Schrödinger's equation—and
- measurement process—in which we get different results with different probabilities.

From this viewpoint, a quantum-type description is a natural way to describe nondeterministic phenomena—and social phenomena are, of course, largely nondeterministic—in the most computationally efficient way.

So, from the computational viewpoint, we also naturally arrive at a quantum-type description of social phenomena.

Comments.

- The main idea behind this explanation was first described in [19].
- In this section, we explained why some probabilistic quantum-like transformations are needed. In principle, we can have different probabilistic transformations. In [2], we explain why the transformations used in quantum physics are the most appropriate in this case.

5 Third Explanation: Quantum Physics Has Many Solved Problems

Finally, let us take into account that in quantum physics, we have a large corpus of solved complex problems.

Why is this important? It is known that all classes of sufficiently complex problems (they are known as *NP-hard*) can be reduced to each other; see, e.g., [21–23]. Thus, once we have come up with an efficient method of solving complex problems from one application area, we can reduce problems from other areas to these problems and thus, get a good algorithm for solving problems from other areas as well.

So, a natural way to solve complex problems in economics and finance is to reduce them to complex problems in other application areas—problems for which solutions are mostly known. One such area is physics—analysis of the physical world. Its equations are often very complex, and still, during several centuries of physics, researchers have found reasonably efficient algorithms for solving many of these equations. Thus, to solve complex problems in economics and finance, it makes sense to reduce them to solvable complex problems from physics.

It is known that to get an adequate description of a physical phenomenon, it is necessary to take quantum effects into account, i.e., to get into the domain of quantum physics; see, e.g., [12, 31]. Because of this, the most complex physical equations are equations of quantum physics. Thus, a natural idea is to reduce complex equations arising in economics and finance to complex quantum equations—namely, to complex quantum equations for solving which we have efficient algorithms. Then, by solving the corresponding quantum equation and translating the solutions back into the language of economics and finance, we can get efficient algorithms for solving complex economic and financial problems.

This is exactly what quantum econometrics is doing! So, we have yet another natural explanation of the empirical success of quantum techniques in economics and in social sciences in general.

Comment. This idea was first described in [22]; see also [20].

6 Beyond Quantum

In the previous sections, we explained why quantum techniques work reasonably well in economics. Of course, economics is different from quantum world. So quantum equations provide only a first approximation to economic situations. A natural question is: how can we get an even more accurate approximation—i.e., how can we modify quantum equations so that they will provide an even more adequate description of social phenomena?

In this section, we show that, based on the above explanations, we can come with some recommendations about these modifications. Indeed, in Sect. 3, we show that a quantum-style complex-valued description of different objects naturally appears if, as an approximation of the general geometric description—with n-dimensional vectors—we consider 2-D vectors. Of course, this is just an approximation. To make this approximation more accurate, the next natural way is to approximate the objects by 3-D, 4-D, etc. vectors.

We hope that this natural generalization of quantum-style techniques will, in particular, lead to a more accurate representation of social and economic phenomena.

Acknowledgements This work was supported in part by the National Science Foundation grants 1623190 (A Model of Change for Preparing a New Generation for Professional Practice in Computer Science), and HRD-1834620 and HRD-2034030 (CAHSI Includes), and by the AT&T Fellowship in Information Technology. It was also supported by the program of the development of the Scientific-Educational Mathematical Center of Volga Federal District No. 075-02-2020-1478. The authors are greatly thankful to Dr. Emmanuel Haven for his encouragement.

References

1. L. Accardi, A. Boukas, The quantum Black-Scholes equation. Glob. J. Pure Appl. Math. **2**(2), 155–170 (2007)
2. K. Autchariyapanitkul, O. Kosheleva, V. Kreinovich, S. Sriboonchitta, Quantum econometrics: how to explain its quantitative successes and how the resulting formulas are related to scale invariance, entropy, and fuzziness, in ed. by V.-N. Huynh, M. Inuiguchi, D.-H. Tran, T. Denoeux, *Proceedings of the International Symposium on Integrated Uncertainty in Knowledge Modelling and Decision Making IUKM'2018*, Hanoi, Vietnam, 13–15 Mar. 2018
3. B. E. Baaquie, Quantum field theory of forward rates with stochastic volatility. Phys. Rev. E **65**, Paper 056122 (2002)
4. B.E. Baaquie, *Quantum Finance: Path Integrals and Hamiltonians for Options and Interest Rates* (Camridge University Press, New York, 2004)
5. B.E. Baaquie, S. Marakani, Finite hedging in field theory models of interest rates. Phys. Rev. E **69**, Paper 036130 (2004)
6. B.E. Baaquie, M. Srikant, M. Warachka, A quantum field theory term structure model applied to hedging. Int. J. Theor. Appl. Financ. **6**(5), Paper 443 (2003)
7. J.R. Busenmeyer, P.D. Bruza, *Quantum Models of Cognition and Decision* (Cambridge University Press, New York, 2012)
8. O. Choustova, Bohmian mechanics for financial process. J. Mod. Opt **51**, Paper 1111 (2004)
9. O. Choustova, Quantum Bohmian model for financial markets. Physica A **347**, 304–314 (2006)

10. O. Choustova, Quantum model for the price dynamics: the problem of smoothness of trajectories. J. Math. Anal. Appl. **346**, 296–304 (2008)
11. O. Choustova, Quantum probability and financial market. Inf. Sci. **179**, 478–484 (2009)
12. R. Feynman, R. Leighton, M. Sands, *The Feynman Lectures on Physics* (Addison Wesley, Boston, MA, 2005)
13. O. Galindo, M. Svitek, V. Kreinovich, Quantum (and more general) models of research collaboration. Asian J. Econ. Bank. **4**(1), 77–86 (2020)
14. I. Goodfellow, Y. Bengio, A. Courville, *Deep Leaning* (MIT Press, Cambridge, Massachusetts, 2016)
15. E. Haven, A discussion on embedding the Black-Scholes option pricing model in a quantum physics setting. Physica A **304**, 507–524 (2002)
16. E. Haven, A Black-Scholes Shrodinger option price: 'bit' versus 'qubit'. Physica A **324**, 201–206 (2003)
17. E. Haven, Pilot-wave theory and financial option pricing. Int. J. Theor. Phys. **44**(11), 1957–1962 (2005)
18. E. Haven, A. Khrennikov, *Quantum Social Science* (Cambridge University Press, Cambridge, UK, 2013)
19. V. Kreinovich, O. Kosheleva, A. Ortiz-Muñoz, Need for simplicity and everything is a matter of degree: how Zadeh's philosophy is related to kolmogorov complexity, quantum physics, and deep learning, in *Recent Developments in Fuzzy Logic, Fuzzy Sets, and Their Applications*, ed. by S. Shahbazova, J. Kacprzyk, V. Kreinovich (Springer, Cham, Switzerland), to appear
20. V. Kreinovich, O. Kosheleva, S. Sriboonchitta, Why Bohmian approach to quantum econometrics: an algebraic explanation, in *Algebraic Techniques and Their Use in Describing and Processing Uncertainty*, ed. by H.T. Nguyen, V. Kreinovich (Springer, Cham, Switzerland, 2020), pp. 65–75
21. V. Kreinovich, A. Lakeyev, J. Rohn, P. Kahl, *Computational Complexity and Feasibility of Data Processing and Interval Computations* (Kluwer, Dordrecht, 1998)
22. V. Kreinovich, H. T. Nguyen, S. Sriboonchitta, Quantum ideas in economics beyond quantum econometrics, in *Econometrics for Financial Applications*, ed. by H. Ly, L.S. Dong, V. Kreinovich, N.N. Thach (Springer, Cham, Switzerland, 2018), pp. 146–151
23. C.H. Papadimitriou, *Computational Complexity* (Pearson, Boston, MA, 1994)
24. P. Pedram, The minimal length uncertainty and the quantum model for the stock market. Physica A **391**, 2100–2105 (2012)
25. R. Penrose, *The Emperor's New Mind: Concerning Computers* (Oxford University Press, Oxford, UK, Minds and The Laws of Physics, 1989)
26. W. Seagal, I.E. Seagal, The Black-Scholes pricing formula in the quantum context. Proc. Nat. Acad. Sci. USA **95**, 4072–4075 (1998)
27. D.J. Sheskin, *Handbook of Parametric and Nonparametric Statistical Procedures* (Chapman and Hall/CRC, Boca Raton, FL, 2011)
28. M. Svítek, *Quantum System Theory: Principles and Applications* (VDM Verlag, Saarbrucken, Germany, 2010)
29. M. Svítek, Towards complex system theory. Neural Networl World **15**(1), 5–33 (2015)
30. M. Svitek, O. Kosheleva, V. Kreinovich, T.N. Nguyen, Why quantum (wave probability) models are a good description of many non-quantum complex systems, and how to go beyond quantum models, in *Beyond Traditional Probabilistic Methods in Economics*, ed. by V. Kreinovich, N.D. Trung, N.N. Thach (Springer, Cham, Switzerland, 2019), pp. 168–175
31. K.S. Thorne, R.D. Blandford, *Modern Classical Physics: Optics, Fluids, Plasmas, Elasticity, Relativity, and Statistical Physics* (Princeton University Press, Princeton, NJ, 2017)

How the Pavement's Lifetime Depends on the Stress Level and on the Dry Density: An Explanation of Empirical Formulas

Edgar Daniel Rodriguez Velasquez, Vladik Kreinovich, Olga Kosheleva, and Nguyen Hoang Phuong

Abstract We show that natural invariance ideas explain the empirical dependence on the pavement's lifetime on the stress level and on the dry density.

1 First Problem: Dependence on Stress

General description of the phenomenon. Road pavements have a limited lifetime. As the vehicles pass over the pavement, eventually, the pavement develops fatigue cracking and needs to be repaired.

Clearly, the larger the load, the larger the stress σ, the larger the strain ε, and thus, the smaller the number of repetitions N before the fatigue cracking.

Empirical fact. To estimate the expected lifetime of the pavement, we need to know how the number of repetitions N to fatigue cracking depends on the stress σ, i.e., we need to know the following dependence:

$$N = f(\sigma). \tag{1}$$

E. D. R. Velasquez
Department of Civil Engineering, Universidad de Piura in Peru (UDEP), Av. Ramón Mugica 131, Piura, Peru
e-mail: edgar.rodriguez@udep.pe; edrodriguezvelasquez@miners.utep.edu

Department of Civil Engineering, University of Texas at El Paso, 500 W. University, El Paso, TX 79968, USA

V. Kreinovich (✉) · O. Kosheleva
University of Texas at El Paso, El Paso, TX 79968, USA
e-mail: vladik@utep.edu

O. Kosheleva
e-mail: olgak@utep.edu

N. H. Phuong
Division Informatics, Math-Informatics Faculty, Thang Long University, Nghiem Xuan Yem Road, Hoang Mai District, Hanoi, Vietnam
e-mail: nhphuong2008@gmail.com

N. H. Phuong and V. Kreinovich (eds.), *Biomedical and Other Applications of Soft Computing*, Studies in Computational Intelligence 1045,
https://doi.org/10.1007/978-3-031-08580-2_7

Empirically, this dependence is described by the following formula; see, e.g., [2, 3, 5]:

$$N = N_0 \cdot \exp(-k \cdot \sigma). \tag{2}$$

This formula applies both to the cases when we analyze how N depends on the stress measured at the top layer of the pavement and on the stress measured at the bottom layer of the pavement.

Natural question. How can we explain this empirical formula?

2 Analysis of the Problem

There are several components of stress. In both top-of-pavement and bottom-of-pavement situations, when we apply the formula (1), we measure the overall stress, and implicitly assume that all this stress is caused by the traffic. In reality, in addition to the traffic-related stress, there are other stresses that also contribute to the eventual deterioration of the pavement. Indeed, even an unused road eventually develops cracks, due, e.g., to weather-induced stress.

The empirical formula relates only to the traffic-related stress. The additional stress—e.g., weather-related stress—follows different cyclic patterns than the traffic-related stress: typically a yearly pattern, or—in case of the rain—a pattern corresponding to each instance of rain.

The number of repetitions of these additional stresses is therefore much smaller than the number of repetitions caused by the traffic passing over the pavement. Thus, in comparison with the number of traffic-related repetitions, the number of weather-related repetitions can be safely ignored.

Strictly speaking, only the traffic-related part σ_t of the stress—the part that repeats many times per day—affects the number of repetitions N. In other words, strictly speaking, we should look for a formula

$$N = f(\sigma_t). \tag{3}$$

How can we estimate the traffic-related part of the stress. All we measure is the overall stress σ, which is equal to the sum $\sigma_t + \sigma_o$ of the traffic-related stress σ_t and the stress σ_o caused by other factors. So, to estimate the traffic-related stress σ_t, we need to subtract, from the measured stress σ, our estimate of the remaining stress σ_o.

Resulting uncertainty in estimating the traffic-related stress. The problem is that this remaining stress can only be estimated rather approximately. If instead of the original estimate σ_0 we used a slightly different estimate σ_o', then, instead of the original estimate $\sigma_t = \sigma - \sigma_o$, we get a somewhat different estimate

$$\sigma_t' = \sigma - \sigma_0' = (\sigma - \sigma_o) + (\sigma_o - \sigma_o') = \sigma_t + \delta,$$

where we denoted $\delta \stackrel{\text{def}}{=} \sigma_o - \sigma_o'$.

As a result, the exact same situation can be described by two somewhat different values σ_t and $\sigma_t' = \sigma_t + \delta$.

3 Invariance Requirement

Natural invariance idea. Since the exact same situation can be described by two somewhat different values σ_t and $\sigma_t = \sigma_t' + \delta$, it is therefore reasonable to require that these two somewhat different values lead to the exact same predictions of the pavement lifetime.

Naive approach to invariance does not work. Of course, we cannot interpret this requitement literally, as saying that we should have $f(\sigma_t) = f(\sigma_t + \delta)$: if we impose this requirement for all σ_t and all δ, this would lead to a physically meaningless conclusion that the function $f(\sigma_t)$ is a constant, i.e., that the pavement's lifetime does not depend on the stress at all.

Solution: using experience of physics. However, the experience of physics shows that we do not need to take this requirement literally; see, e.g., [4, 6]. For example, we know that the formula $v = d/t$ describing the velocity v as a function of distance d and time t does not depend on what measuring unit we select for distance: we can describe the distance d in meters, or we can describe it in centimeters, resulting in a different numerical value $d' = 100 \cdot d$.

In this example, invariance does not mean that we always have $d/t = d'/t$: the formula $v = d/t$ does remain valid if we use a different unit for measuring distance, but for this formula to remain valid, we also need to correspondingly change the unit that we use for measuring velocity: in the above example, from m/sec to cm/sec. In this case, we get $v' = d'/t$, where v' is the numeric value of velocity in the new units.

From this viewpoint, a reasonable idea is to require that for $\sigma_t' = \sigma_t + \delta$, the dependence $N = f(\sigma_t)$ should lead to $N' = f(\sigma_t')$, where N' is the description of the pavement lifetime in correspondingly different units.

But can we use the experience of physics in our case? At first glance, the physics-motivated idea does not seem to work in our case:

- in contrast to quantities like distance or velocity, where we do need to select a measuring unit to get a numerical value,
- the number of repetitions N is simply an integer, no measuring is required.

Yes, we can. However, a more detailed analysis shows that the situation is not that uniquely determined. Indeed, usually, when the vehicle goes over a pavement location:

- we can count it as a single stress cycle,
- or we can consider the situation more accurately and take into account that every time each wheel is passing over, it is a different cycle.

Thus, depending on how we count it, what was a single cycle in one counting becomes several cycles if we consider it differently.

In mathematical terms, this means that we can describe the same traffic history:

- by the number N and
- by a different number $N' = c \cdot N$, where c is the average number of axles per vehicle.

Different estimates of the number c can lead, in general, to different re-scalings of N'.

Resulting formulation of the invariance requirement. From this viewpoint, the above requirement that the dependence $N = f(\sigma_t)$ not change if we change our estimate for σ_o takes the following form.

For every real number δ,
there exists an appropriate value $c(\delta)$—which depends on δ—
such that if $N = f(\delta_t)$, then we should have $N' = f(\sigma_t')$,
where $\sigma_t' = \sigma_t + \delta$ and $N' = c(\delta) \cdot N$.

4 Resulting Explanation

Reduction to a functional equation. Substituting the expressions $\sigma_t' = \sigma_t + \delta$ and $N' = c(\delta) \cdot N$ into the formula $N' = f(\sigma_t')$, we conclude that $c(\delta) \cdot N = f(\sigma_t + \delta)$. Since $N = f(\sigma_t)$, we thus conclude that

$$c(\delta) \cdot f(\sigma_t) = f(\sigma_t + \delta). \tag{3}$$

Resulting explanation. It is known—see, e.g., [1]—that every measurable solution to the functional equation (3) has the form $f(\sigma) = N_0 \cdot \exp(-k \cdot \sigma)$. Thus, we have indeed explained the empirical formula (2).

5 Second Problem: Dependence on Dry Density

Empirical formula. The formula (2) describes how the pavement lifetime depends on the stress provided that all other parameters remain constant. The corresponding values N and k depend on the dry density ρ of the underlying soil. This dry density is called *maximum dry density* since when the pavement is built, we try to maximize the dry density of the compacted soil. The dependence of the lifetime N on stress and dry density is described by the following formula [5]:

$$\ln(N) = k_4 \cdot \ln\left(\frac{\rho}{\omega}\right) \cdot \left(1 - \frac{\sigma}{k_5 \cdot UCS}\right), \tag{4}$$

for some parameters k_4, ω, k_5, and UCS, i.e., equivalently, by the formula:

$$\ln(N) = a_{00} + a_{01} \cdot \sigma + a_{10} \cdot \ln(\rho) + a_{11} \cdot \sigma \cdot \ln(\rho), \tag{5}$$

for some values a_{00}, a_{01}, a_{10}, and a_{11}.

How can we explain this empirical formula?

What we know already. We know, from the previous text, that for each fixed value ρ, the dependence of N on σ has the form (2), i.e., equivalently, that $\ln(N)$ is a linear function of σ:

$$\ln(N) = \ln(N_0) - k \cdot \sigma. \tag{6}$$

Let us analyze how N depends on the dry density ρ.

Corresponding invariance. We can use different units to measure dry density. If we replace the original unit with a new unit which is λ times smaller, then all numerical values get multiplied by λ, i.e., instead of each original numerical value ρ, we get a new numerical value $\rho' = \lambda \cdot \rho$ to describe the exact same physical situation.

There is no reason to prefer one specific measuring unit. It is therefore reasonable to require that the dependence of the lifetime N on the density ρ be the same, no matter what measuring unit we use. Similarly to the case when we analyzed the dependence of N on σ, it is reasonable to formulate this requirement in the following precise terms:

For every real number $\lambda > 0$,
there exists an appropriate value $c(\lambda)$—which depends on λ—
such that if $N = f(\rho)$, then we should have $N' = f(\rho')$,
where $\rho' = \lambda \cdot \rho$ and $N' = c(\lambda) \cdot N$.

Reduction to a functional equation. Substituting the expressions $\rho' = \lambda \cdot \rho$ and $N' = c(\lambda) \cdot N$ into the formula $N' = f(\rho')$, we conclude that $c(\lambda) \cdot N = f(\lambda \cdot \rho)$. Since $N = f(\rho)$, we thus conclude that

$$c(\lambda) \cdot f(\rho) = f(\lambda \cdot \rho). \tag{7}$$

Resulting formula. It is known—see, e.g., [1]—that every measurable solution to the functional equation (7) has the form $f(\rho) = N_0 \cdot \rho^a$ for some values N_0 and a. Thus, $\ln(N) = a \cdot \ln(\rho) + \ln(N_0)$, i.e., $\ln(N)$ is a linear function of ρ. So:

- for each value ρ, the logarithm $\ln(N)$ of the lifetime $N = f(\sigma, \rho)$ is a linear function of σ, and

- for each value σ, the logarithm $\ln(N)$ is a linear function of $\ln(\rho)$.

Thus, the function $\ln(N)$ is a bilinear function of σ and $\ln(\rho)$, i.e., has the desired form (5). So, the empirical dependence (5) is also explained.

Acknowledgements This work was supported in part by the National Science Foundation grants 1623190 (A Model of Change for Preparing a New Generation for Professional Practice in Computer Science), and HRD-1834620 and HRD-2034030 (CAHSI Includes), and by the AT&T Fellowship in Information Technology. It was also supported by the program of the development of the Scientific-Educational Mathematical Center of Volga Federal District No. 075-02-2020-1478, and by a grant from the Hungarian National Research, Development and Innovation Office (NRDI).

References

1. J. Aczél, J. Dhombres, *Functional Equations in Several Variables* (Cambridge University Press, 2008)
2. Applied Research Associates, Inc., *National Cooperative Highway Research Program (NCHRP) Guide for MechanisticEmpirical Design for New and Rehabilitated Pavement Structures. Final Report, NCHRP 1-37A* (Transportation Research Board, National Research Council, 2004)
3. M. De Beer, *Aspects of the Design and Behaviour of Road Structures Incorporating Lightly Cementitious Layers*, Ph.D. Dissertation, Faculty of Engineering, University of Pretoria, Pretoria, South Africa (1990)
4. R. Feynman, R. Leighton, M. Sands, *The Feynman Lectures on Physics* (Addison Wesley, Boston, Massachusetts, 2005)
5. X. Li, J. Wang, H. Wen. and B. Muhunthan, Field calibration of fatigue models of cementitiously stabilized pavement materials for use in the mechanistic-empirical pavement design guide. Transp. Res. Rec. (2019). https://doi.org/10.1177/0361198118821924
6. K.S. Thorne, R.D. Blandford, *Modern Classical Physics: Optics, Fluids, Plasmas, Elasticity, Relativity, and Statistical Physics* (Princeton University Press, Princeton, New Jersey, 2017)

Freedom of Will, Physics, and Human Intelligence: An Idea

Miroslav Svítek, Vladik Kreinovich, and Nguyen Hoang Phuong

Abstract Among the main fundamental challenges related to physics and human intelligence are: How can we reconcile the free will with the deterministic character of physical equations? What is the physical meaning of extra spatial dimensions needed to make quantum physics consistent? and Why are we often smarter than brain-simulating neural networks? In this paper, we show that while each of these challenges is difficult to resolve on its own, it may be possible to resolve all three of them if we consider them together. The proposed possible solution is that human reasoning uses the extra spatial dimensions. This may sound weird, but in this paper, we explain that this solution is much more natural than how it sounds at first glance.

1 Three Fundamental Challenges

What is science about? Science studies the world.

Most of the studies cover different objects. Generally speaking, this is what physics is about—be it physics proper, chemistry (and chemical physics), biology (and biophysics), geosciences (and geophysics), astronomy (and astrophysics), etc.

We humans also study ourselves, not just our own biology, but also the way we think and reason.

M. Svítek
Faculty of Transportation Sciences, Czech Technical University in Prague, Konviktska 20, CZ-110 00 Prague 1, Prague, Czech Republic
e-mail: svitek@fd.cvut.cz

V. Kreinovich (✉)
Department of Computer Science, University of Texas at El Paso, 500 W. University, El Paso 79968, USA
e-mail: vladik@utep.edu

N. H. Phuong
Division Informatics, Math-Informatics Faculty, Thang Long University, Nghiem Xuan Yem Road, Hoang Mai District, Hanoi, Vietnam

N. H. Phuong and V. Kreinovich (eds.), *Biomedical and Other Applications of Soft Computing*, Studies in Computational Intelligence 1045,
https://doi.org/10.1007/978-3-031-08580-2_8

Fundamental challenges. Science is an evolving body, it has many open problems, many challenges. Every year, new discoveries are made, new challenges appear: how to describe a new phenomenon? how exactly will different observed phenomena evolve? These are all serious challenges, and Nobel prizes are awarded to those who solve them.

In addition to such "usual" scientific challenges, there are also fundamental challenges, challenges that deal not with our ability or inability to predict new phenomena, but rather with some perceived fundamental flaws of the current theories. Let us describe the main such challenges: challenges of physics, challenges arising from the relation between physics and study of human reasoning, and challenges related to the study of human reasoning itself.

Fundamental challenges in physics. To describe these challenges, let us start with physics proper. Let us recall why many physicists are not always completely happy with their theories – even when their theories provide a perfect prediction of the corresponding phenomena.

To a naive mathematician, a physical theory may be nothing else but the corresponding set of equations. However, physicists understand that physics is much more than the set of equations: it involved deep understanding of the related phenomena, an understanding that helps physicists make predictions. A classical example of the need to have something beyond equation is the history of General Relativity. It is known that the great mathematician David Hilbert came up with the same equations as Einstein—he submitted his paper two weeks later than Einstein, so Einstein was first. This is a known historical fact. What may be less known is that even if Hilbert was first, Einstein would still deserve his fame. Indeed:

- All Hilbert did was derive the equations—the system of complex non-linear partial differential equations. Hilbert did not provide any solutions to these equations – they were too complex to solve by known mathematical methods.
- In contrast, Einstein not only proposed the equations, he also provided some approximate solutions, solutions that were experimentally tested in a few years.

Einstein came up with these solutions not because he was a better mathematician, no, he came up with these solutions because he had a good physical intuition that enabled him to understand which terms in the original nonlinear system could be safely ignored—and this led to a solvable approximation.

Another Einstein-related example (borrowed from [13]) is that when a guide who showed, to Einstein's wife, a state-of-the-art computer, explained that this computer was busy finding the structure of the Universe as a whole—she replied that her husband did the related calculations on the back of an envelope.

From this viewpoint, it is not enough to come up with mathematical formulas for prediction, we also need to gain a good intuitive understanding of the corresponding phenomena, an intuition that would enable us to provide experimental confirmations of the resulting theory. Whenever such an understanding is missing, we have a fundamental challenge. And in theoretical physics, there is indeed such a challenge.

Specific challenge. This challenge is related to the fact that while current equations of quantum field theory predict many future events with very high accuracy, there is still an important challenge – if we write down seemingly natural equations of quantum field theory in our usual 4-dimensional space-time, then, for some future values, we get meaningless infinite values. In many cases, physicists came up with several tricks to avoid such infinities. A fundamental solution to this problem was found by string theory, according to which a consistent quantum theory is only possible in space-times of dimension 11 and higher; see, e.g., [11, 22, 26].

This is an exciting and useful mathematical result, but from the physical viewpoint, this solution is not fully satisfactory, since:

- in contract to the usual 4 dimensions that have precise physical meaning,
- the additional 7 or more dimensions largely remain purely mathematical constructions.

What is the physical meaning of these extra dimensions is not clear.

Another path to extra dimensions comes from foundations of quantum physics. One of the main interpretations of the stochastic character of quantum predictions is that there are actually many possible worlds, and when we perform an experiment, we branch—with appropriate probabilities—into one of these possible worlds. From the fundamental viewpoint, it is the same idea as extra dimensions—namely, the world we observe is a tiny part of the larger world. In this case, additional dimensions make some physical sense—they correspond to possible worlds.

However, in this case too, all we observe is *one* possible world, a tiny portion of the whole multi-world universe. And physicists understandably do not like theories that heavily rely on unobservable quantities, this sounds too much like speculations rather than physics.

Researchers have been trying to find some physical meaning of the extra dimensions. For example, the Nobelist Andrei Sakharov—best known for his human rights activities—conjectured, in [23], that, in contrast to usual many-world interpretation in which possible worlds do not interact, there is actual interaction between possible worlds, and that, e.g., strong interaction between protons and neutrons can be explained in this way. This idea sounded—and still sounds—promising, it was even cited in a textbook on gravitation [16], but no experimental confirmation of such interaction was found.

Fundamental challenges about the relation between physics and study of human reasoning. Modern physics is largely deterministic. According to physics' equations, once we know the initial state of the Universe—or of any closed system—this initial state uniquely determines the future state at all future moments of time.

To be more precise, before quantum physics, the world's description was fully deterministic in this sense. Quantum physics added an additional level of uncertainty: namely, according to quantum physics, the state does not uniquely predict results of future measurements, but it does uniquely predict the probabilities of different future measurement results; see, e.g., [7, 26].

Many equations of physics have been confirmed on multiple examples with high accuracy. However, their deterministic nature contradicts our intuitive understanding that we have freedom of will, that we can make decisions and select different alternatives—thus changing the future state of our environment (and therefore the state of the Universe); see, e.g., see, e.g., [2–5, 8, 9, 12, 14, 21, 24, 25, 27–30] and references therein.

Some people are so much convinced by physics that they start really believing that freedom of will is an illusion [1]—for example, Einstein, who could not swim, nevertheless liked to go yachting on his own, arguing that whether he drowns or not is pre-determined by the initial state of the Universe and does not depend on his decisions; see, e.g., [13]. To most people, however, this sounds weird, they believe that we can change the state of the universe.

But how can we reconcile this belief with physics? The world consists of elementary particles. So, the fact that we change the state of the Universe means that we can somehow, by our sheer will, change the state of some of these particles. Many experiments tried to see if such "telekinesis"—using willpower to change the state of the material objects—is indeed possible, but no such ability was found. No matter how much people tried, they were never able to affect the results of physical experiments.

Challenges related to the study of human reasoning. We have come a long way towards understanding how we reason and how we think. We know, in many detail, how signals are processes in our brains: the signals go from a neuron to a neuron, and these signals are processed by these neurons in a (largely) known way.

This knowledge has led to development of artificial devices that simulate this activity—artificial neural networks. Current deep neural networks indeed have spectacular successes; see, e.g., [10].

However, in many aspects, these networks are still not as smart as we are—and not as smart as we have hoped them to be. For example, while deep networks perform very well on tasks with which previous computer programs had trouble, such as distinguishing between cats and dogs, it is known that a small change in the input—a change that would not affect human decisions—can lead a neural network to a completely erroneous classification; see, e.g., [10].

The fact that the drastic increase in the number of neurons does not make the artificial neural networks as close to our intelligence abilities as we hoped—this fact has been recognized by many researchers, many of whom speculated that the human brain uses physical phenomena beyond the usual processing of electric signals. For example, Roger Penrose famously conjectured, in [20], that the human brain is actually involved in quantum computing—and it is known that quantum computing can indeed drastically increase the system's computational abilities; see, e.g., [18]. This was a very interesting and very promising hypothesis—but no evidence of such quantum activity was found, so this hypothesis was abandoned.

As a result, the challenge remains: why are we so smart—to be more precise, why are our brains much smarter than similar-size artificial neural networks that try their best to emulate how our brains work?

Let us summarize. In modern science, in addition to the "usual" challenges—how to predict future phenomena, how to explain the observed phenomena—there are several fundamental challenges, and we have mentioned three such challenges:

- What is the physical meaning of extra spatial dimensions?
- How to reconcile freedom of will and physics?
- How to explain that we are often much smarter than computer models that seem to adequate simulate our brains?

What we do in this paper. In this paper, we show that all these three challenges can be resolved if we view them together.

2 How We Can Solve These Challenges: An Idea

Let us start with the first challenge. Let us start our analysis with the very first challenge: how to explain the physical meaning of additional spatial dimensions? What are the physical (= observable) consequences of the existence of possible worlds—worlds that exist in parallel to our world?

Natural idea. To answer this physical challenge, let us take the second challenge into account—the existence of free will. Intuitively, free will means that which possible world we get into is not just randomly determined—*we* can select to which of the possible worlds we move. In other words, free will means that we can move in the space of possible worlds—i.e., that we can move in the extra spatial dimensions corresponding to these possible worlds.

If we assume that we humans can indeed move in these extra dimensions, this provides the desired physical meaning to these dimensions. Namely, there are objects that function in these extra dimensions and that move in these extra dimensions—and we ourselves are these objects.

This also helps with the second challenge. This idea also explain why, in spite of the fact that our will cannot change anything in the 3-D world, we can (and do) make changes by our choices: while we *cannot* change anything in the usual 3-D world, what we *can* do is move between possible worlds—in other words, our free will can select in which of the possible worlds we will be in the future.

This also helps with the third challenge. Let us show that the possibility to move in extra dimensions can also provide a solution to the third challenge—namely, as we will show, the possibility to move in additional spatial dimensions increases our computational abilities—i.e., in particular, our abilities to process information and to reason.

But how do additional dimensions contribute to computational abilities? To answer this question, let us recall that in computer science, the complexity of an algorithm is usually measured by the time T_{seq} needed to perform this algorithm step-by-step on a sequential computer. For many problems, this is too long—for example, if it takes a week to predict tomorrow's weather, then this prediction algorithm is useless, since we will observe tomorrow's weather way before the algorithm finishes its computations. In such situations, it is desirable to speed up computations.

How can we speed up computations? If a human being has a time-consuming task—e.g., cleaning all the offices in a building before an inspection—and it would take too long for one person to clean all the rooms, a natural idea is to have several people cleaning rooms at the same time. Similarly, if some computational task takes too long a time for a single processor, a reasonable idea is to have several processors working in parallel. At first glance, it may seem that the more processors we add, the faster the computation can be. However, in reality, there is a limit to the possible speedup—a limit caused by the fact that, according to modern physics, all the speeds are limited by the speed of light c. Let us show—following [17]—how this limits the computation speed.

Suppose that we have a computational device in which several processes work in parallel, and that this device finishes its computations in time $T_{\rm par}$. In other words, $T_{\rm par}$ is the time between the moment when we present the task to this device and the moment when the device presents us the result. During this time, if any processor is involved in this computation, some signal must reach this processor, and some results from this processor must eventually find their way back to us. So, if this processor is located at some distance r from us, the signal must travel at least a distance $2r$. During the time $T_{\rm par}$, the largest distance that can be covered is the distance $c \cdot T_{\rm par}$. Thus, we must have $2r \le c \cdot T_{\rm par}$ and hence, $r \le (c/2) \cdot T_{\rm par}$. So, all processors that affect the computation result must be located within the sphere of radius $R \stackrel{\rm def}{=} (c/2) \cdot T_{\rm par}$. In the usual 3-D space, the volume V of this inside-sphere region is equal to

$$V = \frac{4}{3} \cdot \pi \cdot R^3 = \frac{4}{3} \cdot \pi \cdot \left(\frac{c}{2}\right)^3 \cdot T_{\rm par}^3.$$

Let us denote by ΔV the size of the smallest possible processor, and by $N_{\rm proc}$ the number of processors involved in the computation. Overall, these processors occupy at least the volume $N_{\rm proc} \cdot \Delta V$. This volume cannot exceed the overall volume V: $N_{\rm proc} \cdot \Delta \le V$, so we get an upper bound on the possible number of processors:

$$N_{\rm proc} \le \frac{V}{\Delta V} = C \cdot T_{\rm seq}^3, \text{ where } C \stackrel{\rm def}{=} \frac{1}{\Delta V} \cdot \frac{4}{3} \cdot \pi \cdot \left(\frac{c}{2}\right)^3.$$

Computations on a parallel computer can be simulated on a sequential computer: for each moment of time, we simulate what the first processor does, then what the second processor does, etc. In this simulation, one time period of the parallel computer is simulated in $N_{\rm proc}$ time periods. Thus, the overall computation time increases by a factor of $N_{\rm proc}$. As a result, we perform the same computations on a sequential computer in time

$$N_{\rm proc} \cdot T_{\rm par} \le C \cdot T_{\rm par}^3 \cdot T_{\rm par} = C \cdot T_{\rm par}^4.$$

So, if we denote by T_{seq}, the smallest possible time needed to solve the corresponding problem on a sequential computer, then we get

$$T_{\text{seq}} \le C \cdot T_{\text{par}}^4.$$

From this inequality, we can conclude that

$$T_{\text{par}} \ge C^{-1/4} \cdot T_{\text{seq}}^{1/4}.$$

In other words, we can speed up computations, but there is a limit to this speedup. The main reason why we need speedup is that many problems are NP-hard; see, e.g., [15, 19]. Unless it turned out that P = NP, solving such problems requires, in the worst case, exponential time $\approx 2^n$, where n is the size of the input. For reasonable-size inputs $n = 300$ and $n = 600$, we will need, correspondingly, $2^{300} \approx 10^{100}$ and $2^{600} \approx 10^{200}$ steps—which would take much larger than the lifetime of the Universe. Let us see what happens if we parallelize these problems.

- If the lower limit for sequential computer is $2^{300} \approx 10^{100}$, which is much larger than the lifetime of the Universe, then the parallel computations can be, potentially, performed in the time of $(2^{300})^{1/4} = 2^{75} \approx 10^{25}$ steps. At the usual Gigabyte speed of 10^9 computational steps per second, this means $10^{25}/10^9 \approx 10^{16}$ seconds. Taking into account that there are approximately $3 \cdot 10^7$ seconds in a year, this means that the parallel computations would require

  $$10^{16}/(3 \cdot 10^7) \approx 3 \cdot 10^8 = 300 \text{ million years},$$

  still not very practical, but at least theoretically feasible.
- On the other hand, if computations on a sequential computer require $2^{600} \approx 10^{200}$ computational steps, then the parallel computations would require

  $$10^{50}/(10^9 \cdot 3 \cdot 10^7) \approx 3 \cdot 10^{33} \text{ years},$$

  still much larger than the lifetime of the Universe.

But what if we take into account that the space is actually multi-dimensional, of dimension $d > 3$? In this case, similar arguments lead to a better bound on T_{par}:

$$T_{\text{par}} \ge C \cdot T_{\text{par}}^{1/(d+1)},$$

for some constant C. In particular, for $d = 10$, the sequential time of $2^{600} \approx 10^{200}$ seconds leads to the parallel time of $(10^{200})^{1/11} = 10^{22}$ computational steps, i.e., to

$$10^{22}/(10^9 \cdot 3 \cdot 10^7) \approx 3 \cdot 10^5 = 300\,000 \text{ years},$$

a more feasible time.

So, additional dimensions do help to speed up computations.

3 Let Us Summarize Our Findings

How can we resolve the first challenge: what is the physical meaning of extra dimensions? According to the proposed idea, extra dimensions—in line with the many-world interpretation of quantum physics—represent possible worlds. The main difference between this interpretation and the proposed idea is that in the proposed interpretation, we can use our free will to select one of the possible worlds—i.e., to move in these additional dimensions.

How can we resolve the second challenge: that we can affect the state of the universe, but no one's will was able to move any material object. The resulting explanation of this challenge is straightforward: yes, according to all the experiments so far, our will cannot change anything in the 3-D world, but probably what we can do is move between possible worlds—in other words, our free will can select in which of the possible worlds we will be in the future.

How we can resolve the third challenge: why are we smart. Why are we smarter than similar-size artificial networks? Maybe because the artificial neural networks live in the 3-dimensional space, while our brain can operate in the higher-dimensional space—and this increases our computational abilities.

Acknowledgements This work was supported in part by the National Science Foundation grants 1623190 (A Model of Change for Preparing a New Generation for Professional Practice in Computer Science), and HRD-1834620 and HRD-2034030 (CAHSI Includes), and by the AT&T Fellowship in Information Technology.

It was also supported by the program of the development of the Scientific-Educational Mathematical Center of Volga Federal District No. 075-02-2020-1478, and by a grant from the Hungarian National Research, Development and Innovation Office (NRDI).

References

1. *Compatibilism*, Stanford Encyclopedia of Philosophy. https://plato.stanford.edu, downloaded on 18 Sept 2021
2. F. Crick, *The Astonishing Hypothesis: The Scientific Search for the Soul* (Charles Scribner & Sons, New York, 1993)
3. P. Davies, *God and the New Physics* (Simon & Schuster, New York, 1983)
4. D. Dennett, *Elbow Room* (MIT Press, Cambridge, Massachusetts, 1984)
5. D. Dennett, *Freedom Evolves* (Viking Penguin, New York, 2003)
6. J. Earman, *A Primer on Determinism D* (Reidel, Dordrecht, Holland, 1986)
7. R. Feynman, R. Leighton, M. Sands, *The Feynman Lectures on Physics* (Addison Wesley, Boston, Massachusetts, 2005)
8. J.M. Fischer, *God, Foreknowledge and Freedom* (Stanford University Press, Stanford, California, 1989)
9. H. Frankfurt, Alternate possibilities and moral responsibility. J. Philos. **66**, 829–839 (1969)
10. I. Goodfellow, Y. Bengio, A. Courville, *Deep Leaning* (MIT Press, Cambridge, Massachusetts, 2016)

11. M.B. Green, J.H. Schwarz, E. Witten, *Superstring Theory*, vols. 1, 2 (Cambridge University Press, 1988)
12. J. Hospers, What means this freedom?, in *Determinism and Freedom in the Age of Modern Science*. ed. by S. Hook (Collier-Macmillan, New York, 1958), pp. 126–142
13. W. Isaacson, *Einstein: His Life and Universe* (Simon and Schuster, New York, 2008)
14. R. Kane, *The Oxford Handbook of Free Will* (Oxford University Press, New York, 2002)
15. V. Kreinovich, A. Lakeyev, J. Rohn, P. Kahl, *Computational Complexity and Feasibility of Data Processing and Interval Computations* (Kluwer, Dordrecht, 1998)
16. C.W. Misner, K.S. Thorne, J.A. Wheeler, *Gravitation* (W. H. Freeman, New York, 1973)
17. D. Morgenstein, V. Kreinovich, Which algorithms are feasible and which are not depends on the geometry of space-time. Geombinatorics **4**(3), 80–97 (1995)
18. M.A. Nielsen, I.L. Chuang, *Quantum Computation and Quantum Information* (Cambridge University Press, Cambridge, U.K., 2000)
19. C. Papadimitriou, *Computational Complexity* (Addison-Wesley, Reading, Massachusetts, 1994)
20. R. Penrose, *The Emperor's New Mind: Concerning Computers, Minds, and the Laws of Physics* (Oxford University Press, Oxford, UK, 2016)
21. R. Penrose, A. Shimony, N. Cartwright, S. Hawking, M. Longair, *The Large, the Small and the Human Mind* (Cambridge University Press, Cambridge, Massachusetts, 2000)
22. J. Polchinski, *String Theory*, vols. 1, 2 (Cambridge University Press, 1998)
23. A.D. Sakharov, Topological structure of elementary charges and CPT symmetry, in *Problems in Theoretical Physics, Dedicated to the Memory of I. E. Tamm*, Nauka, Moscow (in Russian); English translation in *Collective Scientific Works of A. D. Sakharov* (Marcel Decker, New York, 1982), pp. 243–247
24. S. Smilansky, *Free Will and Illusion* (Oxford University Press, New York, 2000)
25. P. Suppes, The transcendental character of determinism. Midwest Stud. Philos. **18**, 242–257 (1993)
26. K.S. Thorne, R.D. Blandford, *Modern Classical Physics: Optics, Fluids, Plasmas, Elasticity, Relativity, and Statistical Physics* (Princeton University Press, Princeton, New Jersey, 2017)
27. F.J. Tipler, *The Physics of Immortality: Modern Cosmology* (God and the Resurrection of the Dead, Doubleday, New York, 1994)
28. J. Trusted, *Free Will and Responsibility* (Oxford University Press, New York, 1984)
29. J.C. Urenda, O. Kosheleva, How to reconcile physical theories with the idea of free will: from analysis of a simple model to interval and fuzzy approaches, in *Proceedings of the IEEE World Congress on Computational Intelligence WCCI'2008* (Hong Kong, China, 1–6 June 2008), pp. 1024–1029
30. P. Van Inwagen, *An Essay on Free Will* (Clarendon Press, Oxford, 1983)

Why Normalized Difference Vegetation Index (NDVI)?

Francisco Zapata, Eric Smith, Vladik Kreinovich, and Nguyen Hoang Phuong

Abstract Plants play a very important role in ecological systems—they transform CO_2 into oxygen. It is therefore very important to be able to estimate the overall amount of live green vegetation in a given area. The most efficient way to provide such a global analysis is to use remote sensing, i.e., multi-spectral photos taken from satellites, drones, planes, etc. At present, one of the most efficient ways to detect, based on remote sensing data, how much live green vegetation an area contains is to compute the value of the normalized difference vegetation index (NDVI). In this paper, we provide a theoretical explanation of why this particular index is efficient.

1 Formulation of the Problem

Empirical fact. Plants play a very important role in ecological systems—they transform CO_2 into oxygen. It is therefore very important to be able to estimate the overall amount of live green vegetation in a given area. The most efficient way to provide such a global analysis is to use remote sensing, i.e., multi-spectral photos taken from satellites, drones, planes, etc.

At present, one of the most efficient ways to detect, based on remote sensing data, how much live green vegetation an area contains is to compute the value of the normalized difference vegetation index (NDVI):

F. Zapata · E. Smith · V. Kreinovich (✉)
University of Texas at El Paso, 500 W. University, El Paso 79968, USA
e-mail: vladik@utep.edu

F. Zapata
e-mail: fcozpt@outlook.com

E. Smith
e-mail: esmith2@utep.edu

N. H. Phuong
Division Informatics, Math-Informatics Faculty, Thang Long University, Nghiem Xuan Yem Road, Hoang Mai District, Hanoi, Vietnam

N. H. Phuong and V. Kreinovich (eds.), *Biomedical and Other Applications of Soft Computing*, Studies in Computational Intelligence 1045,
https://doi.org/10.1007/978-3-031-08580-2_9

$$\text{NDVI} \stackrel{\text{def}}{=} \frac{\text{NIR} - \text{Red}}{\text{NIR} + \text{Red}},$$

where Red and NIR are spectral reflectance measurements corresponding to red and near infrared (NIR) parts of the spectrum; see, e.g., [1–8, 10, 12–14].

Related question. Assuming that the amount of life green vegetation can be uniquely determined by the values Red and NDVI, why this particular combination of these two values is the most adequate?

What we do in this paper. In this paper, we provide a theoretical explanation for the formula (1).

2 Towards an Explanation: General Analysis of the Problem

Different scales. When we process data, we process numerical values of the physical quantities. These numerical values depend not only on the quantity itself, they depends also on what measuring unit we select, and what starting point we select for measurement.

For example, to measure temperature, we can use Celsius (C) and Fahrenheit (F) scales. These scales use different measuring units—a 1-degree difference in the C scale corresponds to 1.8 degree difference in the F scale. These scales also use different starting points: 0 degrees on a C scale correspond to 32 degrees on the F scale.

In general, if we replace the original measuring unit by a new unit which is a times smaller, then all numerical values x get multiplied by a: $x \mapsto a \cdot x$. For example, if we replace meters by centimeters—a 100 times smaller unit—then, e.g., the original height of 1.7 m becomes $1.7 \cdot 100 = 170$ cm. Similarly, if we replace the original starting point with a new point which is b degrees smaller, then to all numerical values we add this value b: $x \mapsto x + b$. Thus, in general, if we change both the measuring unit and the starting point, we get a linear transformation $x \mapsto a \cdot x + b$.

Nonlinear rescalings. In many practical situations, we can also have nonlinear rescalings. For example, instead of describing the electric properties of an object, we can use resistance R or we can use conductivity $1/R$. Which of such nonlinear transformations are natural?

In general, linear transformations are natural. If we have a natural transformation from scale A to scale B, then the inverse transformation—from scale B to scale A—is also natural. If we have a natural transformation from scale A to scale B and another natural transformation from scale B to scale C, then their composition is a natural transformation from scale A to scale C. Thus, the class of all natural transformations is closed under composition and under taking an inverse, i.e., in mathematical terms, this means that this class should be a *transformation group*.

At any given moment of time, we can store and process only finitely many values. Thus, to effectively deal with different natural transformations, we should require that a natural transformation should be uniquely determined by the values of finitely many parameters. In mathematical terms, this means that the class of all natural transformations should be *finite-dimensional*.

It is known—see, e.g., [9] and references therein—that for transformations from real numbers to real numbers, each finite-dimensional transformation group containing all linear transformations contains only fractional linear transformations. In other words, all natural transformations should be fractional linear, i.e., have the form

$$x \mapsto \frac{a \cdot x + b}{c \cdot x + d}, \tag{2}$$

for some constants a, b, c, and d.

Natural functions of two variables. For a function $z = f(x, y)$ of two variables:

- if we fix x, then we get a transformation $y \mapsto f(x, y)$, and
- if we fix y, then we get a transformation $x \mapsto f(x, y)$.

It is reasonable to say that a function $f(x, y)$ is natural if all these transformations are natural—and thus, fractional linear. Let us characterize all such functions.

3 First Result: Characterizing All Natural Functions of Two Variables

Reminder. Let us recall that a function $f(x, y)$ is called *bilinear* if it has the form

$$f(x, y) = a_0 + a_1 \cdot x + a_2 \cdot y + a_{12} \cdot x_1 \cdot x_2. \tag{3}$$

In these terms, we have the following result.

Definition 1 We say that a function $f(x, y)$ is *natural* if the following two conditions are satisfied:

- for every x, the mapping $y \mapsto f(x, y)$ is fractional linear, and
- for every y, the mapping $x \mapsto f(x, y)$ is fractional linear.

Proposition 1 *A function $f(x, y)$ is natural if and only if it is a ratio of two bilinear functions.*

Proof If we fix the value of one of the variables, then a bilinear function becomes linear, and thus, the ratio of two bilinear functions becomes fractional linear. Thus, a ratio of two bilinear functions is natural in the sense of Definition 1.

So, to prove the proposition, it is sufficient to prove that every natural transformation is a ratio of two bilinear functions. Indeed, naturalness means, in particular, that for every y, there exists values $a(y)$, $b(y)$, $c(y)$, and $d(y)$ for which

$$f(x, y) = \frac{a(y) \cdot x + b(y)}{c(y) \cdot x + d(y)}. \tag{4}$$

In the generic case, $d(y) \neq 0$, so we can divide both numerator and denominator by $d(y)$ and get a simpler expression

$$f(x, y) = \frac{A(y) \cdot x + B(y)}{C(y) \cdot x + 1}, \tag{5}$$

where we denoted

$$A(y) \stackrel{\text{def}}{=} \frac{a(y)}{d(y)}, \quad B(y) \stackrel{\text{def}}{=} \frac{b(y)}{d(y)}, \text{ and } C(y) \stackrel{\text{def}}{=} \frac{c(y)}{d(y)}.$$

(The case when $d(y) \equiv 0$ can be treated similarly.)

Since the function $f(x, y)$ is natural, for each x, the expression (5) is a fractional linear function of y. In particular, if we select three generic different values x_1, x_2, and x_3, we conclude that

$$f_i(y) = \frac{A(y) \cdot x_i + B(y)}{C(y) \cdot x_i + 1}, \tag{6}$$

where $f_i(y) \stackrel{\text{def}}{=} f(x_i, y)$ is the corresponding fractional linear function. Multiplying both sides of the formula (6) by the denominator, we conclude that

$$C(y) \cdot x_i \cdot f_i(x) + f_i(y) = A(y) \cdot x_i + B(y), \tag{7}$$

i.e., equivalently, that

$$x_i \cdot A(y) + B(y) - f_i(y) \cdot x_i \cdot C(y) = -f_i(y). \tag{8}$$

So, for each y, we have a system of three linear equations to determine the three unknowns $A(y)$, $B(y)$, and $C(y)$. According to Cramer's rule, the solution to a system of linear equations is a ratio of two determinants—which are polynomials in terms of the coefficients. In mathematics, ratios of polynomials are called rational functions. So, the solution to a system of linear equations is a rational function of all the coefficients. In our case, x_i, 1, and -1 are constants, and $f_i(y) \cdot x_i$ is a rational function. So, each of the coefficients $A(y)$, $B(y)$, and $C(y)$ is a rational function of a rational function—and thus, a rational function itself. In other words:

$$A(y) = \frac{P_A(y)}{Q_A(y)}, \quad B(y) = \frac{P_B(y)}{Q_B(y)}, \quad C(y) = \frac{P_C(y)}{Q_C(y)} \tag{9}$$

for some polynomials $P_A(y)$, $Q_A(y)$, $P_B(y)$, $Q_B(y)$, $P_C(y)$, and $Q_C(y)$. Substituting the expressions (9) into the formula (5), we conclude that

$$f(x, y) = \frac{\dfrac{P_A(y)}{Q_A(y)} \cdot x + \dfrac{P_B(y)}{Q_B(y)}}{\dfrac{P_C(y)}{Q_C(y)} \cdot x + 1}. \tag{10}$$

By using the usual formulas for addition and division of fractions, we conclude that

$$f(x, y) = \frac{\dfrac{P_a(y) \cdot Q_B(y) + x \cdot P_B(y) \cdot Q_A(y)}{Q_A(y) \cdot Q_B(y)}}{\dfrac{x \cdot P_C(y) + Q_C(y)}{Q_C(y)}} =$$
$$\frac{P_A(y) \cdot Q_B(y) \cdot Q_C(y) + x \cdot P_B(y) \cdot Q_A(y) \cdot Q_C(y)}{x \cdot Q_A(y) \cdot Q_B(y) \cdot P_C(y) + Q_A(y) \cdot Q_B(y) \cdot Q_C(y)}. \tag{11}$$

In other words, the function $f(x, y)$ is a ratio of two polynomials which are linear in x. These two polynomials may have a non-constant common divisor, which is either linear in x or does not depend on x at all. Dividing both numerator and denominator by the greatest common divisor of these two polynomials, we conclude that

$$f(x, y) = \frac{a_0(y) + x \cdot b_0(y)}{x \cdot c_0(y) + d_0(y)} \tag{12}$$

for some polynomials $a_0(y)$, $b_0(y)$, $c_0(y)$, and $d_0(y)$.

Similarly, from the fact that for each y, the mapping $x \mapsto f(x, y)$ is fractional linear, we conclude that

$$f(x, y) = \frac{A_0(x) + y \cdot B_0(x)}{y \cdot C_0(x) + D_0(x)} \tag{13}$$

for some polynomials $A_0(x)$, $B_0(x)$, $C_0(x)$, and $D_0(x)$. Equating the right-hand sides of the formulas (12) and (13), we conclude that

$$\frac{a_0(y) + x \cdot b_0(y)}{x \cdot c_0(y) + d_0(y)} = \frac{A_0(x) + y \cdot B_0(x)}{y \cdot C_0(x) + D_0(x)}. \tag{14}$$

Multiplying both sides of this equality by both denominators, we conclude that

$$(a_0(y) + x \cdot b_0(y)) \cdot (y \cdot C_0(x) + D_0(x)) = (x \cdot c_0(y) + d_0(y)) \cdot (A_0(x) + y \cdot B_0(x)). \tag{15}$$

The left-hand side of this equality is divisible by $a_0(y) + x \cdot b_0(y)$, which means that the right-hand side must be divisible by the same expression. By our construction of the expression (12), this sum has no common factors with

$$x \cdot c_0(y) + d_0(y).$$

Thus, it must divide $A_0(x) + y \cdot B_0(x)$.

Similarly, the right-hand side of the equality (15) is divisible by

$$A_0(x) + y \cdot B_0(x).$$

Since by our construction of the expression (13), this sum has no common factors with $y \cdot C_0(x) + D_0(x)$. Thus, it must divide $a_0(y) + x \cdot b_0(y)$.

So, the polynomials $a_0(y) + x \cdot b_0(y)$ and $A_0(x) + y \cdot B_0(x)$ must divide each other and thus, they should differ only by a multiplicative constant c:

$$a_0(y) + x \cdot b_0(y) = c \cdot A_0(x) + y \cdot c \cdot B_0(x) \tag{16}$$

for all x and y. In particular, for two different values x_1 and x_2, we conclude that

$$a_0(y) + x_i \cdot b_0(y) = c \cdot A_0(x_i) + y \cdot c \cdot B_0(x_i). \tag{17}$$

So, we have two linear equations with constant coefficients for determining two unknowns $a_0(y)$ and $b_0(y)$. Thus, the solution is a linear combination of the right-hand sides. Right-hand sides are linear functions of y, so both $a_0(y)$ and $b_0(y)$ are linear functions of y: $a_0(y) = a_{00} + a_{01} \cdot y$ and $b_0(y) = a_{10} + a_{11} \cdot y$, for some coefficients a_{ij}. Thus, the numerator of the expression (12) has the form

$$a_0(y) + x \cdot b_0(y) = a_{00} + a_{10} \cdot x + a_{01} \cdot y + a_{11} \cdot x \cdot y. \tag{18}$$

In other words, this numerator is a bilinear function.

Similarly, we can conclude that the denominator of the expression (12) is a bilinear function:

$$x \cdot c_0(y) + d_0(y) = c_{00} + c_{10} \cdot x + c_{01} \cdot y + c_{11} \cdot x \cdot y \tag{19}$$

for some coefficients c_{ij}. Thus, according to the formula (12), the function $f(x, y)$ is indeed equal to the ratio of two bilinear functions:

$$f(x, y) = \frac{a_{00} + a_{10} \cdot x + a_{01} \cdot y + a_{11} \cdot x \cdot y}{c_{00} + c_{10} \cdot x + c_{01} \cdot y + c_{11} \cdot x \cdot y}. \tag{20}$$

The proposition is proven.

Comment. A similar result can be similarly proven for a function of several variables. Let us recall that a function $f(x_1, \ldots, x_n)$ is called *multi-linear* if it has the form

$$f(x_1, \ldots, x_n) = \sum_{S \subseteq \{1, \ldots, n\}} a_S \cdot \prod_{i \in S} x_i. \tag{21}$$

In these terms, we have the following result.

Definition 2 We say that a function $f(x_1, \ldots, x_n)$ is *natural* if for all i and for all possible values $x_1, \ldots, x_{i-1}, \ldots, x_{i+1}, \ldots, x_n$, the mapping

$$x_i \mapsto f(x_1, \ldots, x_{i-1}, x_i, x_{i+1}, \ldots, x_n)$$

is fractional linear.

Proposition 2 *A function $f(x_1, \ldots, x_n)$ is natural if and only if it is a ratio of two multi-linear functions.*

4 Scale Invariance

Requirement. The values Red and NDVI depend on the solar angle. For a different angle, the same situation leads to the values Red and NDVI both multiplied by the same constant. Thus, to make a characteristic that does not depend on the solar angle, we want to select a characteristic $f(\text{NIR}, \text{Red})$ that should not change if we simply change the angle, i.e., for which

$$f(c \cdot x, c \cdot y) = f(x, y)$$

for all possible values of c, x, and y. Such functions are known as *scale-invariant.*

The following result characterizes all scale-invariant natural functions.

Proposition 3 *A natural function $f(x, y)$ which is not identically constant is scale-invariant if and only if it has the form*

$$f(x, y) = \frac{a_{10} \cdot x + a_{01} \cdot y}{c_{10} \cdot x + c_{01} \cdot y}. \tag{22}$$

Proof One can easily prove that each expression (22) is scale-invariant. Vice versa, let us assume that a natural function is scale-invariant. We know, from Proposition 1, that all natural functions are described by expression (20). Thus, scale invariance means that for all x and y, we have

$$f(x, y) = \frac{a_{00} + a_{10} \cdot x + a_{01} \cdot y + a_{11} \cdot x \cdot y}{c_{00} + c_{10} \cdot x + c_{01} \cdot y + c_{11} \cdot x \cdot y} =$$

$$\frac{a_{00} + c \cdot a_{10} \cdot x + c \cdot a_{01} \cdot y + c^2 \cdot a_{11} \cdot x \cdot y}{c_{00} + c \cdot c_{10} \cdot x + c \cdot c_{01} \cdot y + c^2 \cdot c_{11} \cdot x \cdot y}. \tag{23}$$

To prove the proposition, we need to prove that $a_{00} = c_{00} = 0$ and that $a_{11} = c_{11} = 0$.

Let us prove that $a_{00} = c_{00} = 0$. Indeed, if at least of these two coefficients is different from 0, then for $c \to 0$, the right-hand side of the equality (23) becomes a constant a_{00}/c_{00}, so we conclude that the function $f(x, y)$ is constant.

Similarly, if at least one of the coefficients a_{11} and c_{11} is different from 0, then in the limit $c \to \infty$, the right-hand side of the equality (23) becomes a constant a_{00}/c_{00}, so we conclude that the function $f(x, y)$ is constant.

So, if the function $f(x, y)$ is not constant, all these four coefficients should be equal to 0. Thus, Proposition 3 is proven.

Discussion. We have almost got the desired expression (1). To get even closer to the expression (1), let us take into account that the NDVI index changes from −1 to 1. If we require that the range of the function is exactly [−1, 1], then we get the following result.

Proposition 4 *A natural scale-invariant function $f(x, y)$ whose range of values for $x, y \geq 0$ is $[-1, 1]$ has the form*

$$f(x, y) = \frac{x - c \cdot y}{x + c \cdot y} \text{ or } f(x, y) = -\frac{x - c \cdot x}{x + c \cdot y}. \tag{24}$$

Proof If we divide both the numerator and the denominator of the expression (22) by x, we get

$$f(x, y) = F(z) \stackrel{\text{def}}{=} \frac{a_{10} + a_{01} \cdot z}{c_{10} + c_{11} \cdot z}, \text{ where } a = \frac{y}{x}. \tag{25}$$

The fact that the range of this function is between −1 and 1 means that this function is defined for all z. In this case, the fractional linear function (25) is monotonic, so its range when $z \in [0, \infty]$ is simply an interval bounded by the values of this function at the endpoints $z = 0$ and $z = \infty$. So, the fact that the range is equal to $[-1, 1]$ means that one of the values $F(0)$ and $F(\infty)$ should be equal to 1, and another value to −1.

If $F(0) = -1$, then for its opposite $G(z) \stackrel{\text{def}}{=} -F(z) = -f(x, y)$, we have $G(0) = -F(0) = 1$. So, without losing generality, we can consider the case when $F(0) = 1$. Substituting the value $z = 0$ into the right-hand side of the formula (25), we conclude that $a_{10}/c_{10} = 1$, i.e., that $a_{10} = c_{10}$. Dividing both the numerator and the denominator of the expression (25) by this common value $a_{10} = c_{10}$, we conclude that

$$F(z) = \frac{1 + a \cdot z}{1 + c \cdot z}, \text{ where } a = \frac{a_{01}}{a_{10}} \text{ and } c = \frac{c_{01}}{c_{10}}. \tag{26}$$

This ratio has to be defined for all z, so we must have $c \geq 0$—otherwise, if $c < 0$, this expression would not be defined for $z = -1/c$.

For $z = \infty$, we have $F(\infty) = -1$, so $a/c = -1$, thus $a = -c$. The proposition is proven.

Conclusion. We (almost) explained why NDVI is a relevant characteristics.

Comment. Similar argument can—partly—explain the effectiveness of fractional-linear similarity coefficients like Jaccard index

$$s(A, B) = \frac{\mu(A \cap B)}{\mu(A) + \mu(B) - \mu(A \cap B)},$$

where $\mu(S)$ denotes the measure of the set S; see, e.g., [15].

Acknowledgements This work was supported in part by the National Science Foundation grants 1623190 (A Model of Change for Preparing a New Generation for Professional Practice in Computer Science), and HRD-1834620 and HRD-2034030 (CAHSI Includes), and by the AT&T Fellowship in Information Technology.

It was also supported by the program of the development of the Scientific-Educational Mathematical Center of Volga Federal District No. 075-02-2020-1478, and by a grant from the Hungarian National Research, Development and Innovation Office (NRDI).

References

1. R.E. Crippen, Calculating the vegetation index faster. Remote Sens. Environ. **34**, 71–73 (1990)
2. D.M. Gates, *Biophysical Ecology* (Springer, New York, 1980)
3. B.N. Holben, Characteristics of maximum-value composite images from temporal AVHRR data. Int. J. Remote Sens. **7**(11), 1417–1434 (1986)
4. A.R. Huete, A soil-adjusted vegetation index (SAVI). Remote Sens. Environ. **25**, 53–70 (1988)
5. Y.J. Kaufman, D. Tanre, Atmospherically resistant vegetation index (ARVI) for EOS-MODIS, in *Proceedings of the IEEE International Geosciences and Remote Sensing Symposium'92* (New York, 1992), pp. 261–270
6. F.J. Kriegler, W.A. Malila, R.F. Nalepka, W. Richardson, Preprocessing transformations and their effects on multispectral recognition, in *Proceedings of the Sixth International Symposium on Remote Sensing of Environment* (1969), pp. 97–131
7. S.M. Labib, S. Lindley, J.J. Huck, Scale effects in remotely sensed greenspace metrics and how to mitigate them for environmental health exposure assessment, in *Computers, Environment and Urban Systems*, vol. 82, Paper 101501 (2020). https://doi.org/10.1016/j.compenvurbsys.2020.101501
8. R.B. Myneni, F.G. Hall, P.J. Sellers, A.L. Marshak, The interpretation of spectral vegetation indexes. IEEE Trans. Geosci. Remote Sens. **33**, 481–486 (1995)
9. H.T. Nguyen, V. Kreinovich, *Applications of Continuous Mathematics to Computer Science* (Kluwer, Dordrecht, 1997)
10. H. Nouri, S. Anderson, P. Sutton, S. Beecham, P. Nagler, C.J. Jarchow, D.A. Roberts, NDVI, scale invariance and the modifiable areal unit problem: an assessment of vegetation in the Adelaide Parklands. Sci. Total Environ. **584–585**, 11–18 (2017). https://doi.org/10.1016/j.scitotenv.2017.01.130
11. B. Pinty, M.M. Verstraete, GEMI: a non-linear index to monitor global vegetation from satellites. Vegetatio **101**(1), 15–20 (1992). https://doi.org/10.1007/bf00031911
12. A.J. Richardson, C.L. Wiegand, Distinguishing vegetation from soil background information. Photogramm. Eng. Remote. Sens. **43**, 1541–1552 (1977)
13. J.W. Rouse, R.H. Haas, J.A. Scheel, D.W. Deering, Monitoring vegetation systems in the Great Plains with ERTS, in *Proceedings of the 3rd Earth Resource Technology Satellite (ERTS) Symposium*, vol. 1 (1974), pp. 48–62

14. P.J. Sellers, Canopy reflectance, photosynthesis, and transpiration. Int. J. Remote Sens. **6**, 1335–1372 (1985)
15. P.-N. Tan, M. Steinbach, A. Karpatne, V. Kumar, *Introduction to Data Mining* (Pearson, New York, 2018)

Binary Image Classification Using Convolutional Neural Network for V2V Communication Systems

Hoang-Thong Vo, Ngan-Linh Nguyen, Dao N. Ngoc, Trong-Hop Do, and Quang-Dung Pham

Abstract This research presents a method based on deep learning network architecture for image classification problems for a communication system between autonomous vehicles. Vehicular communication is a research direction of data transmission between autonomous vehicles based on a communication system called a vehicular optical communication camera (VOCC). Thanks to the VOCC system, the vehicles transfer information about the position, direction, speed, and future behavior. In addition, the power of the deep learning network-based proposed algorithm improves the VOCC system to gain the accuracy of image processing from the acquisition of signals from autonomous vehicles. Experimental results show that the proposed algorithm achieves high performance on image signals with difficult conditions.

Keywords Deep neural network · Vehicular communication · Vehicle-to-vehicle · LED image classification

---

H.-T. Vo · N.-L. Nguyen · T.-H. Do
University of Information Technology, Vietnam National University, Ho Chi Minh City, Vietnam
e-mail: 18521462@gm.uit.edu.vn

N.-L. Nguyen
e-mail: 18520989@gm.uit.edu.vn

T.-H. Do
e-mail: hopdt@uit.edu.vn

D. N. Ngoc (✉)
Department of Technology, Dong Nai Technology University, Dong Nai 760000, Vietnam
e-mail: daonhungoc@dntu.edu.vn

Q.-D. Pham
Faculty of Information Technology, Vietnam National University of Agriculture, Hanoi, Vietnam
e-mail: pqdung@vnua.edu.vn

N. H. Phuong and V. Kreinovich (eds.), *Biomedical and Other Applications of Soft Computing*, Studies in Computational Intelligence 1045,
https://doi.org/10.1007/978-3-031-08580-2_10

1 Introduction

Transportation certainly has a great influence on the development of a country. In developed countries, public transport systems are invested heavily to serve the travel needs of people. However, the number of vehicles is still increasing significantly each year leading to overcrowding on infrastructure and a series of problems that need to be solved such as traffic jams, increasing emissions, greenhouse, fuel shortage. To thoroughly solve the problem, many proposals have been raised from scientists, one of the possible and effective solutions proposed by the research community is the intelligent transportation system (ITS) [1–3]. Johann Andersen and Steve Sutcliffe [4] presented an overview and specific study of ITS, comparing the system to a parachute that covers the breadth of transportation systems. In addition, Andersen et al. also show a series of features of ITS User Services such as Traffic Management, Vehicle Systems, Vehicles. The features in ITS User Services to which Andersen et al. contributed are the basic ideas that form ITS. Reference from [4], data acquisition, data processing, data communication, data distribution, information utilization is the core technology to enable ITS. Thanks to the development of ITS, one of the technologies that play an important role in this system in the field of autonomous vehicles. The connection and information transmission between ITS and autonomous vehicles in general and vehicular communication in particular through data communication techniques based on dedicated short-range communications (DSRC) enable highly secure, high-speed direct communication between vehicles and the surrounding infrastructure [5]. In Fig. 1, vehicular communication is one of the information communication protocols between vehicles based on data communication, the main communication components in ITS are vehicle-to-vehicle (V2V), vehicle-to-infrastructure (V2I).

The vehicular communication system is not only an information communication system between vehicles, but also has many other great roles such as developing road safety and reducing dangerous accident situations [6–10]. Fabio Arena and Giovanni Pau [6] argue that vehicle-to-everything (V2X) is a set of concepts covering vehicle-related communications including V2V and V2I. One of the main purposes of V2X technology is to support the connection and sharing of data between vehicles and people to help limit accidents and unnecessary dangerous situations. For example, when a vehicle sees an accident ahead, it can send a signal to the vehicle behind to slow down. When receiving this warning signal, the vehicles can respond to agree or disagree with the above warning signals. In V2V communication networks, we see that the vehicles exchange information about their position, speed, the direction of travel, and even road signs. In a specific situation, the yellow car and the white car behind at the same time want to move into the same position, if there is no communication and information exchange between the two automobiles, there is a risk of a collision. For self-driving cars, depending on the self-driving mechanism of the vehicle, more or less human intervention is required; in the context of completely self-driving cars, vehicle communication will be an effective assistant to assist the

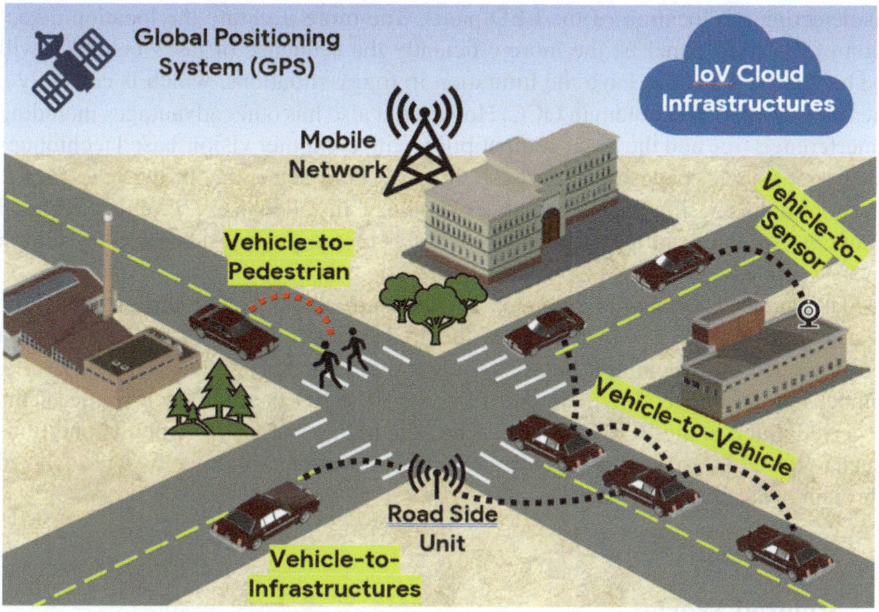

Fig. 1 Overview communication architecture in smart cities

network of autonomous vehicles to keep a safe distance from each other and circulate smoothly in the traffic network in a smart city as Fig. 1.

The development of V2X must go hand in hand with the development of other technologies, especially vehicle-to-vehicle. Over the past decades, there has been a lot of research in the field of vehicular communication for V2V systems in ITS. Radio frequency communication is the most popular type of communication technique. However, it has a big problem of interference in dense traffic. Therefore, the research for the new candidate for V2V communication is still going on. It is possible that the combination of multiple communication techniques is required for a reliable V2V communication channel in all traffic situations. In recent years, optical camera communication (OCC) [11–14] is one of the potential technologies that has received much attention from scientists in recent times. Vehicular optical camera communication (VOCC) is medium-range communication technology and has many benefits compared to other technologies. VOCC utilizes the availability of LED lights and vehicle dashboard cameras for the transmission and reception of signals. The signal is a sequence of binary codes, encoded to the ON and OFF states of the LED chips inside the LEDs, and each LED chip in the panel conveys a single bit corresponding to 1 or 0. The signals are received by capturing video from the camera. Then, image processing techniques are used to extract information from the frames of the video. This system will handle tasks such as determining the position of the LED panel in the image and classifying the signal bits in the LED panel. The accuracy of the task of classifying signal bits in the LED panel will depend on the previous task such as detecting the location of the LED panel. The more accurate the location detection of the LED panel is, the more efficiently the sequence of tasks behind it will be handled. OCC does have the limitation in foggy situations, which is currently a heavily researched problem in OCC. However, it also has other advantages including interference-free and the great compatibility with computer vision-based techniques such as obstacle, pedestrian, traffic light, road lane detection. In the near future, these computer vision-based techniques should be installed in every vehicle. Optical Camera Communication can be one of features of computer vision-based system in every vehicle.

The image classification problem has many approaches, from traditional machine learning algorithms [15–19] to state-of-the-art deep learning models [20–24] has high performance on benchmark datasets. In this research, we propose a deep learning model for the image classification of bits on the LED panel in each frame of the received video in the VOCC system, achieving high accuracy in difficult conditions such as low contrast and image blurriness. The performance of the model is analyzed through the experiments section.

2 Related Work

2.1 Traditional Approaches

Image classification is an important problem in computer vision, there have been many studies on this problem over the years, especially in traffic analysis. Before the appearance of deep learning models, feature extraction and image processing techniques were applied mostly in the interdisciplinary and computer vision fields. Lin et al. [18] presented a new weighted minimum distance classifier that uses the discriminatory power and variance of features. Boser et al. [19] proposed a training algorithm that maximizes the margin between the training patterns and the decision boundary. Lowe [15] introduced a method used to identify and describe features in images. Navneet Dalal and colleagues [16] recommended the grids of Histograms of Oriented Gradient (HOG) descriptors method for feature extraction and human object detection. Ahonen et al. [17] presented an image feature extraction technique based on local binary pattern (LBP) texture features, applied to the face recognition problem.

2.2 Deep Learning Approaches

The Alexnet Deep Neural Networks (DNNs) [24] model won the ImageNet Large Scale Visual Recognition Challenge with the lowest error rate on the Imagenet dataset, setting the stage for the deep learning revolution in the next years. Kaiming

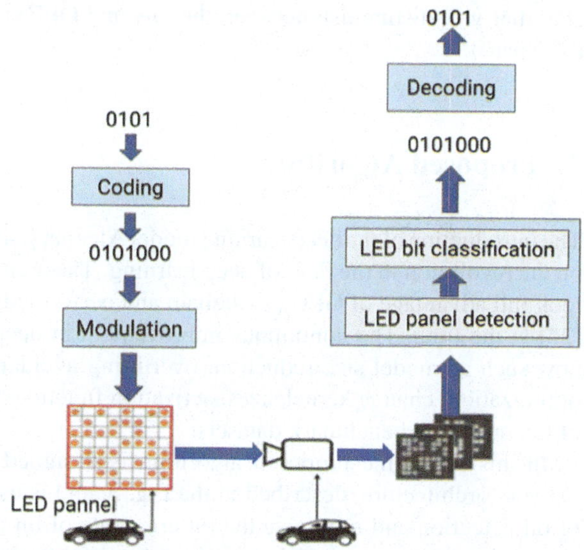

Fig. 2 VOCC system architecture and operational flows

He and colleagues from Microsoft Research developed a new network architecture that uses deep residual learning for image recognition [20]. Ioffe et al. [21] presented a novel mechanism for the training of deep networks. Szegedy et al. [22] proposed a deep convolutional neural network architecture codenamed Inception. Zeiler et al. [23] introduced a novel visualization technique that gives insight into the function of intermediate feature layers and the operation of the classifier.

3 Preliminaries of VOCC Systems

Based on the VOCC system architecture was mentioned in [25], the VOCC system architecture has improved in this research, as described in Fig. 2. LED panel in the vehicle is used to transmit a visible light signal to other vehicles on the road. Each led chip will undertake to transmit a bit, these bits have a value of ON or OFF corresponding to 1 or 0. The collection of bit strings from all led chips in the led panel will form a signal used for transmitting and receiving signal. Dashboard cameras equipped in other vehicles are used to receive this signal in the form of videos. In the video, each frame will be processed to get information after decoding. Autonomous vehicles use this information to communicate with each other to move direction, vehicle identity, relative position, relative speed, future behavior, and warning.

As the Fig. 2, task LED bit classification is processed after being located in the LED panel detection. The LED panel is a 64×64 square matrix. The image processing technique will split the above matrix into LED bits of size 8×8. The classifier will distinguish between the ON and OFF states of the LED bit from the LED panel.

4 Proposed Algorithm

The introduction of the deep learning model Alexnet [24] in 2012 had a great influence on the revolution in the field of deep learning. The architecture of the Alexnet model took full advantage of GPU parallelism and performed well on the Imagenet dataset [26] at the time. The continuous improvement in deep learning architectures until now such as model size reduction, overfitting avoidance techniques, loss function optimization, change kernel sizes, activation functions, have increased the accuracy of the model on benchmark datasets.

In this study, the proposed algorithm is designed based on the deep learning network architecture, described in the Fig. 3, and is used to maximize the accuracy of bit detection and minimize the bit error rate of bit classification. This is an end-to-end deep learning model, taking on two tasks: feature extractor and classifier. With an image input, layers in the same network architecture will have different computational tasks, the final output of the model is a dense layer with an activation function of softmax, a mathematical function that converts a vector of numbers into a vector of probabilities, where the probabilities of each value are proportional to the relative scale of each value in the vector. In the Fig. 4, a dense layer is the implementation of the equation:

$$output = activation(dot(input, kernel) + bias. \tag{1}$$

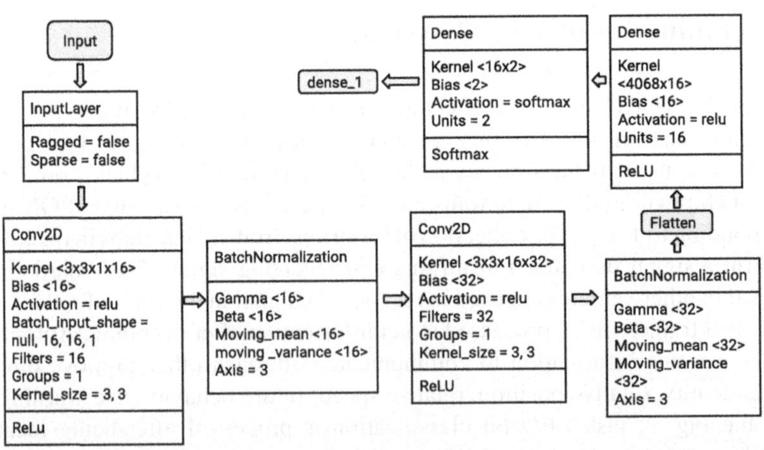

Fig. 3 Neural network architecture for V2V system

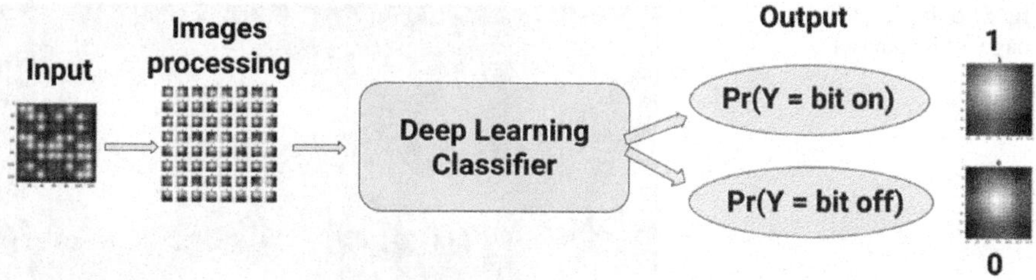

Fig. 4 Architecture of the proposed algorithm

The final output of the model is the probability distribution function of two random variables, a bit on and a bit off. The bit on of LED represents the light on, while the bit off of LED panel represents the off state.

5 Experiment Analyses

In this section, we present detailed information about the dataset, the evaluation metrics for the problem, and the results of the model on the dataset.

5.1 Dataset

The dataset includes 5000 LED panel images, the training, and test sets are divided in a ratio of 5:1, specifically, the test set includes 1000 images and the remaining 4000 images are the train set. The physical size of the sensor is (13.2 × 8.8 mm) (1-inch sensor) with Full-HD (1920 × 1080 pixels) video recording. A wide-angle lens with 18mm of focal length is used. The frame rate of the camera is 100 fps. The fast exposure time is set to 1/1500 s to prevent motion blur. The ratio of LED light to ambient light is from 0.2 to 3.0. LED chips have a circular shape with a diameter of 3 cm. The distance between LED chips is 3 cm. The LED panel consists of 8 × 8 LEDs. The LED illuminance is not fixed but varies from 2000 to 4000 cd/m2. The illuminance of ambient light also ranges from 1000 to 32000 lux. The average reflectance of the background is 0.18. The testing communication distance ranges from 30 to 70 m. As Fig. 5, this is a sample image in the dataset.

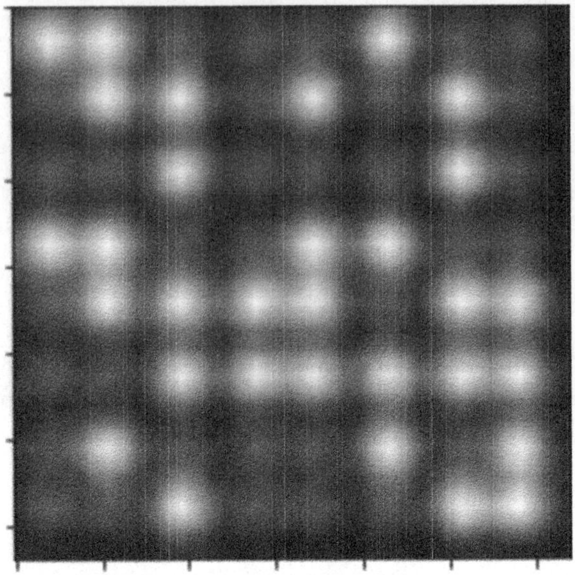

Fig. 5 Sample LED panel image in the dataset

5.2 Evaluation metrics

Bit Error Rate (BER) is calculated based on the following equation:

$$BER = \frac{Number\ of\ wrong\ bit\ LED\ detection}{Total\ number\ of\ bit\ LED}. \quad (2)$$

In this research, we use the attribute Rl2a, the ratio of LED light to ambient light, along with the BER metric to evaluate the performance of the image classification model on the dataset.

5.3 Result

In this subsection, we presented the approach and experiment on the dataset. In the Fig. 6, the detector is based on the Xception backbone network architecture and using bounding box regression to determine the position and size coordinates of the LED region. The results of the LED regions' detection bounding boxes are shown in Fig. 7. Image processing techniques will be applied to remove the background image and the image after the above processing is LED region, also known as LED panels.

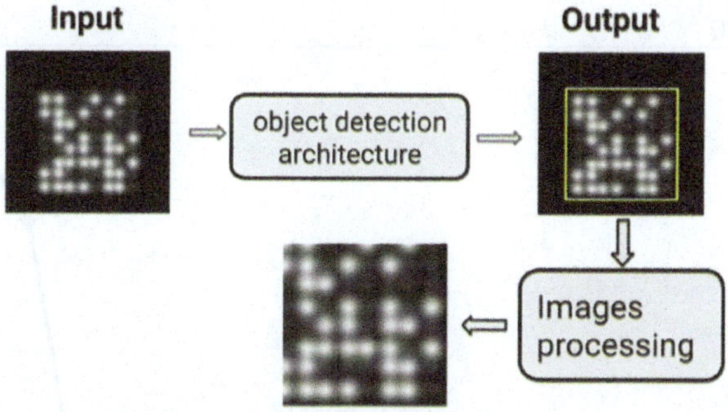

Fig. 6 Flowchart of object detection

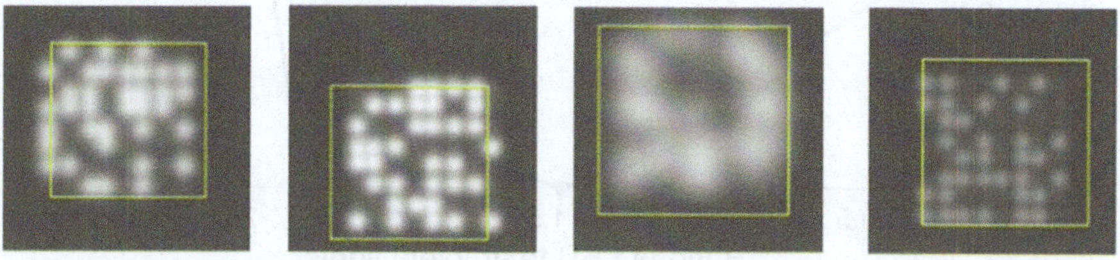

Fig. 7 Detected bounding boxes of LED regions

Then the LED region will be passed through the proposed algorithm to return the result of the binary classification of the LED on or off, shown in Fig. 4.

To evaluate the performance of the proposed algorithms, the classification accuracy at the ratio of LED light to ambient light was obtained and shown in Fig. 8. The metric used for evaluating the classification accuracy is BER (bit error rate). As can be seen in the Figure, BER decreases as the ratio of LED light to ambient light increases in range from 0.2 to 1.5. This is because of the increasing contrast between On and Off LED as the light ratio increases. However, the BER increases after the light ratio exceeds 1.5. This is because both the contrast and blooming effect increases as the light ratio increases. After the light ratio exceed 1.5, the negative effect of the increasing blooming effect surpasses the positive effect of the increasing contrast and makes the increase of BER. This figure reveals that the to achieve the lowest BER, the LED to ambient light ratio should range from 1 to 2. If the ratio is too low, the contrast in the image will be low and the BER will be high. If the ratio is too high, the blooming effect in the image will be too high and the BER will also be high.

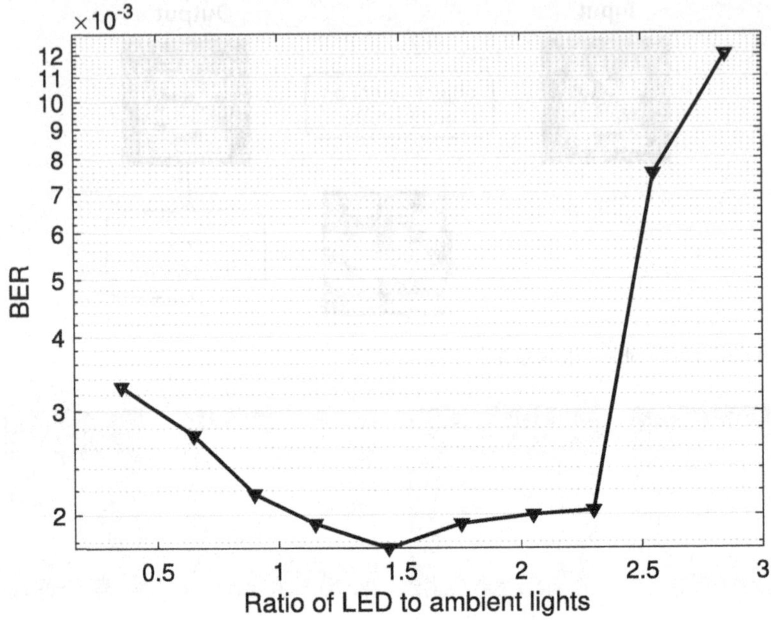

Fig. 8 Bit error rate at different the ratio of LED light to ambient light

6 Conclusion

In this paper, a deep learning-based algorithm is proposed for the bit LED classification task. Experimental results show that the proposed algorithm achieves an approximate bit error rate metric of less than 12×10^{-2}. With BER metrics less than 12×10^{-2}, mid-range communication between vehicles will achieve high efficiency. Furthermore, the performance of the VOCC system based on the proposed algorithm is also significantly improved. As a result, communication signals between autonomous vehicles will be transmitted accurately and increase the efficiency of the VOCC system's medium-range communication.

References

1. A. Haydari, Y. Yilmaz, A survey. IEEE transactions on intelligent transportation systems, Deep reinforcement learning for intelligent transportation systems (2020)
2. M.B. Mollah et al., Blockchain for the internet of vehicles towards intelligent transportation systems: a survey. IEEE Internet Things J. **8**, 4157–4185 (2020)
3. Y. Lee et al., Trustful resource management for service allocation in fog- enabled intelligent transportation systems. IEEE Access **8**, 147313–147322 (2020)
4. J. Andersen, S. Sutcliffe, Intelligent Transport Systems (ITS)–An Overview, in *IFAC Proceedings Volumes 33. IFAC Conference on Technology Transfer in Developing Countries:*

Automation in Infrastructure Creation (DECOM-TT 2000), Pretoria, South Africa, 5–7 July 2000, 99–106. issn:1474-6670 (2000)
5. J.B. Kenney, Dedicated Short-Range Communications (DSRC) Standards in the United States. Proc. IEEE **99**, 1162–1182 (2011)
6. F. Arena, G. Pau, An overview of vehicular communications. Future Internet **11**, 27 (2019)
7. M. Torrent-Moreno, J. Mittag, P. Santi, H. Hartenstein, Vehicle-to-vehicle communication: fair transmit power control for safety-critical information. IEEE Trans. Veh. Technol. **58**, 3684–3703 (2009)
8. S. Biswas, R. Tatchikou, F. Dion, Vehicle-to-vehicle wireless communication protocols for enhancing highway traffic safety. IEEE Commun. Mag. **44**, 74–82 (2006)
9. P. Papadimitratos, A.D. La Fortelle, K. Evenssen, R. Brignolo, S. Cosenza, Vehicular communication systems: Enabling technologies, applications, and future outlook on intelligent transportation. IEEE Commun. Mag. **47**, 84–95 (2009)
10. L. Chisalita, N. Shahmehri, A peer-to-peer approach to vehicular communication for the support of traffic safety applications in proceedings, in *The IEEE 5th International Conference on Intelligent Transportation Systems* (2002), pp. 336–341. https://doi.org/10.1109/ITSC.2002.1041239
11. T. Yamazato et al., Image-sensor-based visible light communication for automotive applications. IEEE Commun. Mag. **52**, 88–97 (2014)
12. Y. Goto et al., A new automotive VLC system using optical communication image sensor. IEEE Photonics J. **8**, 1–17 (2016)
13. I. Takai et al., Optical vehicle-to-vehicle communication system using LED transmitter and camera receiver. IEEE Photonics J. **6**, 1–14 (2014)
14. S. Albayati, An overview of visible light communication systems. Int. J. Comput. Sci. Mob. Comput. **8**, 51–56 (2019)
15. D.G. Lowe, Distinctive image features from scale-invariant keypoints. Int. J. Comput. Vision **60**, 91–110 (2004)
16. N. Dalal, B. Triggs, Histograms of oriented gradients for human detection, in *2005 IEEE computer society conference on computer vision and pattern recognition (CVPR'05)*, vol. 1 (2005), pp. 886–893
17. T. Ahonen, A. Hadid, M. Pietikainen, Face description with local binary patterns: application to face recognition. IEEE Trans. Pattern Anal. Mach. Intell. **28**, 2037–2041 (2006)
18. H. Lin, A. Venetsanopoulos, A weighted minimum distance classifier for pattern recognition, in *Proceedings of Canadian Conference on Electrical and Computer Engineering*, vol. 2 (1993), pp. 904–907. https://doi.org/10.1109/CCECE.1993.332440
19. B.E. Boser, I.M. Guyon, V.N. Vapnik, A training algorithm for optimal margin classifiers, in *Proceedings of the 5th Annual ACM Workshop on Computational Learning Theory*, pp. 144–152
20. K. He, X. Zhang, S. Ren, J. Sun, Deep residual learning for image recognition, in *Proceedings of the IEEE Conference on Computer Vision and Pattern Recognition* (2016), pp. 770–778
21. S. Ioffe, C. Szegedy, Batch normalization: accelerating deep network training by reducing internal covariate shift, in *International Conference on Machine Learning* (2015), pp. 448–456
22. C. Szegedy, et al., Going deeper with convolutions, in *Proceedings of the IEEE Conference on Computer Vision and Pattern Recognition* (2015), pp. 1–9
23. M.D. Zeiler, R. Fergus, Visualizing and understanding convolutional networks, in *European Conference on Computer Vision* (2014), pp. 818–833
24. A. Krizhevsky, I. Sutskever, G.E. Hinton, Imagenet classification with deep convolutional neural networks. Adv. Neural. Inf. Process. Syst. **25**, 1097–1105 (2012)
25. T.-H. Do, M. Yoo, The necessity of LED to ambient light ratio optimization for vehicular optical camera communication. Sensors **20** (2020). issn: 1424-8220. https://doi.org/10.3390/s20010292, https://www.mdpi.com/1424-8220/20/1/292
26. J. Deng, et al., Imagenet: a large-scale hierarchical image database, in *2009 IEEE Conference on Computer Vision and Pattern Recognition* (2009), pp. 248–255

Topic Model—Machine Learning Classifier Integrations on Geocoded Twitter Data

Gillian Kant, Christoph Weisser, Thomas Kneib, and Benjamin Säfken

Abstract Topic Models are unsupervised probabilistic models that are used to explore the hidden semantic structure of a corpus of documents by generating latent topics. Latent topics are discrete distribution over words that need to be interpreted and labeled by humans. The latent topics can be also considered as a summary of the various topics that are discussed in the corpus of documents, so that the topic model can also serve as a dimension reduction method. Latent topics generated by topic models are difficult to measure in their quality [4] and their outputs are commonly used in a descriptive manner in order to gain topical insights into the topic distributions of chosen documents. A so far unexplored way, is the use of labeled data to assess the informational content of latent topics. In this paper, topic models are examined in terms of their informative value regarding classification problems for labeled text data. We use geo-coded social media posts from Twitter, but our approach can be expanded to other labeled documents. The output of Latent Dirichlet Allocation (LDA) [3] models and Structural Topic Models (STM) [29] are used as an input for machine learning classifiers, after pooling tweets based on the hashtag pooling algorithm of [24]. Their predictive power is compared with the performance of state-of-the-art Artificial Neural Networks (ANN) that are trained on a specifically optimized word-embedding and all the available metadata of the tweets. We find that the machine learning classifiers that are trained on topics can compete with the predictive performance of the ANNs, even for out-of-sample predicted topic distributions.

Keywords Topic model · Natural language processing · Twitter · Hashtag pooling · LDA · Latent dirichlet allocation · SVM · Naive bayes · XGBoost · K-NN · Artificial neural networks · GloVe · LSTM

G. Kant
University of Göttingen, Göttingen, Germany

C. Weisser (✉) · T. Kneib · B. Säfken
University of Göttingen, Campus-Institut Data Science (CIDAS), Göttingen, Germany
e-mail: c.weisser@stud.uni-goettingen.de

N. H. Phuong and V. Kreinovich (eds.), *Biomedical and Other Applications of Soft Computing*, Studies in Computational Intelligence 1045,
https://doi.org/10.1007/978-3-031-08580-2_11

1 Introduction

The goal of this work is to discover the generalization capabilities of topic models on new data and to quantify these information to ascertain, if topic models can contribute to Natural Language Processing (NLP) beyond their typical field of application as a purely descriptive tool. This paper therefore provides a blueprint for using the output of LDA Models [3] and STMs for predictions, involving machine learning classifiers. We apply the concept by classifying Twitter social media posts into categories after pooling them based on the hashtag pooling algorithm of [24]. The predictive power of the results is compared with the performance of state-of-the-art ANNs on the same classification task. Using topic model results as input for classifiers has many potential benefits: (i) the topic distribution itself becomes meaningful for the prediction, (ii) certain topics with high predictive power can be identified and different classifiers can be applied based on their accuracy, and (iii) statistical inference can be applied, for instance p-values or confidence intervals can be calculated.

We show, that machine learning classifiers that are trained on a few topics, can match the capabilities and qualities of the classifications of ANNs. Even though our concept can be applied to any kind of labeled text data, the biggest problem with processing the data of our choice, Twitter data, is the sparseness of the tweets text. This lack of information poses major problems for topic models, as they perform poorly when documents are short [37]. We optimize and implement the hashtag-pooling algorithm by [24] for various samples of geocoded tweets.

First, several LDA and STM topic models are trained on the pooled tweets. Secondly, the optimal topic models are selected by their topical coherence score. Thirdly, the predictions of a topic distribution for each of the individual tweets, which were used to generate the pooled tweets (LDA/STM topic model training set) and for the remaining tweets of the corpus (LDA/STM Topic Model test set) are generated using the optimized topic models. Fourthly, we use machine learning procedures with the generated topic distributions of the individual tweets from the LDA/STM topic model training set as input to perform a classification. Hereby, the LDA/STM topic model training set is split again into a classifier training and test set. The goodness of fit for the classifiers that are trained on the LDA/STM topics, which are estimated on the training set, are taken as the baseline result. Next, these steps are repeated for the LDA/STM topic model test set, in order to determine a possible generalization. Here, the LDA/STM topic model test set is also split into a training and test set for the classifiers. Finally, all steps are repeated for all data sets for both LDA and STM. The applied machine learning classifiers are: Gaussian Naive Bayes, K-Nearest Neighbors (K-NN), (linear) Support Vector Machines (SVM) and eXtreme Gradient Boosting (XGBoost). For most models, the parameters have been optimized using *sklearns GridSearch*. Finally, the classifications are compared by their F-score. Unlike for LDA, the tweets were processed unpooled by the STM and the ANN. In case of the ANN, pre-processing was carried out at the level of vectorization and transformation of the metadata to input features. A word embedding was used in pre-trained form based on Global Vectors (GloVe) [28]. The classification results of the ANN were also measured by the F-score. An overview on the workflow can be found in the appendix in Fig. 5.

2 Related Literature

This paper draws on the literature from NLP, Machine Learning and Deep Learning. First, the literature on the problem of estimating LDA topic models on sparse text and potential solutions will be discussed. Secondly, topic model optimization for LDA and STM will be briefly covered. Finally, the Machine Learning and Deep Learning literature for the classification of sparse text will be sketched. Since the development of the LDA topic model by [3] it has been extensively applied in many areas. However, as already stated above, the biggest issue with processing Twitter data with LDA topic models is the sparseness of the tweets text (up to max. 280 characters per tweet). Reference [15] show that pooling tweets, especially when it comes to the textual content, improves the performance of the topic models significantly. Reference [24] provide a solution for the performance problem by comparing different tweet-pooling procedures, with hashtag pooling performing best. They provide an automatic hashtag generation algorithm for unlabeled tweets, which we implement in this paper. Reference [32] conclude that aggregating tweets over hashtags would produce better results than author-pooled tweets. Reference [1] suggest pooling through aggregating tweets by conversation, i.e. tweets, which are answered by other users. They compare their test results with other pooling methods like hashtag and author-pooling as well as to unpooled tweets. Their outcome suggests that pooling by conversation leads to similar results in LDA output quality as hashtag pooling, but comes with a more efficient usage of computational resources. Since our data is streamed, it cannot be filtered for conversations and henceforth this approach has been unsuitable for our application. Some of the mentioned preparations of tweets used in this paper are based on [19].

When using topic models, a decisive parameter is the optimal number of topics. There are several approaches for their selection. Nevertheless measuring the quality of a topic is one of the biggest difficulties in evaluating topic models. One possibility is the hold-out likelihood, based on hold-out documents for verification [4]. Another option is using a Hierarchical Dirichlet Process Topic Model, which generates the optimal number of topics during the variational estimation of the parameters on the data [12, 20, 34]. A third choice is using semantic coherence scores for the topic models [26]. This is the most suitable metric for our analysis, because it returned subjective better topic models in self-implemented tests. A qualitative approach is provided by [6]. They conduct an experiment with results suggesting that quantitative metrics for measuring topic coherence do not correspond to human assessment of semantic coherence. This result suggests, that there are no optimal measures for topic coherence available except human monitoring.

The STM model of [29] has previously been used successfully for Twitter analysis. Reference [27] applied STM on a very small Twitter corpus (about 4000 tweets) with tweets of four different users, whose tweets contents can be differentiated according to their political views. [9] used a STM to analyze tweets in the aftermath of the 2016 Berlin terrorist attack to label tweets of the same topic into different categories, in order to examine the psychological reactions of the users to the attack. [8] used a STM to differentiate the Twitter usage of judges in the USA in years with and without judicial elections. An approach for further research on Twitter data analysis could be Dynamic Topic Models [2]. These models specifically estimate the variation of topics over time and could possibly account very well for the topical changes driven by present events.

Reference [36] have achieved good results in classifying Twitter data with SVM for adapting hashtags to non-hashtagged tweets. Reference [21] have used various machine learning methods in order to classify (short) spam emails, and have accomplished considerable success with Naive Bayes and SVM. Reference [17] have also used Naive Bayes, SVM, and Random Forest for the classification of Twitter data on 19 classes and achieved good results with all classifiers.

Using topic models output as machine learning input is a dimensionality reduction in itself and thus this compression of information for an already sparse input is challenging. Therefore, it is worthwhile to investigate which classifiers perform best using just a few training data and input features. Reference [10] evaluate machine learning classifiers with small, unbalanced data sets for SVM, Naive Bayes, Multinomial Naive Bayes, and logistic regression. In particular, Multinomial Naive Bayes and SVM perform well, logistic regression mediocre, with the Naive Bayes classifiers being less sensitive to changes in the distribution of the training set. SVM and Gaussian Naive Bayes are two of the applied models in this paper.

Presently, ANNs are often used in text classification tasks for Twitter and other short documents. In many cases convolutional networks or combinations of recurrent and convolutional networks are implemented. Reference [31] use a simple n-gram based convolutional model for author detection in short texts. The exceptional point here is, that the model can detect local character-level interactions - a strength of convolutional networks. Ren et al. (2016) create a model that captures and includes contextual information, such as the hashtag or the author of a tweet. Based on their approach, information beyond the text field of the tweets were included in the ANN model. The classification implemented in our approach used both, 1-D convolutional layers, as well as recurrent layers, to combine their advantages and improve the informational value gathered from the tweets texts, similar to [35].

3 Data

The tweets for the analysis performed were obtained using the Twitter API, which has been accessed through a developer account. The Python package Tweepy [30] was used to stream the data. The data used for the analysis consists of 738.439 tweets from Great Britain from November 16, 2019 to January 07, 2020. The classification targets are specific regions which are defined with the geo-data information that is provided by each tweet: London, England (excl. London), Scotland and Wales. Three different data sets are created, namely *Scotland versus Wales* with 68852 versus 21901 tweets, *London versus Scotland* with 178333 versus 68852 tweets and *England versus London* with 469353 versus 178333 tweets. The data preprocessing consisted of stop-word removal, lemmatization and removal of links, mentions, and symbols as well as numbers. An additional preparation step has been the pooling of the tweets by hashtag to produce longer pseudo-documents using the algorithm of Mehrotra et al. (2013). Hereby, the package TTLocVis from [19] was utilized. The particular benefits of Twitter data are the structural variables that are provided with each tweet and that can be incorporated into the STM. We extract the following structural variables for each tweet: hashtags, number of retweets, favorite count (number of likes) and emojis.

4 Methods

In this section, the LDA Topic Model [3] and STM [29] and their outputs transfer to the machine learning classifiers is described. Gaussian Naive Bayes, SVM, K-NN and XGBoost and the implemented ANN are briefly outlined.

4.1 Topic Models Formulation

For the LDA as the generative probabilistic model of a corpus, the corpus is defined by a group of M documents, $\mathbf{D} = \{\mathbf{w}_1, \ldots, \mathbf{w}_M\}$. A document is denoted as a sequence of words, denoted by $\mathbf{w} = (w_1, \ldots, w_N)$. Words are represented by unit-basis vectors and indexed on a fixed vocabulary in $\{1, \ldots, V\}$, see [3]. Furthermore for a document (unobserved) topic variables z_n, $n = 1, \ldots, N$ are associated with the corresponding words w_n, $n = 1, \ldots, N$. The goal of inference in this model is to obtain corpus wide parameters $\boldsymbol{\beta}$ and $\boldsymbol{\alpha}$ as well as $\boldsymbol{\theta}_j$, $j = 1, \ldots, M$, that can be interpreted as the document-specific probabilities of k topics. Hence, the joint probability distribution $p(\boldsymbol{\theta}, \mathbf{z}, \mathbf{w}|\boldsymbol{\alpha}, \boldsymbol{\beta})$ of $\boldsymbol{\theta}$, $\mathbf{z}$ and $\mathbf{w}$ is defined as

$$p(\boldsymbol{\theta}|\boldsymbol{\alpha}) \prod_{n=1}^{N} p(z_n|\boldsymbol{\theta}) p(w_n|z_n, \boldsymbol{\beta}) \tag{1}$$

with the (corpus wide) word probability matrix $\boldsymbol{\beta} \in \mathbb{R}^{k \times V}$, the (corpus wide) parameters of a Dirichlet distribution $\boldsymbol{\alpha}$ and $\boldsymbol{\theta} \sim Dir(\boldsymbol{\alpha})$ is the (document-level) k-dimensional topic mixture, which describes distribution of the topics for each of the corpus documents. This allows for the modelling of the marginal distribution of a document $p(\mathbf{w}|\alpha, \beta)$ and the probability of a corpus $p(\mathbf{D}|\alpha, \beta)$, see Blei et al. (2003). Notice that for the probability of a corpus with M documents we obtain latent document-level variables $\boldsymbol{\theta}_j,\ j = 1, \ldots, M$.

There are different approaches for inference in these models but we only describe the algorithm by [14] that is applied in this paper. The authors' approach, called Online Variational Inference for LDA uses a variational inference attempt to optimize memory usage and computation time. Variational approaches assume that the untraceable posterior can be approximated by other traceable distributions and aim to find the one that is closest to the real posterior. This is mostly done by Kullback-Leibler divergence to measure the distance between two probability distributions [3]. Finally, by using the Expectation Maximization algorithm, the parameters of the variational posterior are approximated.

The theoretical details of the implemented algorithm in the Python package Gensim, which we used, can be found in [14]. Alternatively, model inference can be achieved by a Markov-Chain-Monte-Carlo approach using Gibbs Sampling. Details on that can be found in [4].

STM are an entity of topic models that extends LDA to incorporate metadata related to the documents. The model distinguishes between *topical prevalence covariates*, which influence the proportion of topics occurring in a document, and *topical content covariates*, which influence the variation of words occurring within a topic. Both, one out of the two, or neither of these sets of variables can be included in the model [29].

4.2 From Topic Models to Machine Learning Classifiers

In the second step, documents are classified according to their topic distribution. Therefore, labeled documents belong to a class c_r from class vector $\mathbf{c} = (c_1, \ldots, c_R)$. We assume that the topic probabilities of a document $\mathbf{w}_j$ defined by $\boldsymbol{\theta}_j = \left(\theta_{1j}, \ldots, \theta_{kj}\right)$ influences the probability of the document belonging to a class c_r. Thus, we apply a range of machine learning classifiers to predict the class of a document with latent topic probabilities $\boldsymbol{\theta}_j$ as inputs. In the following the classifiers are outlined according to the notation of the topic model output.
The Naive Bayes classifier is a very simple and fast model for classification built on the premise of independence of every pair of input data $\boldsymbol{\theta}_j = \left(\theta_{1j}, \ldots, \theta_{kj}\right)$. Based on Bayes' theorem the conditional probability of a class c_r is defined by

$$P(c_r \mid \boldsymbol{\theta}_j) = \frac{P(c_r) \prod_{i=1}^{k} P(\theta_{ij} \mid c_r)}{P(\boldsymbol{\theta}_j)} \tag{2}$$

Hence classification is achieved choosing the most probable class

$$\hat{y} =_{r\in\{1,\ldots,R\}} P(c_r)\prod_{i=1}^{k} P(\theta_{ij} \mid c_r) \tag{3}$$

For the Gaussian Naive Bayes approach, $\theta_{ij} \mid c_r$ is assumed to be normally distributed. Hence, the classification is realized by calculating the corresponding likelihood and solving for $\hat{y}$ in a second step [23].

In order to emphasis how our approach translates to further machine learning classifiers we discuss SVMs for a binary classification of documents $\mathbf{w}_j,\ j = 1,\ldots,M$ based on topic probabilities $\boldsymbol{\theta}_j = \left(\theta_{1j},\ldots,\theta_{kj}\right),\ j = 1,\ldots,M$. SVMs maximize the margin between two classes using hyperplanes to build a decision boundary for classification. The following explanation covers the simple case of two classes separated linearly. A data subsample is used to decide on the location of a linear separator for classification in a k-dimensional space. These points are called *support vectors* and halfway in between them lies the hyperplane that maximizes the margin between the classes. $(\boldsymbol{\theta}_1, y_1),\ldots,(\boldsymbol{\theta}_M, y_M)$ refer to the documents and associated topic probabilities used for training, with $y_i \in \{-1, 1\}$ describing the binary classification label. The goal is to separate the k-dimensional points by their respective label $y_i,\ i = 1,\ldots,M$. A hyperplane can be defined by $\{\boldsymbol{\theta} : f(\boldsymbol{\theta}) = \boldsymbol{\theta}^T\boldsymbol{\beta} + \beta_0 = 0\}$ where $\|\boldsymbol{\beta}\| = 1$. The linear classifier is then defined as

$$G(\boldsymbol{\theta}) = (\boldsymbol{\theta}^T\boldsymbol{\beta} + \beta_0). \tag{4}$$

The maximization problem for the optimal margin can be transformed into a minimization problem, since its hyperplane lies in between the two *support vectors*:

$$\min_{\boldsymbol{\beta},\beta_0} \|\boldsymbol{\beta}\| \text{ s.t. } y_i(\boldsymbol{\theta}^T\boldsymbol{\beta} + \beta_0) \geq 1, \tag{5}$$

for $i = 1,\ldots,M$, where the constraint on $\|\boldsymbol{\beta}\|$ is dropped.

Another simple classification model is the K-NN algorithm. Hereby, the data is analyzed regarding its "neighborhood" in the vector space. For a target point $\boldsymbol{\theta}_0$, k training points $(\boldsymbol{\theta}_1,\ldots,\boldsymbol{\theta}_k)$ are chosen closest to $\boldsymbol{\theta}_0$. The target point is then classified as the same class label as the majority of the training points [13].

XGBoost is applied in this paper and takes a slightly different approach by utilizing decision trees using gradient boosting. Gradient Boosting is an algorithm that combines the results from many classifiers (here decision trees) iteratively to increase the overall classification accuracy. It was first introduced by [11]. Based on that, XGBoost provides a modern, enhanced application with many advantages like tweaking with data sparsity and optimized computational time [7]. A detailed explanation of the generic and applied algorithm would go beyond the scope of this paper but can be found in [7, 13].

4.3 Artificial Neural Networks

We use an ANN model for performance comparison, besides the Gaussian Naive Bayes, K-NN, SVM and XGBoost. In order to train an ANN model with text data, it is possible to use pre-trained weights to effectively utilize the documents, if only a limited vocabulary is available. GloVe [28] is one of the biggest and most sophisticated freely available embeddings up to date. This word-embedding was trained on millions of tweets and therefore has a higher accuracy when it comes to weighing the impact of a certain word in relation to the vocabulary. It ships embeddings in different dimensions, but for all data sets the largest one with 200 dimensions is used to ensure that the tweets tokens are maximally utilized. The ANN models for each data set were build with the Keras Python package on the Tensorflow backend. Figure 4 (in the appendix) shows the structure of the model. The tokens and corresponding structural variables from the tweets are processed into *numerical features, categorical features, textual features* and *text of tweets*, which contains the text tokens in padded form.

The *categoricalfeatures* input is piped through five dense layers and the *numerical features* are piped through three dense layers. Subsequently, both are concatenated after passing a batch normalization layer. The resulting layer and the *text of tweets* input are then passed through three dense layers and a batch normalization layer, each. Both are concatenated. This resulting layer is led through four dense layers and a batch normalization layer. The *texts of tweets* input is embedded by an embedding layer, creating the corresponding GloVe word vectors, then passed through two one-dimensional convolutional layers, two LSTM layers, three dense layers and a batch normalization layer. Next, these two loose ends are concatenated and finally passed through two dense layers producing the models' output. Model callbacks are provided. They assure that the learning process is stopped, if the model's accuracy does not increase for three epochs and that the learning rate is reduced if the validation loss is not improving for 10 epochs. The best model was saved for each data set. In order to assess the results of the classification the F-score is calculated.

For evaluating the models we use F-scores. The F-score [16] is defined as $F_\beta = \frac{(\beta^2+1)\pi\rho}{\beta^2\pi+\rho}$ whereas π is defined as precision and ρ is defined as recall, while $\beta > 0$ is a weight for π and ρ, related to the classification goal. $\beta = 1$ if both are weighted equal.

5 Results

First, the performance of the various machine learning classifiers, which are fed with the topic model output will be evaluated. Finally, the ANN performance is shown and compared with the machine learning classifier based results.

5.1 *Topic Models*

An example of the generated topics can be found in Fig. 1.

In Fig. 2 the topic distribution of the LDA topic model training and test set (*London versus Scotland* data set) is shown for a model with 30 topics. Note, that the labeling *trained set* and *test set* in Fig. 2 refers to the LDA topic model training and test set. On the horizontal axis, the topics from 1 to 30 are shown. On the vertical axis, the average prevalence of each topic in the LDA topic models training and test set is displayed. It can be seen, that the distributions of the mean probability per topic are astonishingly similar. This reveals the very high out-of-sample generalization ability of those models.

Fig. 1 The figure shows the 50 words with the highest probability for two topics from *England versus London* data set of a 50 topics LDA model. The topic on the left hand side is *health* related, while the topic on the right hand side is *leisure* related

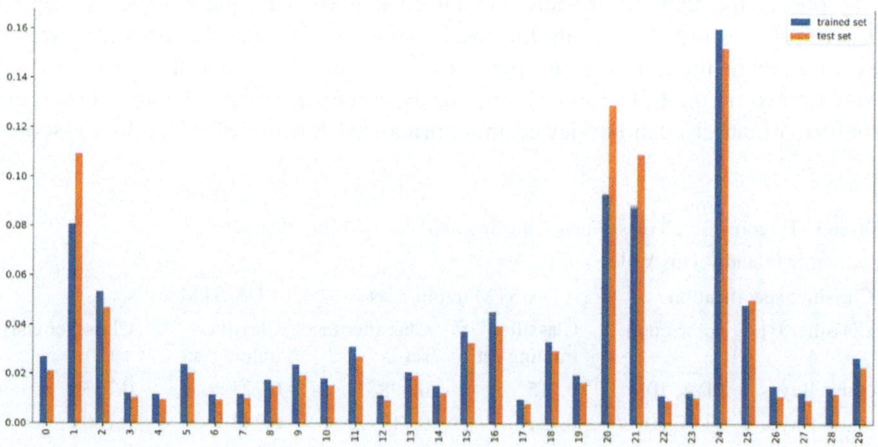

Fig. 2 *London versus Scotland*: mean probability per topic for 30-topics LDA

5.2 Machine Learning Classifiers

The performance of the Naive Bayes, K-NN, SVM and XGBoost classifiers will be discussed for all data sets. For the four machine learning classifiers, that are trained with LDA and STM Models with several different topics, only the best performing classifiers are displayed in the tables below. For each data set, we choose the best three LDA models, measured by the coherence metric, and fed the output into the classifiers. Additionally, for the *Scotland versus Wales* data set, STM models are estimated and used. The F-scores for all data sets are provided. The three LDA models of the *Scotland versus Wales* data set consist of 10, 30 and 50 topics. A STM with 10 topics is implemented.

The F-scores can be found in Table 1, which shows the LDA/STM training set and LDA/STM test set. Further, the split into the classifier training set and classifier test set is displayed. The results show clearly, that the classifiers which were trained on the LDA test set show similarly high F-scores as the classifiers that were trained on the LDA training set. An exception is only the Naive Bayes implementation for the 30 and 50-topics LDA models. This indicates that the LDA models have a very high generalization capability. The STM-based calculations produced very similar results as their LDA counterpart. Only for Naive Bayes, the STM-based model outperforms the Naive Bayes models that was trained with the LDA model output. For the *London versus Scotland* data set, the corresponding F-scores can be found in Table 2. LDA models with 30, 40 and 75 topics were selected based on the coherence score, in order to produce the topic distributions. It can be seen, that Naive Bayes performs very poorly for all LDA models. For all other classifiers, the F-score is ranging between 0.74 and 0.78. Again, the results show clearly that the classifiers which were trained on the LDA test set show a similarly high F-score as the classifiers that were trained on the LDA training set. For the *England versus London* data set, the confusion matrices can be viewed in a normalized form in Fig. 3 and its F-scores

Table 1 F-scores for all classifiers of the *Scotland versus Wales* data set

F-score: Scotland versus Wales					
Classifier specification		LDA/STM training set		LDA/STM test set	
Classifier type	Topic count	Classifier training set	Classifier test set	Classifier training set	Classifier test set
Naive Bayes	LDA: 10	0,715	0,698	0,725	0,713
Naive Bayes	STM: 10	0,721	0,717	0,745	0,760
K-NN	LDA: 50	0,749	0,741	0,757	0,758
K-NN	STM: 10	0,749	0,751	0,746	0,757
SVM	LDA: 50	0,756	0,752	0,764	0,770
SVM	STM: 10	0,759	0,761	0,756	0,764
XGBoost	LDA: 50	0,757	0,758	0,768	0,756
XGBoost	STM: 10	0,759	0,760	0,754	0,750

Table 2 F-scores for all classifiers of the *London versus Scotland* data set

F-score: London versus Scotland					
Classifier specification		LDA training set		LDA test set	
Classifier type	Topic count	Classifier training set	Classifier test set	Classifier training set	Classifier test set
Naive Bayes	LDA: 30	0,530	0,549	0,537	0,529
K-NN	LDA: 75	0,747	0,745	0,731	0,725
SVM	LDA: 75	0,757	0,763	0,743	0,744
XGBoost	LDA: 75	0,771	0,774	0,752	0,750

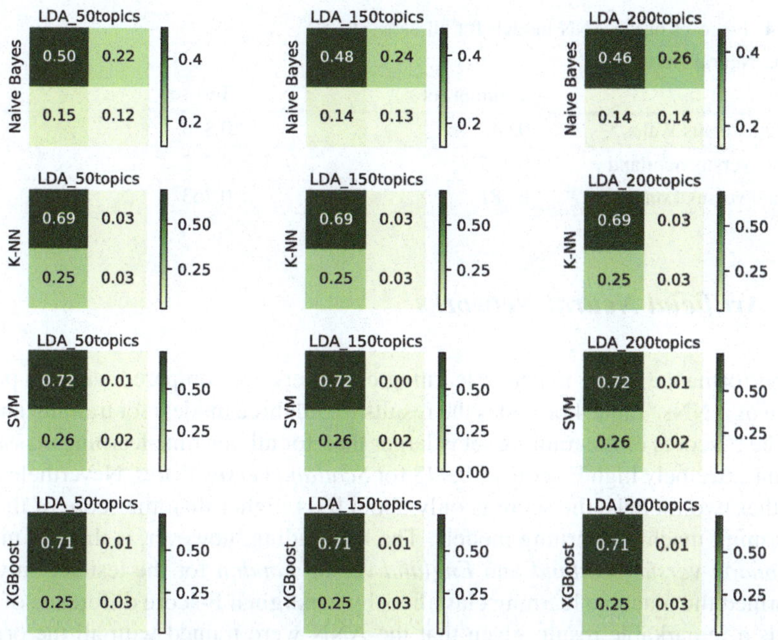

Fig. 3 Confusion matrices of the *England versus London* LDA topic model test set

are presented in Table 3. The overall F-scores are very close to the scores in the previously discussed data sets. It is noteworthy, that the number of topics does not have a strong effect on the classification accuracy. This could be due to the small informational content of a single tweet, which is mostly covered by one to two topics and the temporal relation between the tweets. Hence, the effect of additional topics might be marginal, if the data set contains a lot of tweets covering trending topics.

Table 3 F-scores for all classifiers of the *England versus London* data set

F-score: England versus London					
Classifier specification		LDA training set		LDA test set	
Classifier type	Topic count	Classifier training set	Classifier test set	Classifier training set	Classifier test set
Naive Bayes	LDA: 50	0,609	0,627	0,643	0,630
K-NN	LDA: 50	0,725	0,718	0,728	0,720
SVM	LDA: 150	0,737	0,735	0,741	0,737
XGBoost	LDA: 200	0,748	0,746	0,748	0,743

Table 4 F-scores of the ANN models for all data sets

F-score: Neural Networks		
Model	Training set	Test set
Scotland versus Wales	0.948	0.833
London versus Scotland	0.785	0.762
England versus London	0.781	0.763

5.3 *Artificial Neural Networks*

The performance of the machine learning classifiers are compared with the performance of ANNs. Table 4 provides the results for all three models for training and test set. The F-scores of the training set is larger than for all machine learning classifiers, with an extremely high F-score of 0.948 for *Scotland versus Wales*. Nevertheless, for the other two models, the score is only about 0.04 higher than the score of the best performing machine learning models. The key finding, however, is that the models for *London versus Scotland* and *England versus London* for the test set only outperformed the machine learning classifiers by a marginal F-score difference of 0.01. This is a remarkable result, given that the ANN were trained with all the original tweets, the additional structural variables, and GloVe-based word-embeddings, while the machine learning classifiers were trained only on the topic model output.

6 Conclusion

In this paper, we suggest the use of a new approach to assess the informational content of topics models. We classify geocoded tweets from the social media platform Twitter according to the location of the tweet by feeding machine learning classifiers with the out-of-sample topic predictions of LDA [3] models and STM [29]. Subsequently, the predictive performance is compared with the performance of state-of-the-art ANN. Strong evidence for generalization potential in topic model output is revealed by the fact, that the machine learning classifiers performed equally well on the LDA training set and LDA test set, i.e. on out-of-sample predicted topic distributions. Further, the topic models and machine learning classifiers performance is impressive, since the topic models' poor performance on short and sparse text is commonly known. The ANN used GloVe, one of the best available pre-trained word-embeddings, which was trained especially on Twitter data and all of the available tweets metadata. Nevertheless, the ANN could not perform significantly better than machine learning classifiers that were fed with the output of a LDA model that was only trained with the tweets text.

Possible further improvements in performance could be achieved by using LDA model output as additional input for ANN classifiers. In such a modelling approach, the metadata could be more easily processed on a broader scale as it would be possible with a STM. Other approaches, such as a combination of topic model output with a neural network as the classifier or a symbiosis of neural networks and topic models that include probabilistic elements inside the network [5, 18, 22] as well as alternative word embeddings to GloVe, like Word2vec [25] and Tweet2vec [33], are options for further research on this topic.

7 Appendix

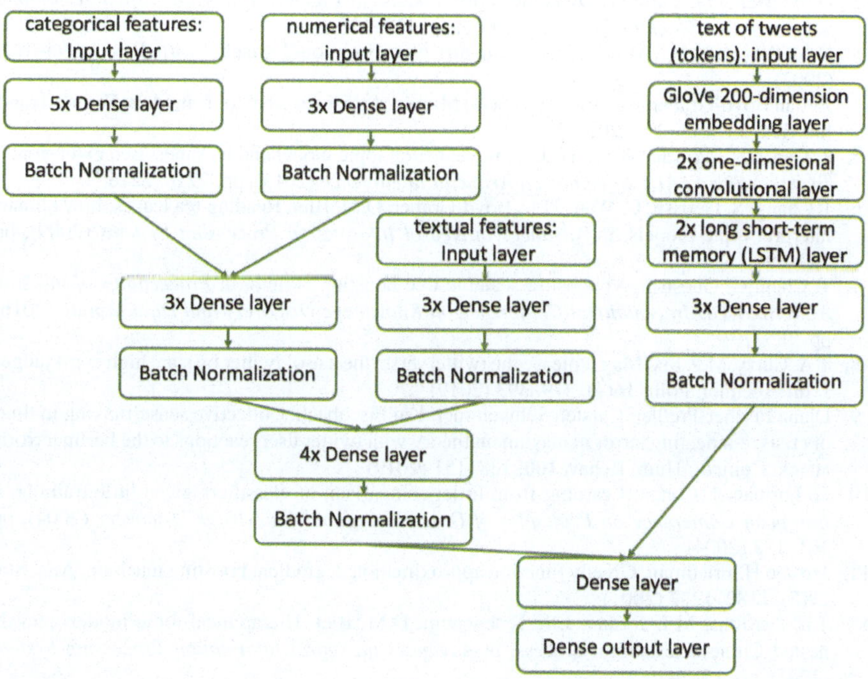

Fig. 4 ANN model

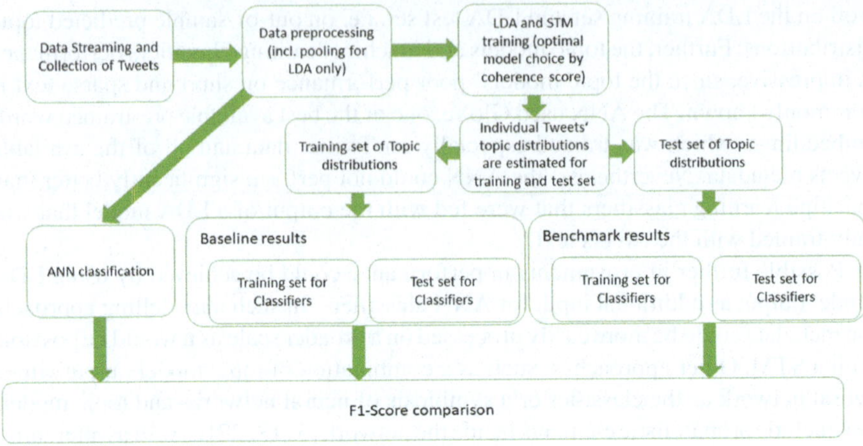

Fig. 5 An overview on the implemented workflow

References

1. D. Alvarez-Melis, M. Saveski, Topic modeling in twitter: aggregating tweets by conversations, in *Tenth International AAAI Conference on Web and Social Media* (2016), pp. 519–522
2. D.M. Blei, J.D. Lafferty, Dynamic topic models, in *Proceedings of the 23rd International Conference on Machine Learning* (2006), pp. 113–120
3. D.M. Blei, A.Y. Ng, M.I. Jordan, Latent dirichlet allocation. J. Mach. Learn. Res. **3**, 993–1022 (2003)
4. Jordan Boyd-Graber, Hu. Yuening, David Mimno, Applications of topic models. Found. Trends Inf. Retr. **11**, 143–296 (2017)
5. Z. Cao, S. Li, Y. Liu, W. Li, H. Ji, A novel neural topic model and its supervised extension, in *Twenty-Ninth AAAI Conference on Artificial Intelligence* (2015), pp. 2210–2216
6. J. Chang, S. Gerrish, C. Wang, J.L. Boyd-Graber, D.M. Blei, Reading tea leaves: how humans interpret topic models, in *Advances in Neural Information Processing Systems* (2009), pp. 288–296
7. T. Chen, C. Guestrin, Xgboost: a scalable tree boosting system, in *Proceedings of the 22nd ACM SIGKDD International Conference on Knowledge Discovery and Data Mining* (2016), pp. 785–794
8. T.A. Curry, M.P. Fix, May it please the twitterverse: the use of twitter by state high court judges. J. Inf. Technol. Polit. **16**(4), 379–393 (2019)
9. Diana Fischer-Preßler, Carsten Schwemmer, Kai Fischbach, Collective sense-making in times of crisis: connecting terror management theory with twitter user reactions to the berlin terrorist attack. Comput. Hum. Behav. **100**, 138–151 (2019)
10. G. Forman, I. Cohen, Learning from little: comparison of classifiers given little training, in *European Conference on Principles of Data Mining and Knowledge Discovery* (2004), pp. 161–172 (2004)
11. Jerome H. Friedman, Greedy function approximation: a gradient boosting machine. Ann. Stat. **29**(5), 1189–1232 (2001)
12. T.L. Griffiths, M.I. Jordan, J.B. Tenenbaum, D.M. Blei, Hierarchical topic models and the nested Chinese restaurant process, in *Advances in Neural Information Processing Systems* (2004), pp. 17–24
13. T. Hastie, R. Tibshirani, J. Friedman. *The Elements of Statistical Learning: Data Mining, Inference and Prediction* (Springer, 2009)

14. Matthew Hoffman, Francis R. Bach, David M. Blei, Online learning for latent dirichlet allocation, in *Advances in Neural Information Processing Systems*, vol. 23 (2010), pp. 856–864
15. L. Hong, B.D. Davison, Empirical study of topic modeling in twitter, in *Proceedings of the First Workshop on Social Media Analytics* (2010), pp. 80–88
16. E. Ikonomakis, S. Kotsiantis, V. Tampakas, Text classification using machine learning techniques. WSEAS Trans. Comput. **4**, 966–974 (2005)
17. M. Imran, P. Mitra, C. Castillo, Twitter as a lifeline: human-annotated twitter corpora for NLP of crisis-related messages, in *Proceedings of the Tenth International Conference on Language Resources and Evaluation* (2016), pp. 1638–1643
18. M. Jin, X. Luo, H. Zhu, H.H. Zhuo, Combining deep learning and topic modeling for review understanding in context-aware recommendation, in *Proceedings of the 2018 Conference of the North American Chapter of the Association for Computational Linguistics: Human Language Technologies (Long Papers)*, vol. 1 (2018), pp. 1605–1614
19. G. Kant, C. Weisser, B. Säfken, Ttlocvis: a twitter topic location visualization package. J. Open Source Softw. **5**(54) (2020)
20. Fedor Krasnov, Anastasiia Sen, The number of topics optimization: clustering approach. Mach. Learn. Knowl. Extr. **1**(1), 416–426 (2019)
21. C.-C. Lai, M.-C. Tsai, An empirical performance comparison of machine learning methods for spam e-mail categorization, in *Fourth International Conference on Hybrid Intelligent Systems* (2004), pp. 44–48
22. J.H. Lau, T. Baldwin, T. Cohn, Topically driven neural language model, in *Proceedings of the 55th Annual Meeting of the Association for Computational Linguistics* (2017), pp. 355–365
23. W. Lou, X. Wang, F. Chen, Y. Chen, B. Jiang, H. Zhang, Sequence based prediction of DNA-binding proteins based on hybrid feature selection using random forest and gaussian Naïve Bayes. PloS One **9**(01), e86703 (2014)
24. R. Mehrotra, S. Sanner, W. Buntine, L. Xie, Improving LDA topic models for microblogs via tweet pooling and automatic labeling, in *Proceedings of the 36th international ACM SIGIR Conference on Research and Development in Information Retrieval* (2013), pp. 889–892
25. T. Mikolov, I. Sutskever, K. Chen, G.S. Corrado, J. Dean, Distributed representations of words and phrases and their compositionality, in *Advances in Neural Information Processing Systems* (2013), pp. 3111–3119
26. D. Mimno, H.M. Wallach, E. Talley, M. Leenders, A. McCallum, Optimizing semantic coherence in topic models, in *Proceedings of the Conference on Empirical Methods in Natural Language Processing* (2011), pp. 262–272
27. A. Mishler, E.S. Crabb, S. Paletz, B. Hefright, E. Golonka, Using structural topic modeling to detect events and cluster twitter users in the Ukrainian crisis, in *International Conference on Human-Computer Interaction* (2015), pp. 639–644
28. J. Pennington, R. Socher, C.D. Manning, Glove: global vectors for word representation, in *Proceedings of the 2014 Conference on Empirical Methods in Natural Language Processing* (2014), pp. 1532–1543
29. M.E. Roberts, B.M. Stewart, D. Tingley, et al., STM: R package for structural topic models. J. Stat. Softw. **10**(2), 1–40 (2014)
30. J. Roesslein, Tweepy: twitter for python! (2020). https://github.com/tweepy/tweepy
31. P. Shrestha, S. Sierra, F.A. González, M. Montes, P. Rosso, T. Solorio, Convolutional neural networks for authorship attribution of short texts, in *Proceedings of the 15th Conference of the European Chapter of the Association for Computational Linguistics: Short Papers*, vol. 2 (2017), pp. 669–674 (2017)
32. A. Steinskog, J. Therkelsen, B. Gambäck, Twitter topic modeling by tweet aggregation, in *Proceedings of the 21st Nordic Conference on Computational Linguistics* (2017), pp. 77–86
33. S. Vosoughi, P. Vijayaraghavan, D. Roy, Tweet2vec: learning tweet embeddings using character-level CNN-LSTM encoder-decoder, in *Proceedings of the 39th International ACM SIGIR Conference on Research and Development in Information Retrieval* (2016), pp. 1041–1044

34. C. Wang, J. Paisley, D. Blei, Online variational inference for the hierarchical dirichlet process, in *Proceedings of the Fourteenth International Conference on Artificial Intelligence and Statistics* (2011), pp. 752–760
35. X. Wang, W. Jiang, Z. Luo, Combination of convolutional and recurrent neural network for sentiment analysis of short texts, in *Proceedings of COLING 2016, the 26th international Conference on Computational Linguistics: Technical Papers* (2016), pp. 2428–2437
36. L. Yang, T. Sun, M. Zhang, Q. Mei, We know what@ you# tag: does the dual role affect hashtag adoption? in *Proceedings of the 21st International Conference on World Wide Web* (2012), pp. 261–270
37. W.X. Zhao, J. Jiang, J. Weng, J. He, E.-P. Lim, H. Yan, X. Li, Comparing twitter and traditional media using topic models, in *European Conference on Information Retrieval* (2011), pp. 338–349

Shop Product Tracking and Early Fire Detection Using Edge Devices

Tuan Linh Dang and Viet Tien Ha

Abstract With the blooming of Internet of Things technology, edge device applications have become very popular in many areas. However, edge devices have constraints on power and resources that make edge device applications different from conventional applications. This paper proposes an architecture for tracking shop products and detecting early fire detection using computer vision that can be implemented in edge devices. Experimental results showed that the proposed architecture could be a feasible application that can be deployed to support staff in retail stores.

Keywords Edge device · Computer vision · Product tracking · Fire detection

1 Introduction

An edge device is any hardware device that is responsible for controlling data flow at the boundary of two different networks. The edge device can also be thought of as a point of entry into company or service provider core networks. Previously, Internet of things (IoT) devices normally sent data to the central server by the edge device [1]. However, the amount of transmission data, especially video-related data, can lead to server overload, increased bandwidth costs, or bottlenecks in data transmission. Therefore, processing at the edge device is gradually becoming a new trend in IoT research. In this situation, edge devices are also equipped with additional computing capabilities to handle machine learning problems.

The use of edge devices and machine learning is becoming more popular thanks to the continuous development of technology. In the retail industry, the applications of computer vision and edge devices are diverse. Some practical applications are

T. L. Dang (✉) · V. T. Ha
School of Information and Communications Technology, Hanoi University of Science and Technology, Hanoi 100000, Vietnam
e-mail: linhdt@soict.hust.edu.vn

V. T. Ha
e-mail: tien.hv167467@sis.hust.edu.vn

N. H. Phuong and V. Kreinovich (eds.), *Biomedical and Other Applications of Soft Computing*, Studies in Computational Intelligence 1045,
https://doi.org/10.1007/978-3-031-08580-2_12

out-of-stock checking, customer behavior analysis, stealing detection, and fire prevention [2–4].

According to some surveys and studies, out-of-stock situations can cause harm to retailers. In this situation, customers may increase their dissatisfaction or do not want to use the service again [5]. On the other hand, customer behavior analysis may help retailers to increase revenue, for example, knowing which products customers pick up but do not purchase. Early fire detection using computer vision may improve the accuracy of the installed fire alarm system inside the store and may help to detect fire outside the store where there are no sensors.

There have been different studies in this research field. However, most of them focus on proposing algorithms [4, 6–15]. In addition, not many studies have effectively implemented these algorithms.

The main contribution of this paper is to propose a suitable approach for shop product tracking and early fire detection implemented on edge devices with computer vision.

This paper is structured as follows. Section 2 describes related work, and Sect. 3 presents our proposed application architecture. The approaches for early fire detection and shop product tracking are also presented in this section. Section 4 presents the results. Section 5 concludes our paper.

2 Related Work

2.1 Real World Applications

The demands for product recognition and product tracking have been increasing dramatically. Hence, various enterprise applications have been invented to handle these tasks. One of the most famous examples is Amazon Go which uses recognition and tracking to identify what the customers are putting into their carts [16]. Another company, Angus, employs product recognition techniques to check out-of-stock products automatically [17].

The fire detection also attracted many researchers [4, 6–8]. Vision-based approaches can be employed to make conventional fire alarm systems, which have only sensors, more accurate. In addition, computer vision may also detect fire outside the store, where there are no sensors.

2.2 Technologies

2.2.1 Product Recognition

One of the earliest studies for grocery products recognition was presented by Merler that used local invariant features [18]. However, experimental results showed that this approach was not suitable for a real-world deployment due to low accuracy and slow speed. After that, different studies have focused on improving recognition capability, such as using the combination between strong features and classification, using the correlation between products on shelves, or using hierarchical multi-stage recognition [9–13]. However, to our knowledge, the previous manuscript solutions were not scalable. Their approaches can recognize a maximum of hundreds of different items, while usually, in a real shop, thousands of products are on sale. By applying a single model image for each product, another paper has proposed a multi-stage system that can scale up to 3400 different products [14]. In 2018, a group of researchers proposed an approach that obtained higher accuracy than previous studies by using neural networks [15].

2.2.2 Object Tracking

Object tracking tries to locate object positions in different video frames which can be used for autonomous driving, camera surveillance, or sports analytics. Object tracking is divided into single object tracking (SOT) and multiple object tracking (MOT). The SOT tracker can track an object in the subsequent frames using a bounding box containing this object without training. The most common and robust tracking algorithms for such tasks are Siamese network-based and correlation filter-based tracking algorithms. On the other hand, MOT trackers track multiple objects. The tracking algorithm of this type is responsible for determining the number of objects in each frame and keeping track of their identities. MOT problem is more difficult because it is hard to find a class of tracking algorithms that would outperform other algorithms [19].

2.2.3 Image Classification

Image classification tries to determine the probability of an object in an image as one "class". The class, considered as a label, can be "people" or "vehicles".

Bag of Feature is an early image classification algorithm that does not use a convolutional network [20]. This algorithm is a lightweight approach that may obtain high inference time. However, the accuracy of this approach is affected by backgrounds, angles, or poses. LeNet, one of the first convolutional networks (CNN) was presented in 1989 [21]. However, the idea of CNN was not popular until the appearance of AlexNet [22] in 2012. Since then, different CNN models have been proposed such as VGGNet [23] in 2014, ResNet [24] in 2015, and EfficientNet [25] in 2019.

3 Method

3.1 Architecture

Figure 1 describes our application architecture. Firstly, the edge device connects to a camera to receive the images of the product shelf in real-time. After that, the edge device asks the server to create a separate room that will be used for clients to connect and to receive video streams and notifications from the edge device. The server can create multiple rooms for different edge devices.

To track products on the shelf, the edge device sends requests to the server to receive information regarding product positions and types. The product detection module is implemented in the server instead of the edge device because for several reasons which will be explained in Sect. 3.2. The early fire detection task is a cron job to check for a fire threat after each time interval.

After the clients ask to join the edge device room, they will make a peer-to-peer (P2P) connection with the edge device to receive shelf images as a stream and fire alarm messages via Web Real-Time Communication (WebRTC) from the edge device to all clients if there is a fire threat.

3.2 Product Recognition Module

This module architecture is based on previous research [15]. Figure 2 describes the process of adding new products into the application and how to query and get product information from the product images.

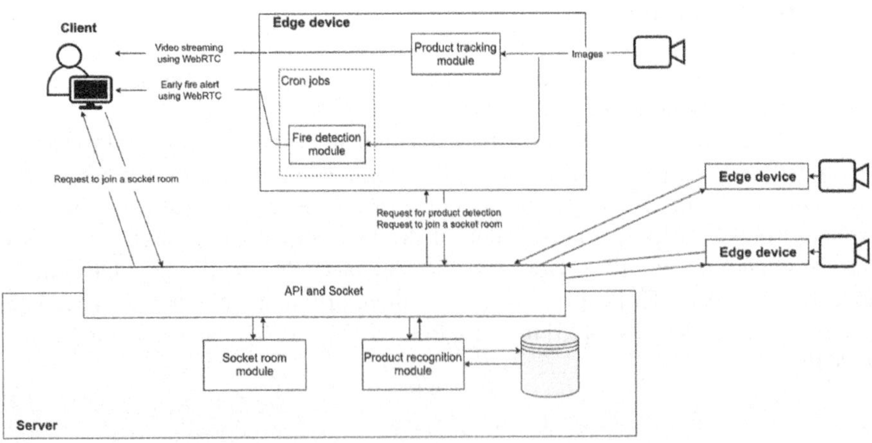

Fig. 1 Application architecture

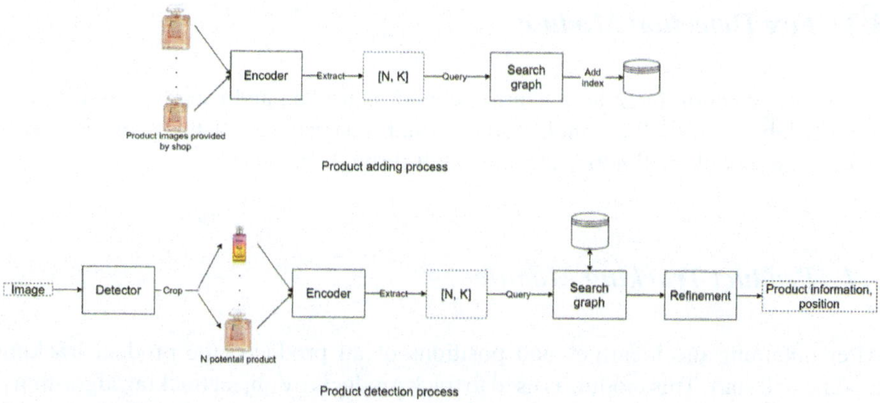

Fig. 2 Product adding and product detection process

This module architecture is based on previous research [15]. Figure 2 describes the process of adding new products into the tracking application (adding process) and how to query and get product information from the product images (detection process).

The adding process uses easy-to-recognize cropped images of one product. These input images are processed and sent through an encoder to get an output matrix of size [N, K], where N is the number of input images and K is the feature vector length. The output matrix is sent to a search graph which is the approximate nearest neighbor search algorithm. The added graph vector indexes are also mapped with the product in the database for a future query.

In the recognition process, the product images are sent to a detector. The detector output is the proposed product positions. According to proposed positions, the product images are cropped and passed to the encoder to get the output matrix with size [N, K], where N is the number of input images and K denotes the feature vector length. Then, each of these feature vectors is queried in the search graph to get the closest vector. If the distance between two vectors is smaller than a threshold, the proposed position is recognized as a product. Finally, the refinement module is used to refine the bounding box positions.

This module should be implemented on the server because several reasons.

- The recognition process does not operate frequently. As a result, putting this module on edge devices is a waste of resources.
- The module needs much access to the database. Thus, directly connecting to a database from a server makes the architecture more efficient.

3.3 Fire Detection Module

In the fire detection module, a neural network is trained and used to classify three product labels called fire, smoke, and normal, respectively. If the result is fire or smoke, the module will notify the shop staff through WebRTC.

3.4 Product Tracking Module

After obtaining the identities and positions of all products, the product tracking module will start. This module is used to track products by object tracking algorithms.

4 Experiments

Although the proposed approach can be used to track different products, this manuscript focuses on perfume as a product in the experiments.

4.1 Datasets

4.1.1 Perfume Shape Dataset

The detector must be trained on an image dataset of product shapes to detect the proposed product positions. The dataset can be images of any given shape, but perfume is the focus of this research. It is necessary to get a dataset containing images of perfume bottles in various shapes. Our "Perfume shape" dataset is crawled from Google Image. Our self-built dataset can be download in [26].

4.1.2 Fire Smoke Dataset

The "fire smoke dataset" provided by Kaggle is used for fire and smoke detection. However, these images are outdoor images, which are not very similar to fire in shops. So, the dataset is customized by adding more indoor fire images. The customized dataset can be downloaded in [27].

4.2 *Choosing Algorithms*

4.2.1 Product Recognition Module

Different state-of-the-art neural network models were evaluated in the product detection module. The comparison results between different models can be observed in Table 1 concerning the accuracies and model sizes. Each model is trained with a batch size of 32, epochs of 100, and Adam optimizer [28]. The loss function for each neural network is the same as their original research.

Experimental results show that the EfficientDet-D2 [31] obtained better accuracy and efficiency than other models. With approximately five times fewer parameters and nine times fewer floating-point operations per second (FLOPS), the EfficientDet-D2 [31] still got the best accuracy among these models.

The algorithms used for the encoder in the product detection module are neural network backbones. This experiment used Resnet 50 as the backbone with the pre-trained weights as encoder [24]. The use of pre-trained models overcomes the challenges of determining how much data is sufficient and how to obtain a proper classification model.

The algorithm used for the Search graph is HNSW [32]. The decision was made according to the researched benchmark [33].

The Refinement algorithm is the same as the original research [15].

4.2.2 Fire Detection Module

The neural networks were used for the fire detection algorithm. To use with edge devices, the lightweight NN models were investigated. The MobileNetV2 [34], MobileNetV3 [35], and SqueezeNetV1.1 [36] were examined to compare and to choose the most suitable model. Table 2 demonstrates the comparison between these lightweight models in terms of accuracies and sizes. Each model was trained with a batch size of 32, epochs of 100, cross-entropy loss, and Adam optimizer as similar to previous research [28].

Table 1 Comparison of the accuracy and size of different models on self-built perfume shape dataset

Model	Average precision	Parameters	FLOPs
RetinaNet-R50 [29]	63.66	34M	97B
RetinaNet-R101 [29]	65.23	53M	127B
YOLOv3 [30]	61.15	65M	71B
EfficientDet-D2 [31]	67.25	8.1M	11B

Table 2 Comparison of the accuracy and size of different image classification models on customized "fire smoke" dataset

Model	Top 1 accuracy	Parameter	FLOPs
MobileNetV3 [35]	0.94	5M	225M
MobileNetV2 [34]	0.91	4M	314M
SqueezeNetV1.1 [36]	0.88	1M	352M

The results show that the MobileNetV3 [35] achieved the best accuracy. In addition, the MobileNetV3 also had the least FLOPs, and the number of parameters was not too high compared to other models.

4.2.3 Product Tracking Module

The tracking algorithm needs to meet the requirements of being lightweight, tracking multiple objects, and fast. But due to the limitation in power and resources of our experiment device and the fact that most MOT algorithms are heavyweight, the Kernelized Correlation Filter algorithm was chosen [37]. This detection-free SOT algorithm was used to apply to every product that was detected.

4.3 Application Testing

The architecture proposed in Sect. 3.1 and the algorithms chosen in Sect. 4.2 were selected to build the application in our experiments.

4.3.1 Hardware Configurations

- Server:
  - CPU: Intel Core i5-5200U, 8 GB RAM
  - GPU: NVidia Geforce GT 940M, 2 GB VRAM
- Edge device:
  - Model: Raspberry Pi 3B+
  - CPU: Quad Core 1.2 GHz Broadcom BCM2837 64bit, 1 GB RAM

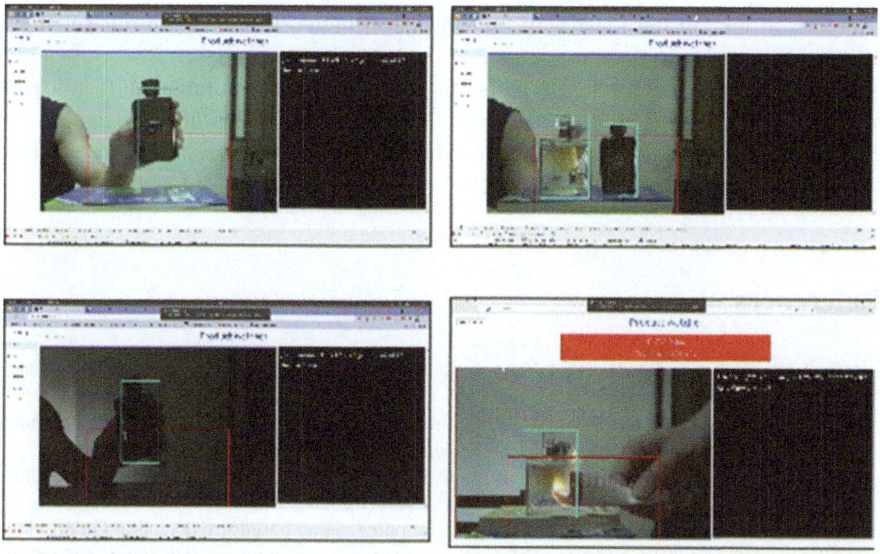

Fig. 3 Some of our test cases

4.3.2 Results

Experimental shows that the speed of image streaming was approximately 12 to 14 frames per second (FPS). With the newer model of edge devices such as the Raspberry Pi 4, which is almost four times faster than the current device used in our experiments, the FPS could potentially reach up to 40 or 50 FPS.

We also tested the accuracy of product detection, product tracking, and fire detection on various backgrounds, light conditions, and products. Experimental results were shown in Fig. 3. With 200 different combinations of cases, the precision was 0.88 while the recall was 0.96.

5 Conclusions

This paper proposes a potential approach for shop product tracking and detecting early fire using edge devices. The server in our architecture not only manages edge devices and clients but also performs compute-intensive tasks. In this situation, the server handles the product detection task. The edge devices manage the tracking of detected products and detecting fire. To make communication between edge devices and clients more efficient, a P2P connection and stream information via WebRTC and socket are employed. In addition, an approach for detecting products using object detection, encoding, and nearest neighbor search algorithms is also proposed. The comparison among the most well-known neural network models on our self-built customized dataset is also investigated.

A possible avenue for future research is to investigate suitable MOT tracking algorithms with more powerful edge devices. It is also necessary to examine the effectiveness of our approach with real-life scenarios. Our proposed architecture could also be tested with other products.

References

1. W.Z. Khan, E. Amed, S. Hakak, I. Yaqoob, A. Ahmed, Edge computing: a survey. Future Gener. Comput. Syst. Elsevier **97**, 219–235 (2019)
2. M.A. Majdi, B.S.B. Dewantara, M.M. Bachtíar, Product stock management using computer vision. in *2020 International Electronics Symposium* (IEEE, 2020), pp. 424–429
3. M. Wang, Consumer behavior analysis in the offline retail stores based on convolutional neural network. J. Phys. Conf. Ser., IOP Publishing **1544**(1) 2020
4. G. Sathyakala, V. Kirthika, B. Aishwarya, Computer vision based fire detection with a video alert system, in *2018 International Conference on Communication and Signal Processing* (IEEE, 2018), pp. 0725–0727
5. https://www.unleashedsoftware.com/blog/consumer-response-stockouts-can-significantly-impact-business [Accessed 29 July 2021]
6. H. Wu, D. Wu, J. Zhao, An intelligent fire detection approach through cameras based on computer vision methods. Process Saf. Environ. Prot., Elsevier **127**, 245–256 (2019)
7. B.U. Töreyin, Y. Dedeoğlu, U. Güdükbay, A.E. Cetin, Computer vision based method for real-time fire and flame detection. Pattern Recognit. Lett., Elsevier **27**(1), 49–58 (2006)
8. T. Celik, K.K. Ma, Computer vision based fire detection in color images, in *2008 IEEE Conference on Soft Computing in Industrial Applications* (IEEE, 2008), pp. 258-263
9. M. Cotter, S. Advani, J. Sampson, K. Irick, V. Narayanan, A hardware accelerated multilevel visual classifier for embedded visual-assist systems, in *2014 IEEE/ACM International Conference on Computer-Aided Design (ICCAD)* (IEEE, 2014), pp. 96–100
10. S. Advani, B. Smith, Y. Tanabe, K. Irick, M. Cotter, J. Sampson, V. Narayanan, Visual co-occurrence network: using context for large-scale object recognition in retail, in *2015 13th IEEE Symposium on Embedded Systems for Real-time Multimedia (ESTIMedia)* (IEEE, 2015), pp. 1–10
11. I. Baz, E. Yoruk, M. Cetin, Context-aware hybrid classification system for fine-grained retail product recognition, in *2016 IEEE 12th Image, Video, and Multidimensional Signal Processing Workshop (IVMSP)* (IEEE, 2016), pp. 1–5
12. A. Tonioni, L. Di Stefano, Product recognition in store shelves as a sub-graph isomorphism problem, in *International Conference on Image Analysis and Processing* (Springer, 2017), pp. 682–693
13. A. Franco, D. Maltoni, S. Papi, Grocery product detection and recognition. Expert Syst. Appl., Elsevier **81**, 163–176 (2017)
14. M. George, C. Floerkemeier, Recognizing products: a per-exemplar multi-label image classification approach, in *European Conference on Computer Vision* (Springer, 2014), pp. 440–455
15. A. Tonioni, E. Serra, L. Di Stefano, A deep learning pipeline for product recognition on store shelves, in *2018 IEEE International Conference on Image Processing, Applications and Systems (IPAS)* (IEEE, 2018), pp. 25–31
16. A. Polacco, K. Backes, The amazon go concept: implications, applications, and sustainability. J. Bus. Manag. **24**(1), 79–92 (2018)
17. https://www.angus.ai/ [Accessed 29 July 2021]

18. M. Merler, C. Galleguillos, S. Belongie, Recognizing groceries in situ using in vitro training data, in *2007 IEEE Conference on Computer Vision and Pattern Recognition* (IEEE, 2007), pp. 1–8
19. https://medium.com/@cindy.trinh.sridykhan/a-tour-of-video-object-tracking-part-i-presentation-8a8aa9da9394 [Accessed 29 July 2021]
20. G. Csurka, C. Dance, L. Fan, J. Willamowski, C. Bray, Visual categorization with bags of keypoints, in *Workshop on Statistical Learning in Computer Vision*, vol. 1, no. 1–22 (ECCV, 2004), pp. 1–2
21. Y. Le Cun, L.D. Jackel, B. Boser, J.S. Denker, H.P. Graf, I. Guyon, D. Henderson, R.E. Howard, W. Hubbard, Handwritten digit recognition: applications of neural net chips and automatic learning, in *Neurocomputing* (Springer, 1990), pp. 303–318
22. A. Krizhevsky, I. Sutskever, G.E. Hinton, Imagenet classification with deep convolutional neural networks. Adv. Neural Inf. Process. Syst. **25**, 1097–1105 (2012)
23. K. Simonyan, A. Zisserman, Very deep convolutional networks for large-scale image recognition (2014). arXiv:1409.1556
24. K. He, X. Zhang, S. Ren, J. Sun, Deep residual learning for image recognition, in *Proceedings of the IEEE conference on computer vision and pattern recognition* (IEEE, 2016), pp. 770–778
25. M. Tan, Q. Le, Efficientnet: rethinking model scaling for convolutional neural networks. Int. Conf. Mach. Learn. PMLR **97**, 6105–6114 (2019)
26. https://drive.google.com/file/d/1SS7lYVVHz8P2DiU2HxCpxVN35U8citkm [Accessed 29 July 2021]
27. https://drive.google.com/file/d/1yI66qHowj4dWH7bOGcsqWyTSU6Rb2NOi [Accessed 29 July 2021]
28. D.P. Kingma, J. Ba, Adam: a method for stochastic optimization, in *International Conference on Learning Representations (ICLR)* (2014). arXiv:1412.6980
29. T.Y. Lin, P. Goyal, R. Girshick, K. He, P. Dollár, Focal loss for dense object detection, in *Proceedings of the IEEE International Conference on Computer Vision* (IEEE, 2017), pp. 2980–2988
30. J. Redmon, A. Farhadi, Yolov3: an incremental improvement (2018). arXiv:1804.02767
31. M. Tan, R. Pang, Q.V. Le, Efficientdet: Scalable and efficient object detection, in *Proceedings of the IEEE/CVF Conference on Computer Vision and Pattern Recognition* (IEEE, 2020), pp. 10781–10790
32. Y.A. Malkov, D.A. Yashunin, Efficient and robust approximate nearest neighbor search using hierarchical navigable small world graphs. IEEE Trans. Pattern Anal. Mach. Intell., IEEE Computer Society **42**(4), 824–836 (2018)
33. https://github.com/erikbern/ann-benchmarks [Accessed 29 July 2021]
34. M. Sandler, A. Howard, M. Zhu, A. Zhmoginov, L.C. Chen, Mobilenetv2: inverted residuals and linear bottlenecks, in *Proceedings of the IEEE conference on computer vision and pattern recognition* (IEEE, 2018), pp. 4510–4520
35. A. Howard, M. Sandler, G. Chu, L.C. Chen, B. Chen, M. Tan, W. Wang, Y. Zhu, R. Pang, V. Vasudevan, Q.V. Le, Searching for mobilenetv3, in *Proceedings of the IEEE/CVF International Conference on Computer Vision* (IEEE, 2019), pp. 1314–1324
36. F.N. Iandola, S. Han, M.W. Moskewicz, K. Ashraf, W.J. Dally, K. Keutzer, SqueezeNet: AlexNet-level accuracy with 50x fewer parameters and< 0.5 MB model size (2016). arXiv:1602.07360
37. J.F. Henriques, R. Caseiro, P. Martins, J. Batista, High-speed tracking with kernelized correlation filters. IEEE Trans. Pattern Anal. Mach. Intell., IEEE Computer Society **37**(3), 583–596 (2014)

SDNs Delay Prediction Using Machine Learning Algorithms

Tuan Linh Dang and Nhat Minh Ngo

Abstract The transmitting time of a packet between two devices is an essential factor in evaluating the network quality. Previous studies have applied machine learning to predict the connection delay value between two devices in traditional networks. However, there is little research using Software-Defined Networks (SDNs) because of the lack of SDNs traffic datasets. A method for collecting SDNs traffic data with delay values is proposed in this manuscript. In addition, this paper also evaluates different learning algorithms for SDNs delay prediction using the collected data. Experimental results showed that the Bidirectional LSTM had the lowest losses and the Recurrent Neural Networks had the shortest inference time.

Keywords SDNs · Machine learning algorithms · Network delay prediction

1 Introduction

A delay in the network between two network devices directly affects the network quality. Thus, the network delay value prediction may improve the network quality by avoiding packet transmission through high delay locations.

Many studies have tried to solve the problem related to the delay on traditional computer networks using different machine learning approaches [1–7]. A network delay can be predicted using regression algorithms. On the other hand, a comparison between classic machine learning algorithms and deep learning algorithms has also been conducted. The Recurrent Neural Networks (RNN) obtained the highest accuracies in predicting traditional network traffic in this experiment [8].

T. L. Dang (✉)
School of Information and Communications Technology, Hanoi University of Science and Technology, Hanoi 100000, Vietnam
e-mail: linhdt@soict.hust.edu.vn

N. M. Ngo
Hanoi University of Science and Technology, Hanoi 100000, Vietnam
e-mail: minh.nn170098@sis.hust.edu.vn

N. H. Phuong and V. Kreinovich (eds.), *Biomedical and Other Applications of Soft Computing*, Studies in Computational Intelligence 1045,
https://doi.org/10.1007/978-3-031-08580-2_13

However, there are still a few studies that predict a delay in Software-Defined Networks (SDNs) that have a separation between network control planes and forwarding planes [10]. The SDNs architecture moves the logic controls to the controller layers. The forwarding planes that do not have logic controls will operate by the rules set by the controllers. Hence, this architecture can be directly programmed and dealt with the high bandwidth. The details can be seen in the previous manuscript [10].

The previous research has only tried to classify SDNs traffic using naive Bayes, support vector machine, and nearest centroid algorithm [9]. Other researchers have examined SDNs traffic classification with RNN and Long-Short Term Memory (LSTM) [11, 12]. To our knowledge, there is limited study on SDNs traffic delay since one of the main reasons is a lack of SDNs dataset.

The main contribution of this paper is to propose a method to collect data from SDNs traffic. A self-built SDNs dataset is presented. In addition, this paper also investigates the SDNs delay in our self-built dataset using different machine learning approaches.

The structure of this manuscript is as follows. Section 2 introduces the related work. Section 3 proposed our data collecting method. The delay prediction is also presented in this section. Section 4 illustrates the experimental results. Finally, Sect. 5 concludes our paper.

2 Related Work

2.1 Machine Learning Algorithms

- Classic regression algorithms have fast inference time and a simple training process even these classic algorithms may not be suitable for complex problems. This paper investigates three different regression algorithms called linear regression, ridge regression, and lasso regression, respectively [13].
- Neural network-based algorithms are also investigated to predict SDNs delay values. RNN is a class of neural networks that uses internal states to process variable-length sequences of inputs. LSTM is an improved version of RNN that enhances the ability to remember information [14]. Previous studies have employed the RNN to predict network traffic on the traditional computer network [4, 5]. This study examines RNN, LSTM, and bidirectional LSTM to predict SDNs delay value.

2.2 Evaluation Metrics

The mean squared error (MSE) and the mean absolute error (MAE) are used for evaluating both the classic machine learning algorithms and neural network-based algorithms with time-series data. The calculations for MSE and MAE are presented in Eqs. (1) and (2).

$$MSE = \frac{1}{n}\sum_{i=1}^{n}(y_i - \widehat{y}_i)^2 \quad (1)$$

$$MAE = \frac{1}{n}\sum_{i=1}^{n}|y_i - \widehat{y}_i| \quad (2)$$

3 Method

3.1 Data Collection

This paper proposes a method to collect SDNs traffic using Mininet [15] and TCPReplay [16].

Configurations

- SDNs Controller: it is necessary to have an SDNs controller to setup all "host to host" flows between all hosts of the Topology. This paper uses **ONOS** with **Reactive Forwarding** which setups all packets going with the shortest path [17].
- Dataplane Simulation: this manuscript employs **Mininet** for simulation software. On the other hand, **BSO Network** is considered as topology. The **BSO Network** has 14 hosts, 14 switches, and 36 links which are supplied by **Topology Zoo** [18].

Conducting

This paper simulates the SDNs traffic using **TCP Replay** to send prepared packets through the hosts to the network. Our SDNs controller will direct all packets to the correct destination.

- Prepare virtual network traffic: it is necessary to have a **pcap** file with all inside IP addresses similar to our simulated topology to create network traffic using **TCP Replay**. By default, the **Mininet** creates the IP addresses of hosts in the format of "**10.0.0.n**" with **n** is the ID of this host. Our research uses a random **pcap** file while **Wireshark** [19] and **Bit-Twist** [20] are used to change the IP addresses. The output of this step is a **pcap** file that contains network traffic.
- Getting network traffic information: this study collects information on every link in the topology. Specifically, for each link, the information is collected as follows.
  - Delay value: this value is the required time to transmit a packet from the source to the destination. The delay value is approximately equal to round trip time (RTT) divided by two. The RTT value is measured using ping in **Mininet**.

Table 1 SDNs dataset feature description

Feature name	Describe	Unit
Timestamp	The timestamp on this link	
Node 1	Identity number of the first node of the link	
Node 2	Identity number of the second node of the link	
Bandwidth	The maximum capacity of this link which is setup by **Mininet**	Gbits/s
FW-Bytes	Average number of bytes transmitted through this link from the first node to the second node per second	bytes/s
BW-bytes	Average number of packets transmitted through this link from the second node to the first node per second	bytes/s
FW-pkts	Average number of packets transmitted through this link from first node to second node per second	packets/s
BW-pkts	Average number of packets transmitted through this link from second node to first node per second	packets/s
Delay	Delay value on this link	ms

– Other network traffic features: this paper also collects other SDNs traffic features as described in Table 1. The **openvswitch** commands can collect packets and bytes transmitted through a switch. The timestamps are also recorded.

3.2 Data Preprocessing

This study uses Pandas, Scikit-learn, Spicy, and Numpy [21–24] to preprocess data. The preprocessing phase includes data cleaning, outliers removal, and data normalization.

- Data cleaning removes all records with non-numeric values or no values.
- Outliers removal calculates the **z-score** of all records using the **Spicy** and the **Pandas** tool. After that, this step removes outlier records that have the **z-score** of any feature larger than three.
- Data normalization uses **MinMaxScaler** to scale the values of one feature to be in a range from 0 to 1.

3.3 Training and Evaluation

This paper investigates the preprocessed data using different machine learning approaches to find a suitable solution to predict the SDNs delay value. Our research uses Sklearn [23] and Keras [25] libraries in the experiments. For a fair comparison, all classic regression algorithms use the same configuration, and all neural network-based algorithms also get a similar configuration. The parameters are as follows.

- Classic regression algorithms
  - Linear regression: no special configuration
  - Lasso regression: alpha = 0.01
  - Ridge regression: alpha = 0.01
- RNN, LSTM, Bidirectional LSTM:
  - Architecture: the hidden layer contains 128 units of the corresponding algorithm (RNN, LSTM, or Bidirectional LSTM). Each unit is fully connected to only one output unit at the output layer.
  - Optimizer: adam
  - Learning rate: 0.001
  - Early stopping

Input for Training

After preprocessing, the data is used in the training phase. Ridge, Lasso, and Linear regression algorithms use directly preprocessed data. On the other hand, some continuous data rows in preprocessed data are combined to create inputs for the network-based algorithms. Details of the data are as follows.

- Classic regression algorithm uses seven columns of the dataset which are Node 1, Node 2, Bandwidth, FW-Bytes, BW-Bytes, FW-pkts, and BW-pkts.
- Neural network-based algorithm employs five columns of dataset which are Bandwidth, FW-Bytes, BW-Bytes, FW-pkts, and BW-pkts. In addition, ten continuous rows in the dataset are combined to make a data sample.

4 Experiments

4.1 Datasets and Data Preprocessing

Our experiment captured 558450 records of network traffic on 14 links on BSO topology by simulating SDNs in three days. There were no missing values in this dataset that had a size of 69.6 MB. The timestamps were also stored as inputs for algorithms that need time-series input data. Table 2 illustrates the summary of our self-built dataset.

In the preprocessing phase, 18001 outliers were removed. In addition, The MinMaxScaler was used to scale all Float64 features.

Table 2 SDNs dataset feature summary

Column	Number of flows	DType
Timestamp	558450	Float64
Node1	558450	Int64
Node2	558450	Int64
Bandwidth	558450	Float64
FW-pkts	558450	Float64
BW-pkts	558450	Float64
FW-bytes	558450	Float64
BW-bytes	558450	Float64
Delay	558450	Float64

4.2 Comparison Between Different Approaches

4.2.1 Hardware Configuration

The experiments were conducted on Google Colab with hardware configuration was as follows.

- CPU: Intel Xeon Processor with two cores @2.30 GHz
- GPU: No
- RAM: 13 GB.

4.2.2 Classic Regression Algorithms

As mentioned, linear regression, ridge regression, and lasso regression were investigated in our research. Each algorithm was trained on our preprocessed data and evaluated by the MAE and MSE. Table 3 shows the experimental results.

As can be seen, all algorithms had similar MAE and MSE in the training and the testing phases. In addition, linear regression had the shortest inference time at 0.075 ms. On the other hand, the inference times for the Ridge regression and Lasso regression were 0.078 ms and 0.085 ms, respectively.

4.2.3 Neural Network-Based Algorithms

Our research investigated three neural network-based algorithms called RNN, LSTM, and Bidirectional LSTM, respectively. Each algorithm was trained on our preprocessed data. The MSE and MAE were selected as two evaluated metrics. Table 4 demonstrates the experimental results.

Table 3 Classic regression algorithms with MAE, MSE and inference time

Model	Training		Testing		Inference time (ms)
	MAE	MSE	MAE	MSE	
Linear regression	0.052915	0.013533	0.052958	0.00327	0.075
Ridge regression	0.052915	0.013533	0.052958	0.00327	0.078
Lasso regression	0.052915	0.013533	0.052958	0.00327	0.085

Table 4 Neural network-based algorithms with MAE, MSE and inference time

Model	Validation		Testing		Inference time (ms)
	MAE	MSE	MAE	MSE	
RNN	0.02955	0.00299	0.02958	0.00328	39.292
LSTM	0.02861	0.00299	0.02875	0.00328	40.163
Bidirection LSTM	0.02856	0.00299	0.02875	0.00315	43.233

All three models had similar MSE errors on the validation dataset. The lowest MSE testing error belonged to Bidirectional LSTM. In addition, with the MAE evaluate metric, the validation error and test error of bidirectional LSTM were still the lowest. Results show that the more complex the model was, the lower the score obtained and the longer the inference time.

5 Conclusions

This paper presents a method to collect network traffic on SDNs that can be used in future research. The SDNs delay prediction is also investigated with not only classic regression algorithms but also neural network-based algorithms in our self-built dataset. The examined regression algorithms are linear regression, ridge regression, and lasso regression. On the other hand, RNN, LSTM, and bidirectional LSTM are selected as neural network-based algorithms.

Experimental results showed that the classic regression algorithms had the shortest inference time at 0.075 ms. However, the MAE and MSE in the delay prediction of the classic regression algorithms were still high. For the neural network-based algorithms, the MAE and MSE in validation and testing datasets were much lower than the error values in classic regression algorithms. The bidirectional LSTM had the lowest MAE and MSE on both validation and testing sets. On the other hand, the inference time of RNN was fastest at 39.292 ms.

The data collecting method proposed in this paper may solve the problem related to the lack of SDNs traffic datasets. This proposed approach can be used in the future to build a new SDNs traffic dataset. Furthermore, the results also showed that the neural network-based algorithms were more suitable for dealing with network traffic delay prediction than the classic models.

A potential area for further research is to broaden the dataset by collecting more features. Another direction for future research is to develop real-time SDNs traffic prediction in a practical application.

References

1. H. Feng, Y. Shu, Study on network traffic prediction techniques, in *Proceedings of 2005 International Conference on Wireless Communications, Networking and Mobile Computing, 2005* (2005), pp. 1041–1044. https://doi.org/10.1109/WCNM.2005.1544219
2. H. Zare Moayedi, M.A. Masnadi-Shirazi, Arima model for network traffic prediction and anomaly detection, in *2008 International Symposium on Information Technology* (2008), pp. 1–6. https://doi.org/10.1109/ITSIM.2008.4631947
3. R. Vinayakumar, K.P. Soman, P. Poornachandran, Applying deep learning approaches for network traffic prediction, in *2017 International Conference on Advances in Computing, Communications and Informatics (ICACCI)* (2017), pp. 2353–2358. https://doi.org/10.1109/ICACCI.2017.8126198
4. R. Madan, P.S. Mangipudi, Predicting computer network traffic: a time series forecasting approach using DWT, ARIMA and RNN, in *2018 Eleventh International Conference on Contemporary Computing (IC3)* (2018), pp. 1–5. https://doi.org/10.1109/IC3.2018.8530608
5. N. Ramakrishnan, T. Soni, Network traffic prediction using recurrent neural networks, in *2018 17th IEEE International Conference on Machine Learning and Applications (ICMLA)* (2018), pp. 187–193. https://doi.org/10.1109/ICMLA.2018.00035
6. W. Shihao, Z. Qinzheng, Y. Han, L. Qianmu, Q. Yong, A network traffic prediction method based on LSTM. ZTE Commun. **17**(2), 19–25 (2019)
7. M.F. Iqbal, M. Zahid, D. Habib, L.K. John, Efficient prediction of network traffic for real-time applications. J. Comput. Netw. Commun. **2019** (2019). https://doi.org/10.1155/2019/4067135
8. T.P. Oliveira, J.S. Barbar, A.S. Soares, Computer network traffic prediction: a comparison between traditional and deep learning neural networks. Int. J. Big Data Intell **3**, 28–37 (2016)
9. M.M. Raikar, M.S. Meena, M.M. Mulla, N.S. Shetti, M. Karanandi, Data traffic classification in software defined networks (SDNs) using supervised-learning. Proc. Comput. Sci. **171**, 2750–2759 (2020). https://doi.org/10.1016/j.procs.2020.04.299
10. https://opennetworking.org/sdn-definition/ [Accessed 29/07/2021]
11. A. Yassine, H. Rahimi, S. Shirmohammadi, Software defined network traffic measurement: current trends and challenges software defined network traffic measurement: current trends and challenges. IEEE Instrum. Meas. Mag. **18**(2), 42–50 (2015). https://doi.org/10.1109/MIM.2015.7066685
12. A.R. Mohammed, S.A. Mohammed, S. Shirmohammadi, Machine learning and deep learning based traffic classification and prediction in software defined networking. IEEE Int. Symp. Meas. Netw. (M&N) **2019**, 1–6 (2019). https://doi.org/10.1109/IWMN.2019.8805044
13. M.A. Golberg, A.C. Hokwon, *Introduction to Regression Analysis* (WIT press, 2004)
14. A. Sherstinsky, Fundamentals of recurrent neural network (RNN) and long short-term memory (LSTM) network. Phys. D: Nonlinear Phenom. **404** (2020). https://doi.org/10.1016/j.physd.2019.132306

15. https://github.com/mininet/mininet/wiki/Documentation [Accessed 29/07/2021]
16. https://tcpreplay.appneta.com/wiki/howto.html [Accessed 29/07/2021]
17. https://opennetworking.org/onos/ [Accessed 29/07/2021]
18. S. Knight, H.X. Nguyen, N. Falkner, R. Bowden, M. Roughan, The internet topology zoo. IEEE J. Sel. Areas Commun. **29**(9), 1765–1775 (2011). https://doi.org/10.1109/JSAC.2011.111002
19. https://www.wireshark.org [Accessed 29/07/2021]
20. http://bittwist.sourceforge.net/doc.html [Accessed 29/07/2021]
21. https://numpy.org/doc/stable/user/index.html [Accessed 29/07/2021]
22. https://pandas.pydata.org/docs/user_guide/index.html [Accessed 29/07/2021]
23. https://scikit-learn.org/stable/user_guide.html [Accessed 29/07/2021]
24. https://docs.scipy.org/doc/scipy/reference/ [Accessed 29/07/2021]
25. https://keras.io/guides/developer-guides [Accessed 29/07/2021]

A Linear Neural Network Approach for Solving Partial Differential Equations on Porous Domains

T. T. V. Le, C.-D. Le, and K. Le-Cao

Abstract This paper presents a typical linear neural network (e.g. radial basis function networks), for solving partial differential equations (PDEs) in a domain $\mathbb{D}_R = \mathbb{D} \setminus \Gamma$, where $\mathbb{D}$ in R^2 and Γ is the set of continuous circular subdomains. The computational domain $\mathbb{D}$ is discretised by a uniform radial basis function grid nodes. The solution of the given PDEs is assumed to vary over interior subdomains according to appropriate analytic functions (e.g. polynomials) that are predefined in order to satisfy the inner boundary conditions and their first and second derivatives (i.e. C^2-continuous). Several boundary-value and initial-value problems, some of which have exact solutions, are considered to validate the proposed technique.

Keywords Partial differential equation · Irregular domains · Artificial neural networks

1 Introduction

Porous domains appear in many engineering applications such as flow, heat transfer in porous materials (fixed inner interfaces) or colloid/particulate suspensions. A general approach to handling such geometrically complex regions in solving PDEs is using boundary fitted unstructured meshes. The unstructured mesh can be locally enhanced in the inner boundary regions to obtain fine information of the numerical solutions.

T. T. V. Le (✉)
Institute of Applied Mechanics and Informatics, Viet Nam Academy of Science and Technology (VAST), Hanoi, Vietnam
e-mail: thuyvanleusq@gmail.com

C.-D. Le
Institute of Mechanical Engineering, Ho Chi Minh City University of Transport, Ho Chi Minh City, Vietnam

K. Le-Cao
Department of Mechanical Engineering, National University of Singapore, Queenstown, Singapore

N. H. Phuong and V. Kreinovich (eds.), *Biomedical and Other Applications of Soft Computing*, Studies in Computational Intelligence 1045,
https://doi.org/10.1007/978-3-031-08580-2_14

However, creating an unstructured mesh is a time consuming and computationally intensive process. It's even worse in the case of moving edges, the computed mesh can be distorted. Hence, a new grid needs to be created and the solutions are then projected onto it, which are also complex tasks. To overcome the above problem, extension domain approaches are employed. The idea is to replace a given problem defined on the real domain $\mathbb{D}_R$, with a new problem defined on a domain of simpler shape (usually rectangular) $\mathbb{D}$ ($\mathbb{D}$ includes $\mathbb{D}_R$) which allows a simple grid and a fast solver to be used. However, special attention should be paid to the handling of boundary conditions at the internal interfaces so that the extended domain solutions $\mathbb{D}$ matches the original domain ($\mathbb{D}_R$) solutions.

To describe the existence of the internal boundaries, there are many extension domain approaches in the literature. For example, fictitious-domain techniques which proposed in [1, 2]. In this method, Lagrange multipliers were employed to impose the physical boundaries associated with Dirichlet boundary conditions. The method has been developed to solve practical problems such as fluid and structures interactions (e.g. [3]), fluid and flexible objects (e.g. [4]), and particle flows (e.g. [5]). Other approaches include finite cell method ([6]) for solving solid mechanics problems, fat boundary method (FBM) ([7]) for solving PDEs in a domain with hole s and implicit FBM ([8]) which combined the classic fictitious domain method and FBM.

RBF approximation schemes have the ability to represent any smooth function to a prescribed degree of accuracy (e.g. [9, 10]). They have rapidly been developed for the approximation of functions and numerical solution of PDEs ([11–16]). Here, we report a new numerical scheme based on RBF networks for the solution of PDEs in porous regions. Unlike other techniques in the literature, the variations of the field variable over interior holes are presently represented by appropriate analytic functions that are constructed to take into account given boundary conditions. Results obtained are compared well with analytic and other numerical solutions.

The remainder of this paper is organised as follows. In Sect. 2, the proposed method is presented for the solutions of elliptic and parabolic equations. Emphasis is placed on the construction of analytic functions representing the solution over interior holes. Details for RBFN discretisations of PDEs in extended (rectangular) regions are also included. Numerical solutions are presented in Sect. 3. Section 4 concludes the paper.

2 The Proposed Radial Basis Function Network Technique

Consider a porous domain such as the one shown in Fig. 1. This domain is decomposed in a network of RBF nodes used for computation. The computational domain, therefore, is composed of two types of subregions: the first is made up of the regions inside the holes and the second by the physical domain. Therefore, one needs to ensure that all boundary conditions on the internal interfaces at the gaps are taken into account.

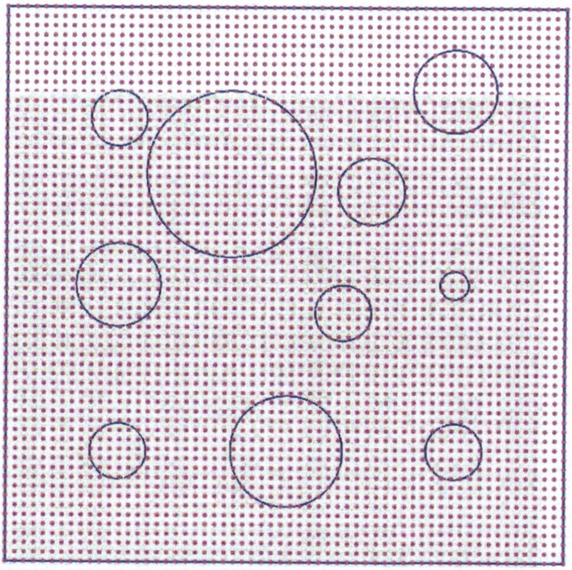

Fig. 1 A typical porous domain. The extension domain is a rectangular which is covered by a uniform RBF network nodes

Let z_f and z_r represent the solution z in the holes and in the real domain, respectively. We assume that (i) the function z is at least C^2-continuous (i.e. the function values and their first derivatives are identical) across the interfaces between the real domain and fictitious regions; and (ii) the solution in the holes can be represented by suitable analytic functions.

2.1 Integrated RBF Network Discretisation for Extended Domain

A network of RBFs allows one to convert general functions representing a physical field in a low-dimensional space (e.g. 1D, 2D and 3D) into a weighted linear combination of RBFs in a very high-dimensional space. It can be mathematically described as

$$\zeta(\mathbf{x}) \approx z(\mathbf{x}) = \sum_{i=1}^{m} w_i g_i(\mathbf{x}), \quad (1)$$

where ζ is the exact function, z the approximate function, $\mathbf{x}$ the position vector, m the number of RBFs, $\{g_i(x)\}_{i=1}^{m}$ the set of RBFs, and $\{w_i\}_{i=1}^{m}$ the set of weights to be found.

The use of integrated RBFs for solving PDEs was reported by Mai-Duy and Tran-Cong [12]. For the integral scheme, firstly, the highest order derivatives in a given PDE are approximated by RBFs, and then lower-order derivatives and the function itself are acquired by integrating RBFs. In this way, one can avoid a decrease in the convergence rate induced by differentiation and obtain a more stable numerical solution [17, 18]. Integrated RBFN-based boundary fitted methods were implemented for solution of the Navier-Stokes equation (e.g. [19, 20]).

Consider a typical term $\frac{\partial^2 z}{\partial x^2}$. As presented in [12], an integrated RBFN-p scheme permits the approximation of a function and its derivatives of orders up to p. One can employ integrated RBFN-2 here to represent z.

$$\frac{\partial^2 z(x)}{\partial x^2} = \sum_{i=1}^{m} w_i g_i(x) = \sum_{i=1}^{m} w_i I_i^{(2)}(x), \tag{2}$$

$$\frac{\partial z(x)}{\partial x} = \sum_{i=1}^{m} w_i I_i^{(1)}(x) + c_1, \tag{3}$$

$$z(x) = \sum_{i=1}^{m} w_i I_i^{(0)}(x) + c_1 x + c_2, \tag{4}$$

where m is the number of RBFs on the grid line. The present technique implements the multiquadric (MQ) function whose form is

$$I_i^{(2)}(x) = \sqrt{(x - c_i)^2 + a_i^2}, \tag{5}$$

$$I_i^{(1)}(x) = \frac{(x - c_i)}{2} A + \frac{a_i^2}{2} B, \tag{6}$$

$$I_i^{(0)}(x) = \left(\frac{-a_i^2}{3} + \frac{(x - c_i)^2}{6} \right) A + \frac{a_i^2 (x - c_i)}{2} B, \tag{7}$$

where c_i and a_i are the centre and the width of the ith MQ, respectively; $A = \sqrt{(x - c_i)^2 + a_i^2}$; and $B = \ln\left((x - c_i) + \sqrt{(x - c_i)^2 + a_i^2} \right)$. A set of collocation points $\{x_i\}_{i=1}^m$ is chosen to be a set of centres $\{c_i\}_{i=1}^m$. Such a set is comprised of two subsets. The first subset consists of the interior nodal points ($\{x_i\}_{i=1}^q$) that are also the grid nodes (regular nodes). The values of the field variable at the interior points are unknown. The second subset is formed from the boundary nodes ($\{x_{bi}\}_{i=1}^2$) that do not generally coincide with the grid nodes (irregular nodes).

The integrated RBFN approximations of derivatives are expressed in terms of the nodal values of z. As presented in [12, 19, 21], one will gather these approximations together to form the global matrices. It is straightforward to implement the boundary conditions on the outer boundary. However, special treatments are required when imposing the boundary conditions on the inner interfaces as the RBF nodes do not generally lie on these boundaries.

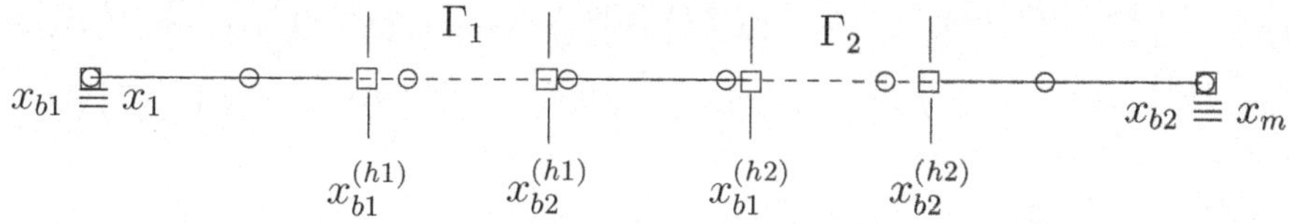

Fig. 2 Points on a grid line consist of interior points x_i (○) and boundary points x_{bi} (□)

Two types of equations, namely elliptic and parabolic, with Dirichlet boundary conditions are considered. Using integrated RBFNs, the PDE reduces to a set of algebraic equations that can be written in the following form

$$\mathcal{A}\widetilde{z} = \widetilde{b}, \tag{8}$$

where $\mathcal{A}$ is the system matrix, $\widetilde{z}$ the vector consisting of nodal values of z over the whole 2D computational domain and $\widetilde{b}$ the known vector generated by the driving function.

This system can be rearranged for unknown values of z in the real domain as

$$\mathcal{A}(i_r, i_r)\widetilde{z}(i_r) = \widetilde{b}(i_r) - \mathcal{A}(i_r, i_b)\widetilde{z}(i_b) - \mathcal{A}(i_r, i_f)\widetilde{z}(i_f), \tag{9}$$

where i_r, i_b and i_f are the sets whose elements are the indices of nodes in the real domain $\mathbb{D}_R$, on the outer boundary and in the interior holes Γ, respectively. We assume that the solution in inner holes is known and can be described by polynomials of fifth order in one dimension. The construction of these polynomials is presented in detail for a typical case: an $x-$grid line that crosses two inner holes as shown in Fig. 2.

Let $x_{b1}^{(hi)}$ and $x_{b2}^{(hi)}$ (e.g. $i = (1, 2)$ for Fig. 2) be the coordinates of the boundary points of the ith hole on the $x-$grid line. We seek a fictitious function $z_f(x)$ over the interval $x_{b1}^{(hi)} \leq x \leq x_{b2}^{(hi)}$ in the form

$$z_f^{(hi)}(x) = ax^5 + bx^4 + cx^3 + dx^2 + ex + f, \tag{10}$$

where $\{a, b, c, d, e, f\}$ are unknown values are can be calculated by using the boundary conditions information and their derivatives. The coefficients in (10) can be found through

$$\begin{pmatrix} a \\ b \\ c \\ d \\ e \\ f \end{pmatrix} = \mathcal{P}^{-1} \begin{pmatrix} z_{b1}^{(hi)} \\ z_{b2}^{(hi)} \\ \frac{\partial z_{b1}^{(hi)}}{\partial x} \\ \frac{\partial z_{b2}^{(hi)}}{\partial x} \\ \frac{\partial^2 z_{b1}^{(hi)}}{\partial x^2} \\ \frac{\partial^2 z_{b2}^{(hi)}}{\partial x^2} \end{pmatrix}, \tag{11}$$

where $z_{b1}^{(hi)} = z\left(x_{b1}^{(hi)}\right)$, $z_{b2}^{(hi)} = z\left(x_{b2}^{(hi)}\right)$, $\partial z_{b1}^{(hi)}/\partial x = \partial z\left(x_{b1}^{(hi)}\right)/\partial x$, $\partial z_{b2}^{(hi)}/\partial x = \partial z\left(x_{b2}^{(hi)}\right)/\partial x$, $\partial^2 z_{b1}^{(hi)}/\partial x^2 = \partial^2 z\left(x_{b1}^{(hi)}\right)/\partial x^2$, $\partial^2 z_{b2}^{(hi)}/\partial x^2 = \partial^2 z\left(x_{b2}^{(hi)}\right)/\partial x^2$, and

$$\mathcal{P} = \begin{bmatrix} \left(x_{b1}^{(hi)}\right)^5 & \left(x_{b1}^{(hi)}\right)^4 & \left(x_{b1}^{(hi)}\right)^3 & \left(x_{b1}^{(hi)}\right)^2 & x_{b1}^{(hi)} & 1 \\ \left(x_{b1}^{(hi)}\right)^5 & \left(x_{b1}^{(hi)}\right)^4 & \left(x_{b2}^{(hi)}\right)^3 & \left(x_{b2}^{(hi)}\right)^2 & x_{b2}^{(hi)} & 1 \\ 5\left(x_{b1}^{(hi)}\right)^4 & 4\left(x_{b1}^{(hi)}\right)^3 & 3\left(x_{b1}^{(hi)}\right)^2 & 2x_{b1}^{(hi)} & 1 & 0 \\ 5\left(x_{b1}^{(hi)}\right)^4 & 4\left(x_{b1}^{(hi)}\right)^3 & 3\left(x_{b2}^{(hi)}\right)^2 & 2x_{b2}^{(hi)} & 1 & 0 \\ 20\left(x_{b1}^{(hi)}\right)^3 & 12\left(x_{b1}^{(hi)}\right)^2 & 6\left(x_{b1}^{(hi)}\right) & 2 & 0 & 0 \\ 20\left(x_{b1}^{(hi)}\right)^3 & 12\left(x_{b1}^{(hi)}\right)^2 & 6\left(x_{b2}^{(hi)}\right) & 2 & 0 & 0 \end{bmatrix}.$$

The values of $z_f^{(hi)}$ at the interior nodes x_i are thus computed by

$$z_f^{(hi)}(x_i) = [x_i^5 \; x_i^4 \; x_i^3 \; x_i^2 \; x_i \; 1]\mathcal{P}^{-1}\left(z_{b1}^{(hi)} \; z_{b2}^{(hi)} \; \frac{\partial z_{b1}^{(hi)}}{\partial x} \; \frac{\partial z_{b2}^{(hi)}}{\partial x} \; \frac{\partial^2 z_{b1}^{(hi)}}{\partial x^2} \; \frac{\partial^2 z_{b2}^{(hi)}}{\partial x^2}\right)^T, \quad (12)$$

where $x_{b1}^{(hi)} < x_i < x_{b2}^{(hi)}$.

Replacing the boundary conditions with given values, the right hand side of (12) becomes known. By collecting $z_f^{(hi)}(x_i)$ for all holes to form $\tilde{z}(i_f)$ and substituting it into (9), one can now apply linear algebra to solve the system (9) for the solution values of z_r.

3 Numerical Solutions

3.1 Elliptic Partial Differential Equation

This problem is governed by

$$\nabla^2 \mathbf{z} = b(x, y), \quad (13)$$

where $b(x, y)$ is a given function. Two cases are considered in this section. For case 1, a domain of interest $\mathbb{D}_R$ is a square region $\mathbb{D}$, $[-1, 1] \times [-1, 1]$, with Γ the two circular holes of radius $R = 0.25$ and their centres located at position $(-0.4, 0.4)$ and $(0.4, -0.4)$. This physical domain is extended to a square computational one that can be conveniently represented by a uniform RBF grid node. The solution accuracy is measured through the discrete relative L_2 norm of the error defined as

$$Ne = \frac{\sqrt{\sum_{i=1}^{n_{ip}} (z_i^e - z_i)^2}}{\sqrt{\sum_{i=1}^{n_{ip}} (z_i^e)^2}}, \tag{14}$$

where n_{ip} is the number of interior points in the physical domain, and z^e and z are the exact and approximate solutions, respectively.

$$b(x, y) = -\pi^2 \cos(\pi x) \cos(\pi y). \tag{15}$$

The exact solution to this case is available

$$z^e(x, y) = \cos(\pi x) \cos(\pi y), \tag{16}$$

from which Dirichlet boundary conditions can be derived analytically. As mentioned earlier, one needs to have the values of $\partial z/\partial x$, $\partial^2 z/\partial x^2$ at $x_{b1}^{(hi)}$ and at $x_{b2}^{(hi)}$ to construct the polynomials. An iterative procedure is employed here for this purpose. In the beginning, these derivative values are set to zero. Through (11), a function z_f is obtained. By substituting nodal values of z_f into (9), one is able to acquire the solution z_r. These z_f and z_r constitute a solution z along a grid line, from which the values of $\partial z/\partial x$ and $\partial^2 z/\partial x^2$ at $x_{b1}^{(hi)}$ and at $x_{b2}^{(hi)}$ are updated. These steps are repeated a few times to pick up one among many possible fifth-order polynomial representations for the interior hole solution z_f. In terms of stability, the proposed method yields relatively low condition numbers (e.g. about 1.0×10^3 for a grid of 66×66). In terms of accuracy, errors obtained are relatively small, varying from 2.09×10^{-2} (grid of 22×22) to 3.73×10^{-3} (grid of 66×66). Figure 3 displays the computed solution using a grid of 42×42. It can be seen that the present scheme is able to capture a complex function.

For case 2, this problem is governed by Poisson equation with the driving function $b(x, y) = -1$ and homogeneous Dirichlet boundary conditions. The domain of interest $\mathbb{D}_R$ is a unit square with Γ, the sum of 10 circular domains of several radii $\{0.05, 0.1, 0.15, 0.25, 0.3\}$. This example provides a good means of testing the capability of the proposed technique in dealing with geometrically complex-domain problems. It is noted that multi-hole regions occur in many practical modelling situations such as thermal conductivity for composite materials, fluid flow in porous medium and particulate flows. The pre-processing for the present technique is much more economical than that for boundary-fitted approaches. Since there is no exact solution available here, we present a visual distribution of z_r of the present technique (grid of 80×80) at some random positions of the inner interfaces (Fig. 4).

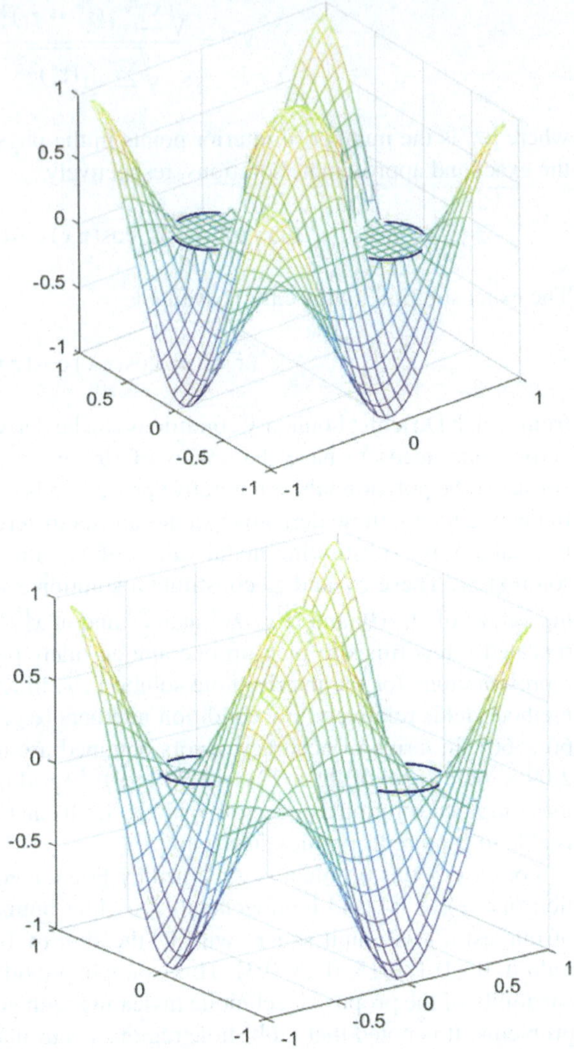

Fig. 3 The RBF solution of elliptic partial differential equations—Case 1: A plot of the analytical solution with discontinuity at inner interfaces (Top) and the solution reconstructed using a RBF network of 42 × 42 nodes (Bottom)

3.2 *Parabolic Partial Differential Equation*

The domain $\mathbb{D}_R$ is the region lying between $\mathbb{D}$, a square of 1 × 1 and Γ a circle of radius 0.2 which is located at the origin.This problem is governed by the following parabolic PDE

$$\frac{\partial z}{\partial t} - \left[\frac{\partial^2 z}{\partial x^2} + \frac{\partial^2 z}{\partial y^2}\right] = (1 + 2\pi^2 l^2 t)\sin(l\pi x)\sin(l\pi y), \tag{17}$$

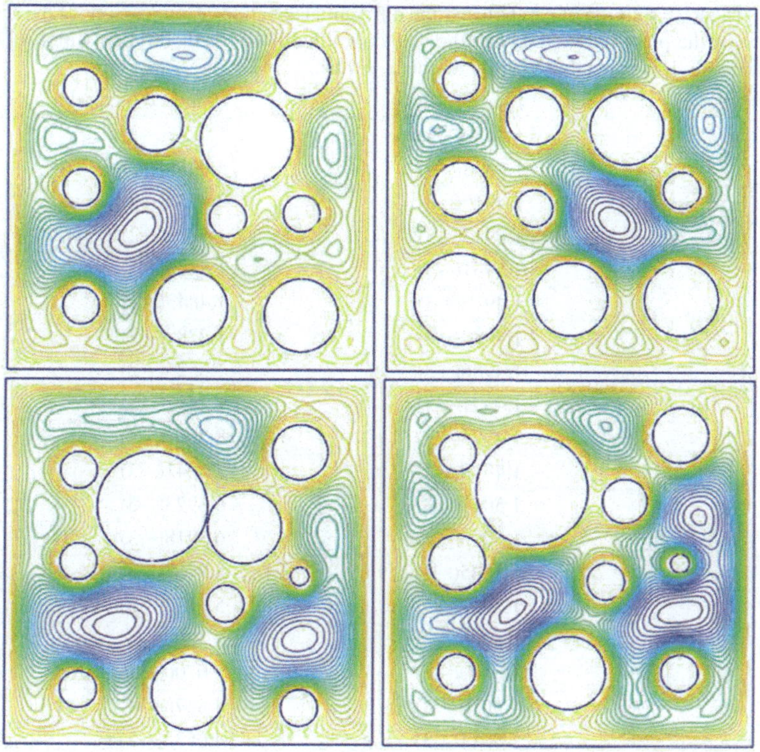

Fig. 4 The RBF solution of elliptic partial differential equations—Case 2: A contour plot of z_r by the present method using grid node of 80×80. The positions of the inner boundaries are randomly arranged

where l is a given number. The exact solution is given by

$$z^e(x, y, t) = \sin(l\pi x)\sin(l\pi y)t, \tag{18}$$

from which the initial solution and Dirichlet boundary conditions are derived analytically. The spatial discretisation is carried out using a uniform RBF grid nodes of 40×40. The values of $\partial z/\partial x, \partial^2 z/\partial x^2$ at $x_{b1}^{(hi)}$ and at $x_{b2}^{(hi)}$ are simply calculated from the previous time level. Results of Ne for $l = 7$ using a time step of 0.1 are presented in Table 1. It appears that the present method produces a solution slightly less accurate than that of boundary fitted grid IRBF because of the cost of simple pre-processing. It is noted that the matrix condition number is only 3.8×10^2 for a grid of 40×40. Plots for z with two values of l at $t = 1$ are shown in Fig. 5.

Table 1 The RBF solution of parabolic partial differential equations: Comparisons of the accuracy of z between the proposed method and the boundary fitted IRBF method ([12]) with $l = 7$ using grid of 40×40. It is noted that $a(b)$ represents $a \times 10^b$

Time	$IRBF$([12])	Proposed method
0.2	1.4690(−3)	5.6225(−2)
0.4	1.4858(−3)	4.0033(−2)
0.6	1.4915(−3)	2.6139(−2)
0.8	1.4943(−3)	1.4765(−2)
1.0	1.4961(−3)	1.8876(−2)
1.2	1.4972(−3)	1.4843(−2)
1.4	1.4980(−3)	1.2288(−2)
1.6	1.4986(−3)	1.0525(−2)
1.8	1.4991(−3)	9.2411(−3)
2.0	1.4995(−3)	8.2667(−3)
2.2	1.4998(−3)	7.5041(−3)
2.4	1.5001(−3)	6.8923(−3)
2.6	1.5003(−3)	6.3918(−3)
2.8	1.5005(−3)	5.9757(−3)
3.0	1.5007(−3)	5.6253(−3)
3.2	1.5008(−3)	5.3268(−3)
3.4	1.5009(−3)	5.070(−3)
3.6	1.5010(−3)	4.847(−3)
3.8	1.5011(−3)	4.6533(−3)
4.0	1.5012(−3)	4.3317(−3)

4 Concluding Remarks

In this work, an integrated RBF network numerical solver for solving elliptic and parabolic equations of second order defined in porous domains is reported. The preprocessing is simple as the physical domain is converted into a computational rectangular one. Unlike other extension domain approaches, no interpolation at the inner interfaces is required here. Polynomials of the fifth order are exploited to represent the fictitious solution within the inner holes. The polynomials are analytically constructed by boundary conditions, 1st and 2nd derivative values at the inner interfaces. The technique is validated successfully through several boundary-value and initial-value problems. In comparison with the integrated RBFN-based boundary fitted methods, the proposed method is seen to be less accurate, but appears to be more efficient (in terms of simple pre-processing) for solving porous domain problems with many moving inner boundaries.

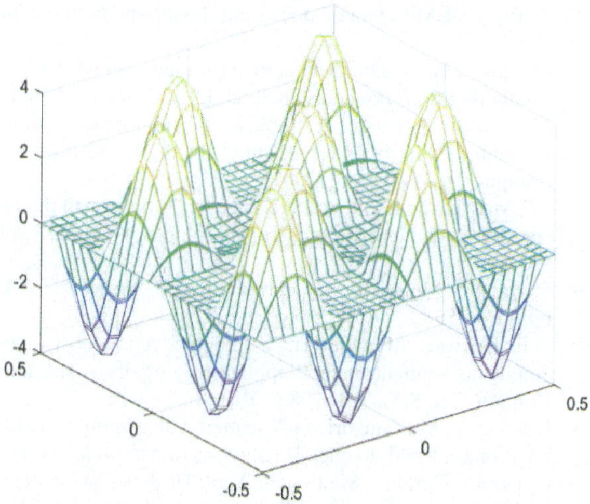

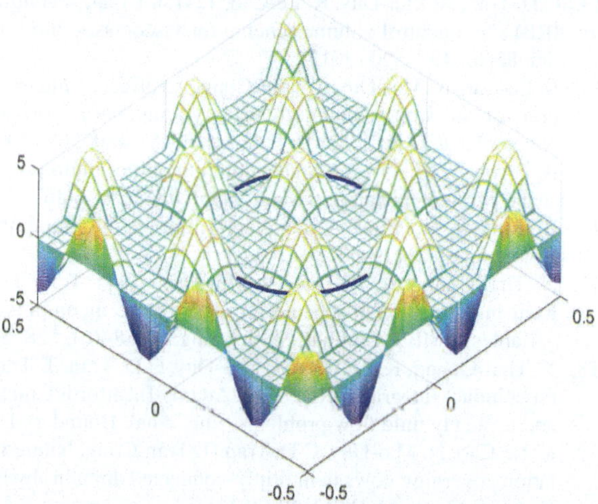

Fig. 5 The RBF solution of parabolic partial differential equations: Plots of the approximate solutions $l = 4$(Top) and $l = 5$(Bottom) using a grid of 40×40

References

1. R. Glowinski, T.W. Pan, J. Periaux, Distributed Lagrange multiplier methods for incompressible viscous flow around moving rigid bodies. Comput. Methods Appl. Mech. Eng. **151**(1–2), 181–94 (1998)
2. R. Glowinski, T.W. Pan, J. Periaux, A fictitious domain method for external incompressible viscous flow modeled by Navier-Stokes equations. Comput. Methods Appl. Mech. Eng. **112**(1–4), 133–48 (1994)
3. N.A. Patankar, P. Singh, D.D. Joseph, R. Glowinski, T.W. Pan, A new formulation of the distributed Lagrange multiplier/fictitious domain method for particulate flows. Int. J. Multiph. Flow **26**(9), 1509–24 (2000)

4. Z. Yu, A DLM/FD method for fluid/flexible-body interactions. J. Comput. Phys. **207**(1), 1–27 (2005)
5. D. Wan, S. Turek, Direct numerical simulation of particulate flow via multigrid FEM techniques and the fictitious boundary method. Int. J. Numer. Meth. Fluids **51**(5), 531–66 (2006)
6. J. Parvizian, A. Duster, E. Rank, Finite cell method. Comput. Mech. **41**(1), 121–33 (2007)
7. B. Maury, A fat boundary method for the Poisson problem in a domain with holes. J. Sci. Comput. **16**(3), 319–39 (2001)
8. P.E. Vos, R. Van Loon, S.J. Sherwin, A comparison of fictitious domain methods appropriate for spectral/hp element discretisations. Comput. Methods Appl. Mech. Eng. **197**(25–28), 2275–89 (2008)
9. S. Haykin, Multilayer perceptrons. Neural Networks: A Comprehensive Foundation (1999), pp. 156–255
10. D.F. Trevisan, J.P. Gois, H.C. Batagelo, A low-cost-memory CUDA implementation of the conjugate gradient method applied to globally supported radial basis functions implicits. J. Comput. Sci. **5**(5), 701–708 (2014)
11. E.J. Kansa, Multiquadrics - A scattered data approximation scheme with applications to computational fluid-dynamics-II solutions to parabolic, hyperbolic and elliptic partial differential equations. Comput. Math. with Appl. **19**(8–9), 147–61 (1990)
12. N. Mai-Duy, K. Le-Cao, T. Tran-Cong, A Cartesian grid technique based on one-dimensional integrated radial basis function networks for natural convection in concentric annuli. Int. J. Numer. Meth. Fluids **57**(12), 1709–30 (2008)
13. C.D. Tran, N. Mai-Duy, K. Le-Cao, T. Tran-Cong, A continuum-microscopic method based on IRBFs and control volume scheme for viscoelastic fluid flows. CMES-Comput. Model. Eng. Sci. **85**(6), 499–520 (2012)
14. K. Le-Cao, N. Mai-Duy, T. Tran-Cong, An effective integrated-RBFN cartesian-grid discretization for the stream function-vorticity-temperature formulation in nonrectangular domains. Numer. Heat Transf. Part B: Fundam. **55**(6), 480–502 (2009)
15. B. Fornberg, G. Wright, Stable computation of multiquadric interpolants for all values of the shape parameter. Comput. Math. Appl. **48**(5), 853–867 (2004)
16. B. Fornberg, C. Piret, A stable algorithm for flat radial basis functions on a sphere. SIAM J. Sci. Comput. **30**(1), 60–80 (2017)
17. N. Thai-Quang, K. Le-Cao, N. Mai-Duy , C.D. Tran, T. Tran-Cong, A high-order compact local integrated-RBF scheme for steady-state incompressible viscous flows in the primitive variables. CMES: Comput. Model. Eng. Sci. **84**(6), 528–557 (2012)
18. N. Thai-Quang, K. Le-Cao, N. Mai-Duy, C.D. Tran, T. Tran-Cong, A numerical scheme based on compact integrated-RBFs and Adams-Bashforth/Crank-Nicolson algorithms for diffusion and unsteady fluid flow problems. Eng. Anal. Boundary Elem. **37**(12), 1653–67 (2013)
19. K. Le-Cao, N. Mai-Duy, C.D. Tran, T. Tran-Cong, Numerical study of stream-function formulation governing flows in multiply-connected domains by integrated RBFs and Cartesian grids. Comput. Fluids **44**(1), 32–42 (2011)
20. T.T.V. Le, N. Mai-Duy, K. Le-Cao, T. Tran-Cong, A time discretisation scheme based on integrated radial basis functions (IRBFs) for heat transfer and fluid flow problems. Numer. Heat Transf. Part B. Part B: Fundam. **74**(2), 498–518 (2018)
21. T.T.V. Le, K. Le-Cao, H. Duc-Tran, A radial basis neural network approximation with extended precision for solving partial differential equations, in *Soft Computing: Biomedical and Related Applications. Studies in Computational Intelligence* ed. by N.H. Phuong, V. Kreinovich (Springer, Cham, 2021), p. 81

Accuracy Measures and the Convexity of ROC Curves for Binary Classification Problems

Le Bich Phuong and Nguyen Tien Zung

Abstract In this work, we offer a theoretical explanation of the convexity of the ROC (receiver operating characteristic) curves for rational binary classifiers, and show some natural important inequalities relating different measures of accuracy.

1 Introduction

In this work, we are interested in the accuracy measures of binary classifiers in the context of artificial intelligence. More specifically, we will be interested in the ROC (receiver operating characteristic) curves and their corresponding AUC (area under the curve), the maximal balanced accuracy and the maximal cost-weighted accuracy, and inequalities relating these accuracy measures to each other.

The main results presented in this paper may be summarized as follows:

- (Section 2) ROC curves of optimal machines are convex.
- (Section 3) Natural inequalities relating different measures of accuracy. In particular, if the machine is rational then the AUC is greater than the balanced accuracy.

Convexity of the ROC curve is not something new, and many research papers and monographs already discussed this convexity property in an empirical way, see, e.g., [1–6] and references therein. However, the only place we we found a rigorous theorem on convexity of the ROC curve is [2, Theorem 3], where the authors showed that the convexity of the ROC curve is equivalent to another natural condition, namely "the conditional event probability (or the likelyhood ratio) is nondecerasing". (The higher the "sigmoid value" is, the higher the probability of the event being true is). In this paper, we study *optimal* machines (see Definition 1), and show that if a machine

L. B. Phuong (✉)
Hanoi University of Mining and Geology, Bac Tu Liem, Vietnam
e-mail: lebichphuong@humg.edu.vn

N. T. Zung
Université de Toulouse and Torus Actions SAS, Toulouse, France
e-mail: tienzung@math.univ-toulouse.fr; ntzung@torus-actions.fr

N. H. Phuong and V. Kreinovich (eds.), *Biomedical and Other Applications of Soft Computing*, Studies in Computational Intelligence 1045,
https://doi.org/10.1007/978-3-031-08580-2_15

is optimal then its ROC curve is automatically convex. Our proof is based on an explicit construction of optimal machines. We then observe that for such machines the conditional event probability is automatically nondecreasing, which implies (as in [2, Theorem 3]) that the ROC curve is convex. To make our paper self-contained, we also write down explicitly a proof for this last implication, which takes only a few lines.

Since, by machine learning, most reasonable constructed machines tend to become "nearly optimal", our result gives an explanation why most ROC curves met in practice are "nearly convex".

The inequalities that we present in Sect. 3 are elementary, but they are important, and we have not seen them written down explicitly anywhere in the literature, and that's why we want to present them in this paper.

Our motivation for studying the ROC curves and the "most rational classifiers" comes from the following voting question: given several different binary classifiers for the same problem, what is the best arithmetical voting method for ensembling their results together to obtain highest possible accuracy. This will be the topic of a separate research paper.

2 Convexity of the ROC Curves

Let us fix some notations:

Ω denotes the space of all inputs, which comes together with some natural probability measure P.

A given binary decision problem on Ω is denoted by a function

$$Y : \Omega \to \{0, 1\},$$

that we want to replicate by a "machine"

$$M : \Omega \to [0, 1]$$

which takes values in the interval $[0, 1]$: We can fix a threshold $\sigma \in [0, 1]$ and declare that $Y_{predict}(x) = 1$ if $M(x) \geq \sigma$ and $Y_{predict}(x) = 0$ otherwise, and hope that $Y_{predict}$ is a good approximation of Y.

The prediction function M is constructed as the composition of two maps,

$$M = \Sigma \circ \phi,$$

where

$$\phi : \Omega \to \Phi$$

is a projection from Ω to a certain "features space" Φ (which can also be called the "information space"—it contains all the information that we can use for predicting Y), and

$$\Sigma : \Phi \to [0, 1]$$

is a "sigmoid function": for each input $x \in \Omega$, we first extract its features, $\varphi = \phi(x)$, and then calculate its sigmoid value, $\sigma = M(x) = \Sigma(\varphi)$. The probability measure on Φ is the push-forward of the probability measure from Ω.

The reason why Σ is called a "sigmoid function" is that sigmoid functions (such as the standard function $\sigma(z) = \exp(z)/(\exp(z) + \exp(-z))$ and similarly shaped functions that map $\mathbb{R}$ to $]0, 1[$) are often used as the last layer in a neural network for binary classification problems. Here we use the words "sigmoid function" in an abstract way, without referring to any specific function, and Φ can be multi-dimensional, doesn't have to be $\mathbb{R}$.

Remark that, in the AI literature, the value σ is very often called the probability, but it is not true in general. In fact, we can define the probability function $p : [0, 1] \to [0, 1]$,

$$p(\sigma) := P\left(Y(x) = 1 \mid \Sigma(\phi(x)) = \sigma\right),$$

which is *not* the identity function in general.

For each threshold value $\sigma \in [0, 1]$, one defines the corresponding *sensitivity* (= true positive) rate $TP(\sigma)$ and *specificity* (= true negative = 1 minus false positive) rate $TN(\sigma)$ by the following formulas:

$$TP(\sigma) = \frac{P(M(x) \geq \sigma | Y(x) = 1)}{P(Y(x) = 1)},$$

$$TN(\sigma) = 1 - FP(\sigma) = \frac{P(M(x) < \sigma | Y(x) = 0)}{P(Y(x) = 0)}.$$

The curve $ROC : [0, 1] \to [0, 1] \times [0, 1]$ given by the formula

$$ROC(\sigma) = (FP(\sigma), TP(\sigma))$$

is called the ROC curve in the literature, and is very widely used in many fields, see, e.g., [1–6] and references therein. See Fig. 1 for some examples of ROC curves (a test of AI for the detection of different classes of skin lesions, courtesy of Torus Actions SAS).

The ROC curves goes "backward" from the point $ROC(0) = (1, 1)$ to the point $ROC(1) = (0, 0)$ in the unit square, and the higher the curve the more accurate the machine. The so called AUC (area under the curve) is the area of the region under the ROC curve in the unit square, and is a popular measure for the accuracy of the machine.

Remark that, if we change the sigmoid function Σ by composing it with another function, $\Sigma' = \theta \circ \Sigma$, where $\theta : [0, 1] \to [0, 1]$ is a strictly increasing bijective function, then Σ and Σ' give the same ROC curve up to a reparametrization by θ. In other words, we can change a σ value to any other value by composing it with

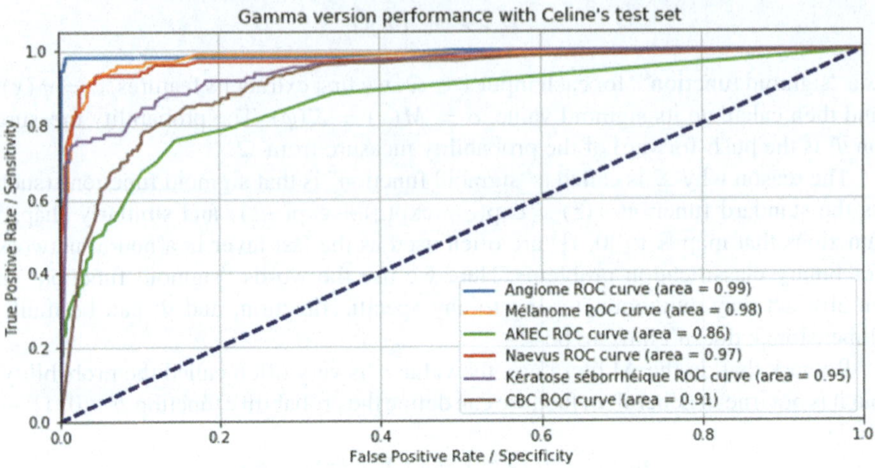

Fig. 1 Examples of ROC curves from a test on skin cancer images

a function, without changing the accuracy of the system, and this is one more reason why the σ values should not be called "probabilities" in general.

All the ROC curves in Fig. 1 are convex or almost convex (i.e. the regions under them almost coincide with their convex hulls). As a matter of fact, it has been observed since long time ago that "reasonable" ROC curves are nearly convex, and moreover, non-convexity is a kind of defect that could be corrected by randominizing certain things, see, e.g., [2]. Philosophically, machine learning aims at creating rational predictors (that are better than just random functions), and that's why ROC curves are mostly convex, but in practice our predictors are not completely rational, and so there are ROC curves that are not completely convex (another reason for non-convexity is lack of data).

Definition 1 We will say that machine $M = \Sigma \circ \phi$ (or its corresponding sigmoid function Σ), for a given binary classifier $Y : \Omega \to \{0, 1\}$ and features map ϕ, is *optimal* or *most rational*, if its ROC curves lies above the ROC curves of all the other sigma functions for the same problem (ϕ is fixed).

Proposition 1 *Among all the (measurable) sigmoid functions $\Sigma : \Phi \to [0, 1]$ (for a given binary classifier $Y : \Omega \to [0, 1]$ and a given features projection map $\phi : \Omega \to \Phi$) there is an optimal sigmoid function (called the "probability sigmoid function"), and its ROC curve is convex.*

As a consequence, if a machine learning algorithm is efficient enough so that it creates a "nearly optimal" machine, then the ROC curve of that machine is convex or nearly convex.

In order to prove the above proposition, we will construct the optimal machine "explicitly". For each $\varphi \in \Phi$, denote by $p(\phi)$ the conditional probability of $Y = 1$ given that the features value is φ:

$$p(\varphi) = P(Y(x) = 1 \mid \phi(x) = \varphi).$$

Then this probability function $\varphi \mapsto p(\varphi)$ is actually the best sigmoid function, i.e., if we put $\Sigma(\varphi) := p(\varphi)$, then this sigmoid function has the best ROC curve.

Indeed, denote by $\Sigma' : \Phi \to [0, 1]$ any other sigmoid function. Fix an arbitrary false positive level $\alpha \in]0, 1[$. Denote by $\sigma \in]0, 1[$ (resp., $\sigma' \in]0, 1[$) the threshold value for the probability sigmoid function $p(\varphi)$ (resp., the function $\Sigma'(\varphi)$ to get this level of false positive. Then we have the following formula,

$$\alpha = \frac{\int_{\{\varphi \in \Phi | p(\varphi) \geq \sigma\}} (1 - p(\varphi)) d\varphi}{\int_{\Phi} (1 - p(\varphi)) d\varphi} = \frac{\int_{\{\varphi \in \Phi | \Sigma'(\varphi) \geq \sigma'\}} (1 - p(\varphi)) d\varphi}{\int_{\Phi} (1 - p(\varphi)) d\varphi},$$

(where $\int_{\Phi} (1 - p(\varphi)) d\varphi = P(Y = 0)$ is the probability measure of the negative set $\{x \in \Omega \mid Y(x) = 0\}$), which implies that

$$\int_{\{\varphi \in \Phi | p(\varphi) \geq \sigma\}} (1 - p(\varphi)) d\varphi = \int_{\{\varphi \in \Phi | \Sigma'(\varphi) \geq \sigma'\}} (1 - p(\varphi)) d\varphi.$$

To simplify the notations, put $A = \{\varphi \in \Phi | p(\varphi) \geq \sigma\}$ and $B = \{\varphi \in \Phi | \Sigma'(\varphi) \geq \sigma'\}$. Then we have $\int_A (1 - p(\varphi)) d\varphi = \int_B (1 - p(\varphi)) d\varphi$, which implies that

$$\int_{A \setminus B} (1 - p(\varphi)) d\varphi = \int_{B \setminus A} (1 - p(\varphi)) d\varphi.$$

Since $(1 - p(\varphi)) \leq 1 - \sigma$ on $A \setminus B$ while $(1 - p(\varphi)) > 1 - \sigma$ on $B \setminus A$, we must have that $P(A \setminus B) \geq P(B \setminus A)$, which implies that

$$\int_{A \setminus B} p(\varphi) d\varphi \geq \sigma P(A \setminus B) \geq \sigma P(B \setminus A) \geq \int_{B \setminus A} p(\varphi) d\varphi,$$

which implies that $\int_A p(\varphi) d\varphi \geq \int_B p(\varphi) d\varphi$. This last inequality means exactly that the true positive level of p at the false positive level α is greater or equal to the true positive level of Σ' at the same false positive level. In other words, the ROC curve of the probability sigmoid function p lies above the ROC curve of Σ' everywhere, i.e., p is the optimal sigmoid function. The two ROC curves coincide if and only if, in the above formulas, B coincides with A (up to a set of measure zero) for every false positive level α, and it basically means that Σ' is obtained from p by composing it with a monotonous function. In other words, up to a reparametrization of the sigmoid values, the probability sigmoid function is the only optimal sigmoid function.

The convexity of the ROC curve of the optimal sigmoid function p follows directly from its construction, which assures that the conditional event probability is nondecreasing (the higher the sigmoid value σ, the higher the conditional conditional probability value, which is obvious because this value is equal to σ in our construction).

See Theorem 3 of [2]. Indeed, denote by

$$\alpha(\sigma) = \frac{\int_{\{\varphi\in\Phi|p(\varphi)\geq\sigma\}}(1-p(\varphi))d\varphi}{\int_\Phi(1-p(\varphi))d\varphi} \quad \text{and} \quad \beta(\sigma) = \frac{\int_{\{\varphi\in\Phi|p(\varphi)\geq\sigma\}}p(\varphi)d\varphi}{\int_\Phi p(\varphi)d\varphi}$$

the false negative and false positive levels at threshold σ for the probability sigmoid function p, then we have

$$\frac{d\beta}{d\alpha} = \frac{\int_\Phi(1-p(\varphi))d\varphi}{\int_\Phi p(\varphi)d\varphi} \cdot \frac{\sigma}{1-\sigma},$$

which is an increasing function in σ, but a decreasing function in α, because α itself is a decreasing function in σ. Hence β is a concave function in α, which means that the ROC curve is convex.

Remark 1 (*Under the above notations*) A machine may be called *consistent* if it satisfies the following natural monotonicity condition: the higher the sigmoid value, the higher the probability of $Y = 1$. In other words, a machine is consistent if $p(\sigma)$ is a strictly monotonous increasing function in σ, with $p(0) = 0$ and $p(1) = 1$. We have shown that optimal machines are automatically consistent (and hence their ROC curves are automatically convex).

Remark 2 Given an original probability distribution P on the data space Ω, which is imbalanced in the sense that $P(Y = 0) \neq P(Y = 1)$ (imbalance between negative and positive cases), we may change it to a new, balanced, probability distribution $\hat{P}$, defined by the following formula:

$$\hat{P}(A) = \frac{1}{2}\left[\frac{P(A\cap\{Y=0\})}{P(Y=0)} + \frac{P(A\cap\{Y=1\})}{P(Y=1)}\right].$$

It is easy to verify that the ROC curve (for a given machine $M = \Sigma \circ \phi : \Omega \to [0, 1]$) with respect to $\hat{P}$ coincides exactly with the ROC curve with respect to P. Indeed, for any $\sigma \in [0, 1]$, the true positive level at σ is $TP(\sigma) = \frac{P(\Sigma > \sigma, Y = 1)}{P(Y = 1)} = 2\hat{P}(\Sigma > \sigma, Y = 1) = \frac{\hat{P}(\Sigma > \sigma, Y = 1)}{\hat{P}(Y = 1)} = \widehat{TP}(\sigma)$, and similarly for $TN(\sigma)$. Thus in the study of accuracy, without loss of generality, one may suppose that the probability distribution is balanced with respect to Y in the sense that $P(Y = 0) = P(Y = 1) = 0.5$.

3 Inequalities Relating Different Measures of Accuracy

Besides AUC, another popular measure of accuracy is the so called *balanced accuracy* defined as follows (for a given machine, i.e., a given sigmoid function):

$$BA(\sigma) = \frac{TP(\sigma) + TN(\sigma)}{2}$$

and (the maximal possible BA when one varies the threshold)

$$MBA = \max_{\sigma \in [0,1]} BA(\sigma)$$

Remark that $BA(0) = BA(1) = 0.5$, so we always have $MBA \geq 0.5$. If $MBA = 0.5$ then the sigmoid function is "useless" (worse than random). Both AUC and MBA are of course smaller or equal to 1, and they are equal to 1 if the machine is "perfect". The following proposition gives relations between these two measures of accuracy, which implies that if one of them tends to 1 then the other also tends to 1.

Proposition 2 *For any given sigmoid function we have*

$$1 - 2(1 - MBA)^2 \geq AUC \geq 2MBA - 1$$

If, moreover, the sigmoid function is rational in the sense that its ROC curve is convex, then we have

$$AUC \geq MBA$$

Instead of the balanced accuracy, one can also use *cost-based weighted accuracy* (see, e.g., [3]): given weight $w \in]0, 1[$ (determined by a cost function), we put

$$WA(\sigma) = w.TP(\sigma) + (1 - w).TN(\sigma) = w.TP(\sigma) - (1 - w).FP(\sigma) + (1 - w),$$
$$MWA = \max_{\sigma \in [0,1]} WA(\sigma).$$

and we have the following similar inequalities relating MWA to AUC:

Proposition 3 *For any given weight $w \in]0, 1[$ and any given sigmoid function, we have*

$$1 - \frac{(1 - MWA)^2}{2w(1 - w)} \geq AUC$$

If, moreover, the sigmoid function is rational in the sense that its ROC curve is convex, then we have

$$AUC \geq 1 - \frac{(1 - MWA)}{2\min(w, 1 - w)}$$

When $w = 0.5$ then MWA coincides with MBA and Proposition 3 basically becomes Proposition 2. Remark also that both AUC and MBA are *balanced* measures of accuracy, in the sense that true positive and true negative ratios are equally important in their formulas, while MWA ($w \neq 0.5$) is not balanced. This fact is important for optimal arithmetical voting methods to be discussed in the next section.

All of the above inequalities are direct consequences of the following elementary proposition:

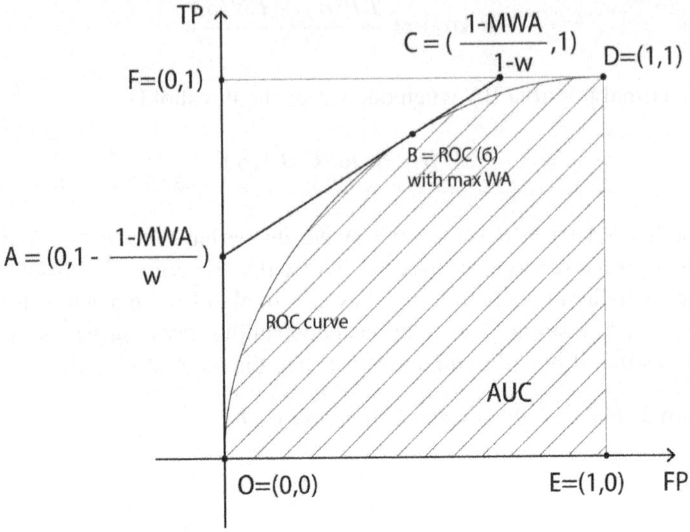

Fig. 2 The ROC curve and the tangent line at a maximal weighted average point

Proposition 4 *Give a sigmoid function* $\Sigma : \Omega \to [0, 1]$ *and a weight* $w \in]0, 1[$, *a number* $\sigma \in [0, 1]$ *is a threshold where* Σ *attains highest weighted accuracy if and only if the straight line through the point* $ROC(\sigma)$, *consisting of the points* $(FP(\sigma) + wt, TP(\sigma) + (1 - w)t)$, $t \in \mathbb{R}$, *lies above the ROC curve.*

The above simple propositions show that, the three different measures of accuracy AUC, MBA and MWA are correlate well with each other, and if one of them tends to 1 (maximal accuracy) then the other two also tend to 1.

Let us indicate here the proof of the above propositions. See Fig. 2 for the illustration.

The lines $\ell = \{(FP(\sigma) + wt, TP(\sigma) + (1 - w)t), t \in \mathbb{R}\}$ of slope $(w, 1 - w)$ in Proposition 4 are simply "lines of constant weighted accuracy". If the point $(FP(\sigma), TP(\sigma))$ of the ROC curve give maximal weighted accuracy, then no point of the ROC curve can lie above its corresponding line of slope $(w, 1 - w)$, because lying above means higher weighted accuracy.

The line $\ell = \{(FP(\sigma) + wt, TP(\sigma) + (1 - w)t), t \in \mathbb{R}\}$, where σ gives the maximal weighted accuracy for the sigmoid function Σ, cuts the boundary of the unit square at two points $A = (0, 1 - \frac{1 - MWA}{w})$ and $C = (\frac{1 - MWA}{1 - w}, 1)$. The triangle ΔACF, where $F = (0, 1)$, is disjoint from the region under the ROC curve, which implies that $AUC + \text{area}(\Delta ACF) \leq 1$. Since $\text{area}(\Delta ACF) = \frac{FA \cdot FC}{2} = \frac{(1 - MWA)^2}{2w(1 - w)}$, we get the inequality

$$AUC \leq 1 - \frac{(1 - MWA)^2}{2w(1-w)}.$$

On the other hand, the region under the ROC curve contains the rectangle whose vertices are $(FP(\sigma), 0)$, $(FP(\sigma), TP(\sigma)$, $(1, TP(\sigma))$, $(1, 0)$. The surface area of this rectangle is $TP(\sigma) \cdot (1 - FP(\sigma)) = TP(\sigma) \cdot TN(\sigma) = TP(\sigma) + TN(\sigma) - 1 + (1 - TP(\sigma))(1 - TP(\sigma)) \geq TP(\sigma) + TN(\sigma) - 1 = 2BA(\sigma) - 1$ (for every σ). Hence we get the inequality

$$AUC \geq 2MBA - 1.$$

If the ROC curve is convex, then the region below it contains the quadrilateral $OBDE$, where $O = (0, 0)$, $B = (FP(\sigma), TP(\sigma))$, $D = (1, 1)$, $E = (1, 0)$ (for any σ). The surface area of this quadrilateral is exactly equal to $BA(\sigma)$, i.e., to MBA, hence we get the inequality $AUC \geq BA(\sigma)$ for any σ, i.e., we have

$$AUC \geq BA(\sigma).$$

Finally, the inequality $AUC \geq 1 - \dfrac{(1 - MWA)}{2\min(w, 1-w)}$ in the case when the ROC curve is convex and the weight w is arbitrary is a direct consequence of the inequalities

$$AUC \geq area(OBDE) \geq \min(area(OADE), area(OCDE))$$

and the equalities

$$area(OADE) = 1 - \frac{(1 - MWA)}{2w}, \quad area(OCDE) = 1 - \frac{(1 - MWA)}{2(1-w)}.$$

(See Figure 2).

References

1. Tom Fawcett, An introduction to ROC analysis. Pattern Recogn. Lett. **27**(8), 861–874 (2006)
2. T. Gneiting, Peter Vogel Receiver Operating Characteristic (ROC) Curves (2018). https://arxiv.org/abs/1809.04808
3. J. Hernandez-Orallo, P.A. Flach, C. Ferri, A unified view of performance metrics: translating threshold choice into expected classification loss. J. Mach. Learn. Res. **13**, 2813–2869 (2012)
4. L.L. Pesce, Ph.D. 1, C.E. Metz, Ph.D. 1, K.S. Berbaum, Ph.D. 2 On the convexity of ROC curves estimated from radiological test results, Acad Radiol. **17**(8), 960–968.e4 (2010). https://doi.org/10.1016/j.acra.2010.04.001
5. J.A. Swets, R.M. Dawes, J. Monahan, Psychological science can improve diagnostic decisions, Psychological Science in the Public Interest, VOL. 1, NO. 1 (2000)
6. Wikipedia page on ROC: https://en.wikipedia.org/wiki/Receiver_operating_characteristic
7. Z.-H. Zhou, *Ensemble Methods: Foundations and Algorithms* 1st edn. (Chapman Hall/CRC, 2012)

Stochastic Simulations of Airborne Particles in a Fibre Matrix

T. T. V. Le, K. Le-Cao, T. Ba-Quoc, and Y. Nguyen-Quoc

Abstract Airborne particles are extremely small (under 0.1 μm) which can be suspended in the air from 30 min to hours. At such scale, these particles' displacement is affected by rapid collisions due to density fluctuation of the air and also by air flows. It is necessary to have a stochastic scheme for numerical simulations to take into account these random movements. In this research work, a stochastic differential equation (e.g. Langevin equation) is incorporated into the velocities field to simulate the movement of airborne particles immersed in air flows and see how it interacts with the fibre matrices to improve air filtration performance.

Keywords Stochastic differential equation · Brownian motion · Simulation · Langevin equation · Filter efficiency simulation · Filter lifetime

1 Introduction

Breathing fresh air is a basic human need especially while the covid epidemic is spreading around the world. New research indicates that the minimum size of a respiratory particle that can contain SARS-CoV-2 is about 4.7 μm. This minimum size of the particles can be further reduced because of the water evaporation on the

T. T. V. Le (✉)
Institute of Applied Mechanics and Informatics, Viet Nam Academy of Science and Technology (VAST), Ho Chi Minh City, Vietnam
e-mail: thuyvanleusq@gmail.com

K. Le-Cao · T. Ba-Quoc
Department of Mechanical Engineering, National University of Singapore, Singapore, Singapore
e-mail: khoa@u.nus.edu

T. Ba-Quoc
e-mail: thaibaquoc@u.nus.edu

Y. Nguyen-Quoc
Department of Engineering, Van Lang University, Ho Chi Minh City, Vietnam
e-mail: y.nq@vlu.edu.vn

N. H. Phuong and V. Kreinovich (eds.), *Biomedical and Other Applications of Soft Computing*, Studies in Computational Intelligence 1045,
https://doi.org/10.1007/978-3-031-08580-2_16

outer surface. Furthermore, if the respiratory particles are less than 9 μm in size, they can be suspended in the air for up to hours and do not follow air streamlines but continuously diffuse across them as they travel around a fibre. Some of the particles will collide with fibre as they diffuse and be removed from the airflow by the Van der Waals attractive force. Because the random motion increases as the particle diameter decrease, the capture of particles by diffusion become more efficient as the particle size becomes smaller. To simulate the airborne particle and fibre interaction, a stochastic approach needs to be investigated.

Stochastic Differential Equations (SDEs) play an important role in simulations of extremely complex physical systems. An understanding of SDE theory requires a fundamental concept of stochastic calculus including random variables, a convergence of sequences of random variables, Hilbert space of random variables, stochastic process and Wiener process, Ito integral, etc. The first detailed account of Brownian motion was carried out by the botanist Robert Brown in 1927 while studying plant life. He experimented with observing pollen grains immersed in a fluid under a microscope and he found out that in all cases the pollen grains were in rapid oscillatory motion. The motion consists of translations and rotations is incredibly erratic. The trajectory seems to have no tangent. Two pollen grains appear to move independently even when they are pretty close to one another. The composition and density of the particles (i.e. pollen grains) have no influence on the motion. The smaller the particles is, the more active the motion. The less viscous the fluid makes the motion larger. The higher temperature will lead to more active motion. The radius of the Brownian particle (i.e. the random motion of particles are large enough) is typically 10^{-9} m $< a < 5.10^{-7}$ m. The motion of a particle is the result of random and rapid collisions due to density fluctuation in the fluid. In the past, computational fluid approaches were often described using ordinary differential equations because random factors were ignored. Thus, the model will be incomplete for small scale particles. In fact, these factors are quite important in studying the filtration of airborne particles. In this study, we investigate stochastic differential equations (SDEs), especially Langevin's equation, which allows us to simulate the motion of extremely small particles in the air and interact with the filter fibre matrix.

The remainder of the paper is organised as follows. In Sect. 2, the equation of a random motion of a particle is described by the Langevin equation. The formula for the mean square displacement of the particle is then derived. It is known that, generally, exact solutions of SDEs are difficult to obtain. Section 3 will introduce two approximated methods namely Euler Maruyama (EM) method and Milstein method (MM). In Sect. 4, The motions of particles in a filter fibre matrix are simulated with emphasis placed on the interaction between particles and the fibre matrix. Numerical solutions in 2D and 3D are also reported. Section 5 concludes the paper.

2 Langevin Equation

Consider a particle immersed in a fluid with a viscosity μ. Langevin equation in a simple form is the equation of motion of the Brownian particle according to Newton's laws under the assumption that the Brownian motion is a sum of two forces, namely (i) a systematic force (viscous drag), which represents dynamical friction experienced by the particle (ii) a rapidly fluctuating force, which is the impact of molecules of the liquid on the particle, called white noise. Newton equation of motion for the particle $m\frac{dv(t)}{dt} = F(t)$, where m is particle mass, $F(t)$ the total instantaneous force on the particle at time t. This force includes a friction force $-\gamma v(t)$ and random force $\xi(t)$

$$dv(t) = -\frac{\gamma}{m}v(t)dt + \frac{1}{m}\xi(t)dt.$$

Let $dW(t) = \xi(t)dt$, $W(t)$ satisfies all requirements for a Wiener process. One can now write (3) as

$$dv(t) = -\frac{\gamma}{m}v(t)dt + \frac{1}{m}dW(t).$$

The friction force is assumed to be governed by Stokes' law which claims that the frictional force decelerating a spherical particle of radius a is $\gamma v = 6\pi\mu a v$. The fluctuation part $\xi(t)$ varies extremely rapid compared to the variation of the position of the particle. This implies that each collision is practically instantaneous. This rapid variation can be presented by $\overline{\xi(t)\xi(t')} = 2\gamma KT\delta(t-t')$, where δ is the Dirac function, t and t' are distinct times and $2\gamma KT$ is called the spectral density. It is also assumed that the average $\overline{\xi(t)} = 0$ since $\xi(t)$ is so irregular. The over-bar means that the statistical average over an ensemble of particles, each particle in the ensemble starts with the same initial conditions.

Analytical Solution of Langevin Equation
In simple case, an analytical solution of Langevin equation can be derived. One can write the Langevin equation as

$$\int_a^b dv(t) = -\frac{\gamma}{m}\int_a^b v(t)dt + \frac{1}{m}\int_a^b dW(t).$$

Multiplying both sides of Eq. (6) by a continuous function $f(t)$ of the time t

$$\int_a^b f(t)dv(t) = -\frac{\gamma}{m}\int_a^b f(t)v(t)dt + \frac{1}{m}\int_a^b f(t)dW(t).$$

All these integrals are well defined for $f(t)$ a continuous function of time. Setting $f(t) = e^{\frac{\gamma}{m}t}$ with $[a, b] = [0, 1]$. We have

$$\int_0^t e^{\frac{\gamma}{m}s} dv(s) = -\frac{\gamma}{m}\int_0^t e^{\frac{\gamma}{m}s} v(s)ds + \frac{1}{m}\int_0^t e^{\frac{\gamma}{m}s} dW(s).$$

And since integration by parts, we have

$$e^{\frac{\gamma}{m}t}v(t) - e^{\frac{\gamma}{m}0}v(0) - \frac{\gamma}{m}\int_0^t e^{\frac{\gamma}{m}s} v(s)ds = -\frac{\gamma}{m}\int_0^t e^{\frac{\gamma}{m}s} v(s)ds + \frac{1}{m}\int_0^t e^{\frac{\gamma}{m}s} dW(s)$$

or

$$e^{\frac{\gamma}{m}t}v(t) - v(0) = \frac{1}{m}\int_0^t e^{\frac{\gamma}{m}s} dW(s).$$

Hence, the solution of Langevin equation is

$$v(t) = v(0)e^{-\frac{\gamma}{m}t} + \frac{1}{m}\int_0^t e^{-\frac{\gamma}{m}(t-s)} dW(s).$$

Mean Square Displacement
One has

$$m\ddot{x}(t) = -\gamma\dot{x}(t) + \xi(t). \tag{1}$$

On multiplying Eq. 1 by $x(t)$, one gets

$$m\ddot{x}(t)x(t) = -\gamma\dot{x}(t)\ddot{x}(t) + \xi(t)x(t).$$

Rewriting as follow

$$\frac{m}{2}\frac{d}{dt}\left(\frac{dx^2(t)}{dt}\right) - m\dot{x}^2(t) = -\frac{\gamma}{2}\frac{dx^2(t)}{dt} + \xi(t)x(t).$$

Using the averaged equation of motion

$$\frac{m}{2}\frac{d}{dt}\left(\frac{d\overline{x^2}}{dt}\right) - m\overline{\dot{x}^2} = -\frac{\gamma}{2}\frac{d\overline{x^2}}{dt} + \overline{\xi x}$$

$$\frac{m}{2}\frac{d}{dt}\left(\frac{d\overline{(x^2)}}{dt}\right) + \frac{\gamma}{2}\frac{d\overline{(x^2)}}{dt} = kT$$

Let $\frac{d\overline{x^2}}{dt} = u$, then $\frac{m}{2}\frac{du}{dt} + \frac{\gamma}{2}u = kT$.
The solution of equation (1) is

$$u = Ce^{-\gamma t/m} + 2KT/\gamma$$

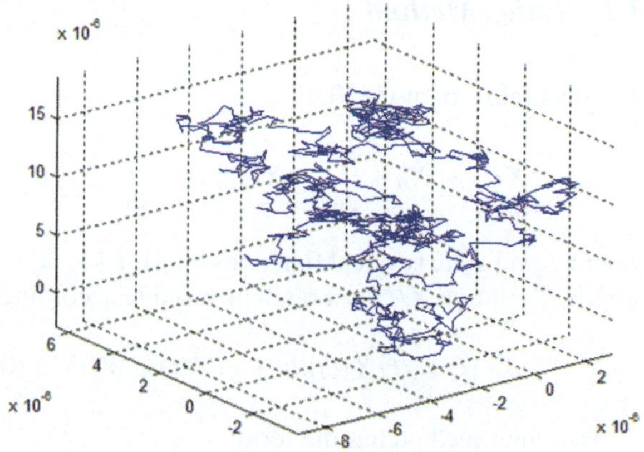

Fig. 1 A typical random motion of a particle under the influence of air density fluctuation

For very large time t, only the first term of the right hand side of equation (the exponential term) is neglected, implying ignoring the effect of the inertia of the Brownian particle, then one can obtain $u = 2KT/\gamma$. Integrating equation from $t = 0$, $\overline{x^2} - \overline{x_0^2} = (2KT/\gamma)t$.

Let $x_0 = 0$ and write $\overline{(\Delta x)^2}$ instead of $\overline{x^2}$, $\overline{(\Delta x)^2} = (2KT/\gamma)t$, which is a formula derived by Langevin. $\overline{(\Delta x)^2}$ has the following meaning: we observe a single particle at a time. During the time interval $(0, t)$ it has undergone a displacement Δs, whose projection on the x-axis is Δx. The same particle is observed at later times $2t$, $3t$,..., and Δx is determined for each interval. According to theory, the mean squared displacement of the particle is proportional to the time interval. The formula for the mean square displacement of the Brownian particle

$$\left\langle \left| \vec{x}(t_0 + t) - \vec{x}(t_0) \right|^2 \right\rangle = (2KT/\gamma)t,$$

where $x(t)$ is position, t is time interval. Consider a particle has radius $a = 10^{-6}$ m, viscosity $\mu = 10^{-3}$ m^2/s. A random movement of the particle can be plotted as in Fig. 1.

3 SDE Solver

For most SDEs, it is tough to find an analytical solution. Instead, we investigate some common approximation methods. Among the simple and effective methods for approximating solutions in SDEs are the Euler (or Euler–Maruyama) method and the Milstein method. The feature of these methods is to find approximations of the real solution at the discrete points $t_0, t_1, t_2, \ldots, t_n$.

3.1 Euler Method

The integral form of the SDE

$$X(t) = X_0 + \int_0^t f(X(s))ds + \int_0^t g(X(s))dW(s), 0 \leq t \leq T,$$

where f and g are two real functions on $[0, T] \times \Omega$, initial condition X_0 is a random variable, solution $X(t)$ is a random variable according to t. In differential form:

$$dX(t) = f(X(t))dt + g(X(t))dW(t), X(0) = X_0, 0 \leq t \leq T$$

The Euler method has the form:

$$X_{i+1} = X_i + f(t_i, X_i)\Delta t + g(t_i, X_i)\Delta W_{i+1}, X_0 = X(0)$$
with $i = 0, 1, 2, \ldots, N-1, t_i = i\Delta t, \Delta t = T/N$ and $\Delta W_{i+1} = W(t_{i+1}) - W(t_i)$.
Apply Euler's method to the Black–Scholes equation:

$$\begin{cases} dX = \mu X dt + \sigma X dW \\ X(0) = X_0 \end{cases}$$

with $\mu = 2, \sigma = 1, X(0) = 1$.

The exact solution of this equation is:

$$X(t) = X_0 e^{(\mu - \frac{1}{2}\sigma^2)t + \sigma W_t},$$

$$X_{i+1} = X_i + \mu X_i \Delta t + \sigma X_i \Delta W_{i+1}.$$

Let consider the Black–Scholes equation with $\mu = 2, \sigma = 1, X_0 = 1$. In Fig. 2, the upper line represents the exact solution $(X(t) = X_0 e^{(\mu - \frac{1}{2}\sigma^2)t + \sigma W_t})$. The lower line represents the approximate solution using the Euler–Maruyama method. The error between the exact solution and the approximate solution at time $t = T$ is err = 0.6907.

Convergence of Euler's Method
A numerical method is said to be strongly convergent of order γ if there exists a constant C such that

$$E|X_n - X(\tau)| \leqslant C\Delta t^{\gamma},$$

with $\tau = n\Delta t \in [0, T]$. Euler's method converges strongly with $\gamma = 1/2$, notice if the function $g \equiv 0$ with X_0 constant then becomes the ordinary differential equation and the method converges to order 1. The error at time is calculated $t = T$, set: $e_{\Delta t}^{strong} = E|X_L - X(T)|$, with $L\delta t = T$.

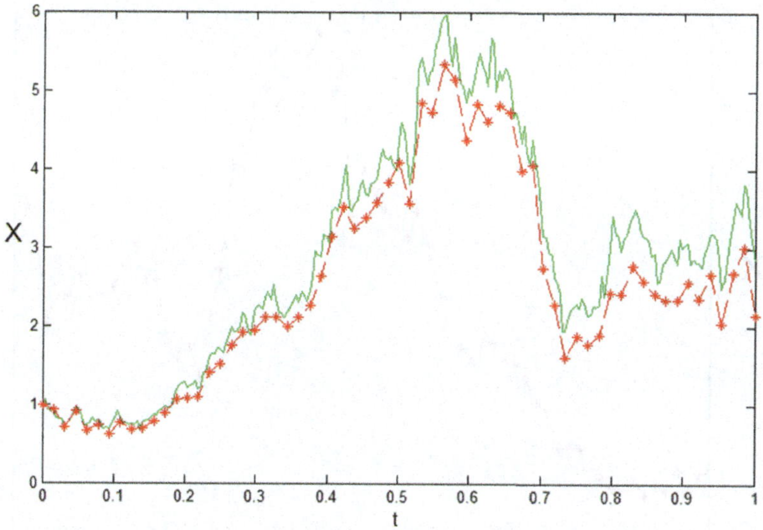

Fig. 2 The exact solution and the approximated solution of Euler's method

If Euler's method converges to $\gamma = 1/2$ at every point $t \in [0, T]$ then it also converges to $\gamma = 1/2$ at $t = T$

$$e_{\Delta t}^{strong} \leqslant C\Delta t^{\frac{1}{2}}.$$

We examine the strong convergence of the Euler method for the Black–Scholes equation with μ, σ and X_0 as above. We calculate 1000 Brownian paths on [0, 1] with $\delta t = 2^{-9}$. For each path, EM is calculated with 5 different divisions :$\Delta t = 2^{p-1}\delta t$ with $1 \leq p \leq 5$. The error at the end of the sth sample curve with the pth division is stored, $q \approx \gamma \approx 0.5384$.

3.2 *Milstein Method*

In the previous section, Euler's method has a degree of convergence of $\gamma = 1/2$. The order of convergence of Euler's method can be improved to 1 by adding a random quantity to the Milstein method, which is due to the Ito–Taylor expansion. The Milstien method has the form:

$$X_{i+1} = X_i + \Delta t f(X_i) + g(X_i)\Delta W_{i+1} + \frac{1}{2}g(X_i)g'(X_i)((\Delta W_{i+1})^2 - \Delta t),$$

with $i = 0, 1, 2, \ldots, N-1$, $t_i = i\Delta t$, $\Delta t = T/N$ and $\Delta W_{i+1} = W(t_{i+1}) - W(t_i)$. Apply to the above Black Scholes equation, the Milstein method has the form:

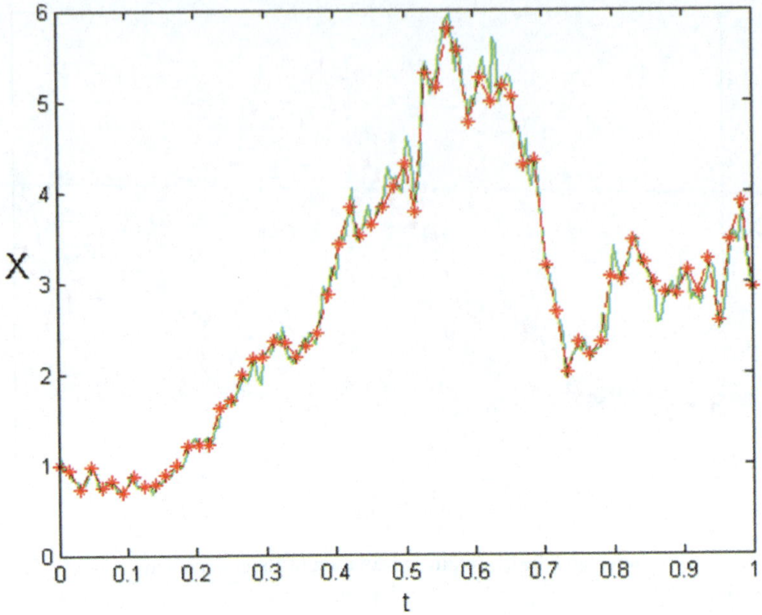

Fig. 3 The exact solution and the approximated solution of Milstein's method

$$X_{i+1} = X_i + \mu X_i \Delta t + \sigma X_i \Delta W_{i+1} + \frac{1}{2}\sigma^2 \left((\Delta W_{i+1})^2 - \Delta t\right).$$

Consider a Black–Scholes equation with $\mu = 2$, $\sigma = 1$, $X_0 = 1$. In Fig. 3, the upper line represents the exact solution ($X(t) = X_0 e^{(\mu - \frac{1}{2}\sigma^2)t + \sigma W_t}$). The lower line represents the approximate solution using the Milstein method.

Milstein Method Convergence

Considering the equation of population change and applying Milstein's method

$$dX(t) = rX(t)(K - X(t))dt + \beta X(t)dW(t), X(0) = X_0,$$

where r, K and β are constants. We take $r = 2$, $K = 1$, $\beta = 0.25$ and $X_0 = 0.5$ and use Brownian paths on [0, 1] with $dt = 2^{-11}$. We calculate the 500 Brownian path on [0, 1] with $\delta t = 2^{-11}$ with $\Delta t = dt$, $\Delta t = 16dt$, $\Delta t = 32dt$, $\Delta t = 64dt$, $\Delta t = 128dt$. For each path, EM is calculated with 5 different divisions $\Delta t = dt$, $\Delta t = 16dt$, $\Delta t = 32dt$, $\Delta t = 64dt$, $\Delta t = 128dt$. The strong convergence of the Milstein method for the population transformation equation with r, K, β and X_0 as above is around $q \approx \gamma \approx 1.0316$.

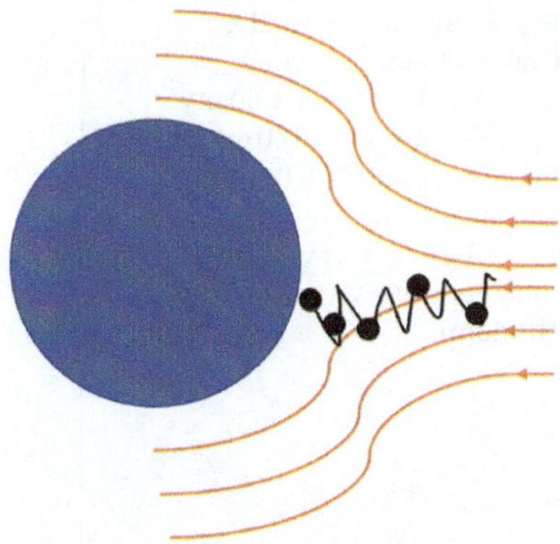

Fig. 4 Diffusion in porous media occurs for extremely tiny particles. Particles move randomly about within the air, due to thermal gradients. Particle motion is determined by the superposition of the Brownian (probabilistic) motion and the inertial (deterministic) motion over the flow field

4 Langevin Equation for Particles' Zig-Zac Motion

Particles from 10 μm are visible to the naked eye in the presence of scattered light. Smaller particles, which have a diameter of about 0.5 μm, are invisible to the naked eye even at high concentrations such as cigarette smoke. Virus particles are in the range of 0.005–0.1 μm. In the fine filtration mechanism, the filter material is commonly made of glass fibre with a filter fibre diameter of about 10 μm, the distance between the fibres about 10 μm. The air velocity across the membrane is usually 2–10 cm/s. The primary particle retention of fine filtration is diffusivity. For particles smaller than 1 μm, there will be Brownian motion like molecules. The particles move in a zig-zag pattern and when the airflow passes through the filter they will cease oscillating and are retained on the fibre surfaces by van der Waals forces (Fig. 4). In the simulation, the air velocity vector field throughout a defined region containing the fibre matrix is first obtained. The particle trajectory through fibre matrix elements is then computed by superimposing the Langevin particle equation of motion over the flow velocity field.

The resulting particle dynamics is given by

$$\frac{d\vec{v}}{dt} = -\gamma \times (\vec{v}(t) - \vec{v}_0(t)) + \sigma \times \frac{d\vec{W}(t)}{dt}$$

and

$$\frac{d\vec{x}}{dt} = \vec{v},$$

where $\vec{v}(t)$ is particle velocity vector and $\vec{v}_0(t)$ air velocity vector.

Two coefficients γ and σ are defined as

$$\gamma = \frac{6\pi\rho\mu R}{C_s m},$$

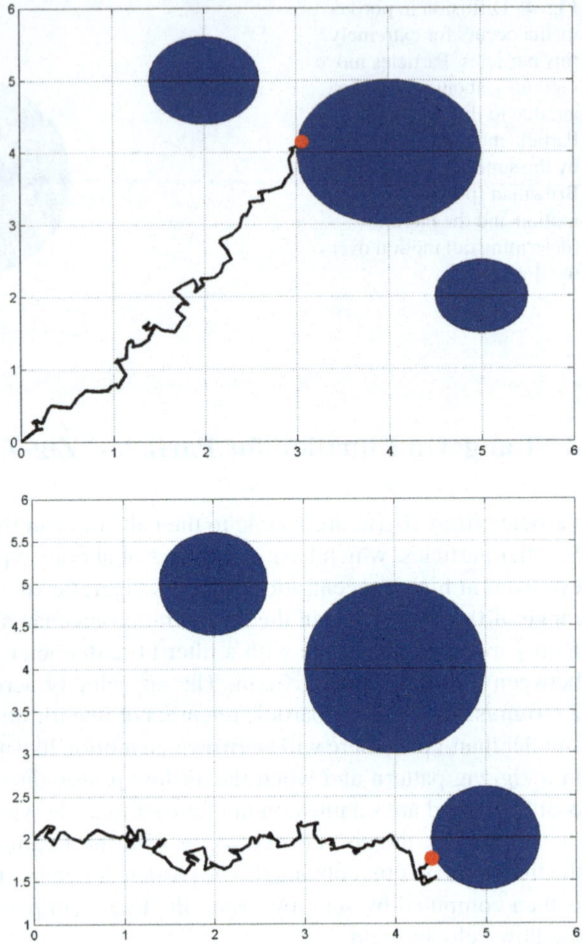

Fig. 5 Simulation in 2D: Consider a particle with 1.5 μm diameter, air velocity $v_x = 0.2\,\text{m/s}$, $v_y = 0.0\,\text{m/s}$, initial position $A(0, 0)$ (Top) and air velocity $v_x = 0.2\,\text{m/s}$, $v_y = 0.3\,\text{m/s}$, initial position $A(0, 2)$ (Bottom)

$$\sigma^2 = \frac{2k_B T \gamma}{m},$$

where T ambient temperature, k_B Boltzman constant, and ρ air density.

The principle of superimposing particle motion over the flow field is illustrated in Fig. 4, where the net particle motion is the sum of a deterministic term and probabilistic term. Numerical solutions of Langevin equation 4 is calculated by using Milstein method for two and three dimensional trajectory Figs. 5 and 6, respectively.

Filter Efficiency
Filter efficiency is defined by the ratio of upstream particle concentration compared to the downstream concentration that has passed through and multiplied by 100

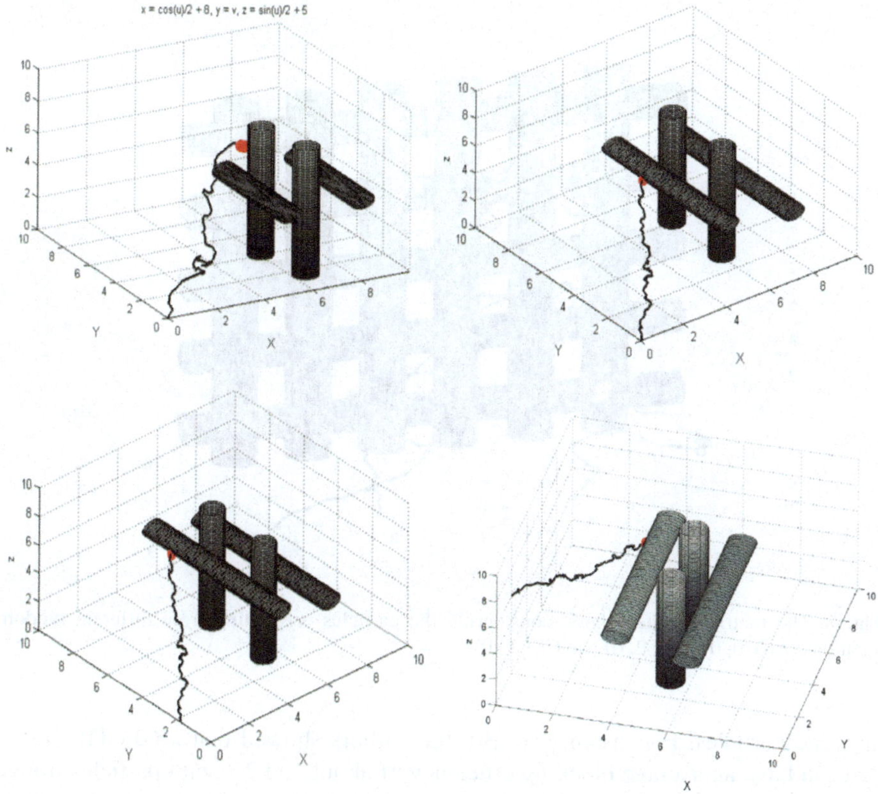

Fig. 6 Simulation in 3D: Consider a particle with 1 μm diameter, air velocity $v_x = 0.2\,\text{m/s}$ $v_y = 0.3\,\text{m/s}$, $v_z = 0.4\,\text{m/s}$, initial position $A(0, 1.8, 0)$, air velocity $v_x = 0.2\,\text{m/s}$ $v_y = 0.0\,\text{m/s}$, $v_z = 0.0\,\text{m/s}$, initial position $A(0, 3.5, 6.5)$, respectively

$$E = \left[1 - \frac{DC}{UC}\right] \times 100,$$

where DC: downstream concentration, UC: upstream concentration. Computing the filter efficiency requires three steps: (1) estimate the airflow field, (2) compute the trajectory of the particle by solving the Langevin equation, and (3) compute the trajectories of particles at random positions at filter inlet to get an average efficiency. For filter efficiency as a function of particle size, step 2 and step 3 must be repeated for each particle size. A typical calculation of particle movement in a matrix of filter fibres is shown in Fig. 7. In the preliminary findings, we have conducted about 1000+ simulations with different initial positions of the particles and the results show that about 60–70% of the particles stick to the fibres of the filter matrix (due to zig-zac motions) although the size of the gaps between the fibres is very large compared to the size of the particles. This result is also quite consistent with the recent study reveal that stochastic diffusion and electrostatic force play an important role in filtering

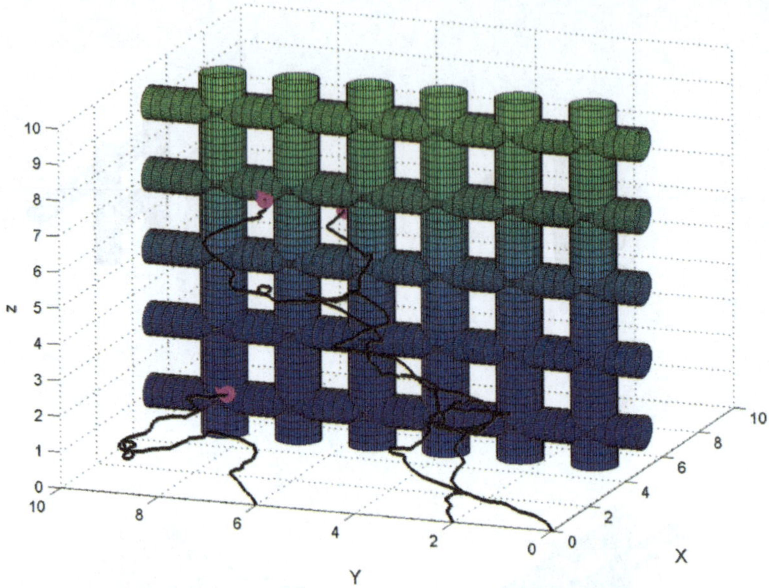

Fig. 7 The matrix of filter fibers: considering the particles originating from different random locations: A(0, 0, 0), B(0, 2, 0) vaø C(0, 6, 0)

airborne particles. For instance, in [9], the authors showed that a 600 TPI cotton pad can have an average blocking efficiency of about 79±2% with particles' range $0.01 - 0.3\,\mu\text{m}$.

5 Conclusion

In this study, we have presented a simulation of the random motion of airborne particles and how they interact with the filter fibre matrix by solving SDEs. Equation of motion for the particle is Langevin equation which is a Newton second law for motion of an airborne particle, where force includes both viscous drags due to the surrounding air and the fluctuations caused by the individual collisions with air molecules. Trajectories of particles are computed in two dimensions and as well as three dimensions. The research results are quite reasonable and explain the filtering mechanism when the filter matrices are much larger than the particle size but still effective in blocking particles. In future works, we will conduct simulations with different particle sizes as well as variable distances between filters. The improvement of the numerical accuracy of SDEs will also be investigated. The differential deterministic component defined in the Langevin equation can be improved by using a higher-order approximation such as IRBF [1].

Acknowledgements This research is supported by Computational Engineering and Science Research Centre (CESRC), University of Southern Queensland, and Institute of Applied Mechanics and Informatics (IAMI), HCMC Vietnam Academy of Science and Technology (VAST).

References

1. T.T.V. Le, N. Mai-Duy, K. Le-Cao, T. Tran-Cong, A time discretisation scheme based on integrated radial basis functions (IRBFs) for heat transfer and fluid flow problems. Numerical and Heat Transfer, Part B (2018)
2. B. Oksendal, *Stochastic Differential Equation* (Springer, Berlin, 1998), pp. 43–72
3. L.C. Evans, *Introduction to Stochastic Differential Equations-Berkeley Lecture Notes* (2000), pp. 55–99
4. E. Allen, *Modelling with Ito Stochastic Differential Equations*, vol. 22. (Springer, Berlin, 2007), pp. 89–127
5. P.E. Kloeden, E. Platen, H. Schurz, *Numerical Solution of Stochastic DE Through Computer Experiments*. (Springer, 2003), pp. 139–194
6. W.T. Colley, *The Langevin Equation with application to stochastic Problems in Physics, Chemistry and Electrical Engineering*, vol. 14 (World Scientific, 2004), pp. 1–60
7. S. Rief, A. Latz, A. Wiegmann, Computer simulation of air filtration including electric surface charges in three-dimensional fibrous microstructures (2005)
8. B.U. Lee, Minimum sizes of respiratory particles carrying SARS-CoV-2 and the possibility of aerosol generation. Int. J. Environ. Res. Public Health **17**(19), 6960 (2020)
9. A. Konda, A. Prakash, G.A. Moss, M. Schmoldt, G.D. Grant, S. Guha, Aerosol filtration efficiency of common fabrics used in respiratory cloth masks. ACS Nano **14**, 6339–6347 (2020)

Disease Diagnosis Based on Symptoms Description

Huong Hoang Luong, Phong Cao Nguyen, and Hai Thanh Nguyen

Abstract When one feels unwell, it is crucial to arrange a time as soon as possible to meet a doctor for early detection of potential health-related problems. However, a relatively large number of Vietnamese people usually avoid going to the hospital as they are afraid of long waits at such crowded places, while the current COVID-19 pandemic means being at those places poses a higher risk of contracting the disease. For simpler health problems, people would prefer a solution that, given their symptoms, provides a reliable diagnosis in a shorter time. This study presents an approach in building a deep-learning-based disease predictor of health conditions conducted from given symptoms in Vietnamese. The proposed method combines a tokenizer and bi-directional recurrent neural networks and achieved an accuracy of 98.96% (compared to a certified doctor's diagnosis) in selected test cases, demonstrating its promising capabilities in the task. The application is expected to easily be integrated into a mobile application and open the way for other deep-learning-based solutions which analyze people's symptoms to help them have their health conditions diagnosed at home.

1 Introduction

In a country with a developed healthcare system, citizens would have easy access to doctors and treatments (with little delay, little transportation needed, and preferably at little cost). However, the situation is far less ideal in Vietnam. The number of medical doctors per 1,000 people is only 0.828 (2016), much lower than the East Asia–Pacific (1.665, 2017) and world (1.566, 2017) averages (see [1, 2]). As a result, when people go to hospitals, they face long queues and cursory examinations while

H. H. Luong · P. C. Nguyen
FPT University, Can Tho 900000, Vietnam

H. T. Nguyen (✉)
College of Information and Communication Technology, Can Tho University, Can Tho 900100, Vietnam
e-mail: nthai.cit@ctu.edu.vn

N. H. Phuong and V. Kreinovich (eds.), *Biomedical and Other Applications of Soft Computing*, Studies in Computational Intelligence 1045,
https://doi.org/10.1007/978-3-031-08580-2_17

doctors have too many patients to handle. These are not to mention the surge of COVID-19, which increases the risk of cross-infection in medical facilities. These factors, all together, account for the widespread reluctance of the people to go to hospitals. Instead, people would generally go to drugstores and ask the pharmacist to diagnose and prescribe simultaneously. This action can have detrimental effects: pharmacists are not trained as doctors, so it is not guaranteed that their diagnoses are correct, or they may prescribe unnecessary antibiotics to increase sales, thus worsening the antibiotic resistance problem. As a response to this, the Vietnamese government is encouraging *telehealth* solutions—systems that allow consultation with licensed doctors without having to meet them in person.

In our attempt to help the government deal with this, we developed a portable, lightweight solution that helps Vietnamese diagnose mild medical conditions that do not yet require expert consultation with doctors, reducing the need to go to hospitals and thus ease the current overload on doctors and the health system as a whole. By conducting this research, we aim to make the following contributions:

- Developing an authentic dataset from real patients input and real diagnosis. We would like to offer our sincere thanks to Dr. Trinh Hoang Thanh for assisting us with this task
- Demonstrating how to apply some Vietnamese language processing tools to solve some specific real-world problem
- Developing a deep-learning model that can be integrated into a mobile solution that helps Vietnamese diagnose some simple medical problems from home, thus easing the pressure on the nation's health system
- Reducing the number of inaccurate prescriptions from non-licensed practitioners that negatively impact people's health

The rest of this paper is organized as follows. In Sect. 2, we review some works that are related to our study. The details of our method are included in Sect. 3, and the results are presented in Sect. 4. The paper then ends, in Sect. 5, with a brief conclusion and our expectations for future works.

2 Overview of Related Works

The problem of predicting health problems based on text input can be divided mainly into two classes based on the input: specialized (which means the input is standard with medical terms, and the sources are mainly medical journals, clinical data, or doctors' input), and non-specialized (which means the text is far from standard with non-technical terms that are synonymous with terms professionals usually use). There has been quite a lot of research into the technical input problem. For example, N. Dragu et al. mines text from an online news article to predict possible outbreaks of diseases [3]. The problem of non-specialized input is, however, much more complex. The variety of synonymous phrases may cause difficulties that cannot efficiently process natural language. Unfortunately, most proposed models do not have an element of deep learning. Instead, they would use full-text databases, and therefore, if the user enters something not found in the databases, the results may be undesirable. One notable proposal is the Automated Disease Prediction System (ADPS), developed by T. R. Laskar et al., which utilized a wide range of techniques (a novel Relevant Attribute data structure, synonym tree, decision tree, etc.) to achieve an average accuracy of 14.35% higher than that of a chosen symptom matching system [4]. Another work tackled the problem with an approach of natural language processing [5], which also allows for linguistic diversity. However, the dataset used was taken from the UK NHS website, considered somewhat specialized. Other attempts at the problem usually focus on one specific class of diseases, like heart diseases in [6, 7].

The number of related researches in Vietnamese is even lower. Most medical-related deep-learning solutions focus on diagnostic imaging. For example, T. T. Tran et al. applied convolutional neural networks (CNNs) to diagnose diseases from pediatric chest radiographs [8]. Another example would be [9], in which H. H. Pham et al. also used CNNs to predict fourteen thoracic diseases from chest X-rays. These models are mainly aimed to assist professional health practitioners by offering them a supplementary opinion in predicting diseases from medical images; few are directed at patients. Linguistic differences between the two languages also make taking a well-established model in English and "translating" it into Vietnamese far from straightforward. One system proposed in recent years is [10], which is based on knowledge bases and logic and requires users to answer Yes/No questions rather than taking in text input. Because of the severe lack of similar works, there were hardly any figures to bring up for comparison purposes against our work.

3 Methodology

3.1 Background

3.1.1 Tokenizer

Word segmentation in Vietnamese is more complicated than English because white spaces do not necessarily mean separating two words. In Vietnamese, most words consist of two syllables, and white spaces can either separate these two syllables or separate one word from another. Deliberately separating two syllables within the same word may lead to significant changes in meaning or render the sentence incomprehensible at all. For example, *dạ dày* is the stomach, but *dạ* is an interjection, while *dày* means thick. Therefore, we decided to employ the word segmented proposed by Nguyen et al. [11]. The segmenter makes use of a variant of a binary tree—the Single Classification Ripple Down Rules (SCRDR) tree [12]—which decides whether a syllable stands alone as a complete word or it is part of a larger word. The result of the tokenizer would be a properly segmented phrase or sentence where syllables of the same word/phrase are joined by an underscore (for example, *Tôi* (I) *bị* (have) *chảy_nước_mũi* (a running nose)).

3.1.2 Bi-directional Recurrent Neural Networks (Bi-RNNs)

Before RNNs, all neural networks were essentially feedforward ones. That is, input is fed into the network and goes straight through to produce an output. Then, the following input is processed similarly, utterly independent of the previous input. Because of this characteristic, feedforward networks only focus on the current information exposed, have no notion of time/place, and are therefore unfit for problems that require context.

RNNs, developed based on the idea of back-propagation introduced by Rumelhart et al. [13], address this issue. Nodes in RNNs can store their previous values, and these values play a certain role in determining the values of the nodes for the current iteration. In certainly simplified diagrams, this feature is represented by a loop indicating output is fed back to the network as part of the input for the next iteration. Figure 1, inspired by [14], compares how feedforward networks ((a)—compressed and (b)—unfolded) and RNNs ((c)—compressed and (d)—unfolded) work. In this figure, x_i, y_i, and A represent input, output, and the network's collection of nodes and connections, respectively.

Bi-RNNs are very similar to RNNs, but another information flow travels from the last cell to the first. Because of this characteristic, Bi-RNN would not be suitable for streaming input (input is appended while the model is learning), but it poses no such problem for our application. We decided to employ Bi-RNN because, in most cases, phrases at the end of a sentence help better understand the meaning of

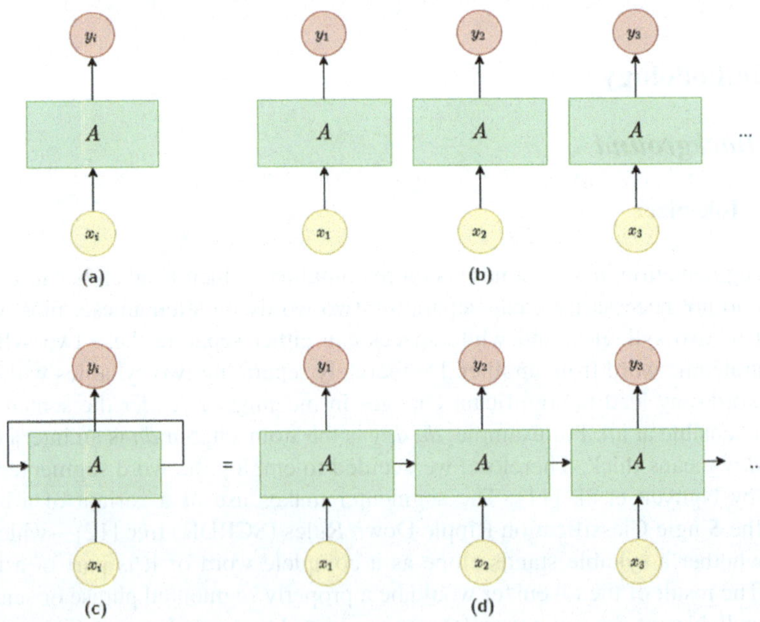

Fig. 1 Feedforward networks and recurrent networks

previous words and because after experiments with several other algorithms, Bi-RNN produced the highest results (see Sect. 4.2).

3.2 *Datasets, Input Pre-processing, Model Selection and Training*

3.2.1 Datasets

Three main datasets were used during the development of our solution. The first dataset, used to train the Tokenizer, was obtained from Association for Vietnamese Language and Speech Processing (VLSP) [15] and Vietnamese Corpus [16]. The sources for these datasets are from dictionaries and news articles. For the Synonym Tree, medical newspapers and journals are used instead.

We decided to develop our dataset because of the lack of related datasets and our wish to include a more authentic (versus specialized) data source. Therefore, we asked Dr. Thanh to provide us with a list of symptoms, together with his diagnosis.

The original dataset, collected from 121 Military Hospital in Can Tho, Vietnam, contains symptoms displayed in a 9-row-2-column table where each row represents the symptoms of the specified disease. To account for the seemingly too small dataset, we select two, three, or more symptoms of the same disease and make each combination into a new line. We ended up with a 105-line dataset used for training. For testing, we used manually prepared symptom combinations that did not appear in the training dataset, and the number of samples in the testing set is 96. Figures 2 and 3 shows some illustrations on some symptoms appearing in samples. The total

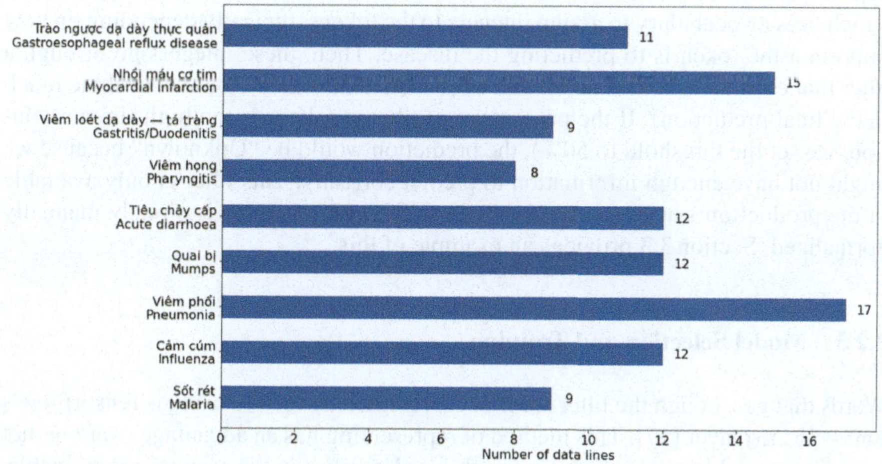

Fig. 2 Distribution of symptom samples in the training set in both English and Vietnamese

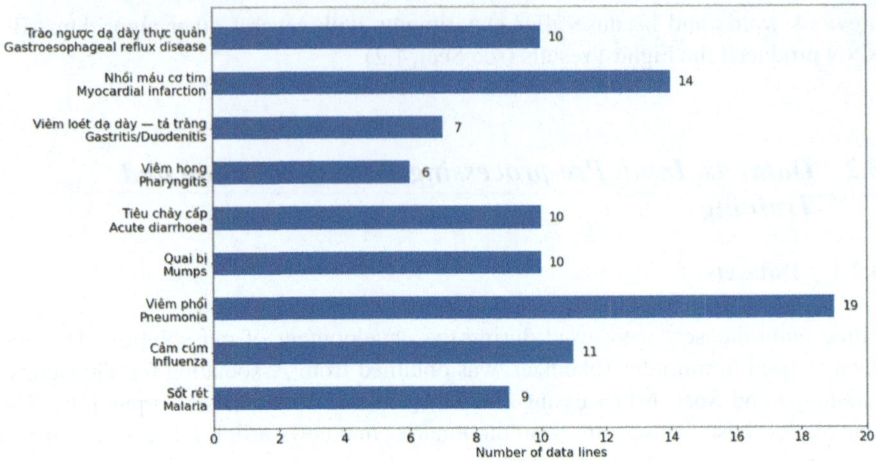

Fig. 3 Distribution of symptom samples in the testing set in both English and Vietnamese

samples for the training and testing sets are 105 (52%) and 96 (48%), respectively. The nine diseases included in this research, as shown in Fig. 2, are Malaria, Influenza, Pneumonia, Mumps, Acute diarrhea, Pharyngitis, Gastritis/Duodentis, Myocardial infarction, and Gastroesophageal reflux.

3.2.2 Input Pre-processing

User raw input first goes through the Tokenizer described in Sect. 3.1. Then, tokens are converted into integers using TensorFlow's `TextVectorization` [17], which uses a vocabulary to assign integers to the tokens, typically depending on how important the token is to predicting the disease. Then, these integers go through a filter that eliminates low-ranking tokens (which means they do not contribute much to the final prediction). If the percentage of filtered tokens is too high (in our solution, we set the threshold to 50%), the prediction would be "Unknown" because we might not have enough information to predict correctly. This filter is only available in the production stage because the input to the training stage is already manually normalized. Section 3.3 provides an example of this.

3.2.3 Model Selection and Training

Words that go through the filter are now represented as vectors using a TensorFlow's `Embedding` layer [17]. This method of representing has an advantage over one-hot encoding or deliberately assigning each word a value as the vectors are trainable, which means the weights can be adjusted during training, and similar vectors will represent similar words if the dataset is large enough. Then comes a Bi-RNN layer as described in Sect. 3.1 with `softmax` activation function [18], `cross-entropy` loss function [19] and `adam` optimizer [20], and a `Dropout` layer [17] with a rate of 0.2 is also used to avoid overfitting, considering the relatively small size of our dataset. Finally, a `Dense` layer [17] does the final processing and returns a classification result. Section 3.3 also includes a diagram of the described model.

3.3 The Proposed Architecture for Diagnosis Based Symptom Descriptions

In this section, we describe one iteration of our model using an example. Much of the details have already been discussed in Sect. 3.2 above. Where appropriate, refer to Fig. 4, which visualizes the process.

1. We start with a tokenized input of *Tôi* (I) *bị* (have) *đau_bụng* (a bellyache) *sốt* (a fever) *nghẹt_mũi* (a stuffy nose) *và* (and) *ho* (a cough).

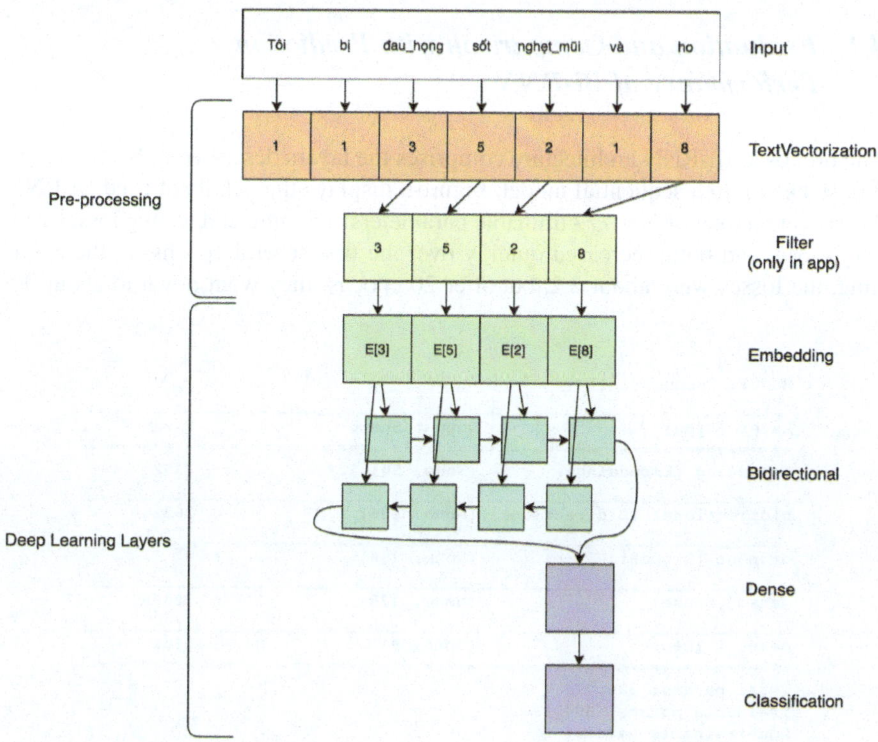

Fig. 4 The proposed architecture for the predictor based on symptoms description

2. The two-part pre-processing stage takes place. First, the tokens go through a `TextVectorization` layer, which scores the words accordingly. We can see that "main" words like *bellyache* or *fever* have higher scores than insignificant words like *I* or *have*. Then, a filter removes words with scores of 1. Only content words remain.
3. The remaining tokens are fed as input to the deep learning model, which consists of an initial `Embedding` layer, a Bi-RNN layer with the dropout rate of 0.2, and a `Dense` layer that returns the final classification. Thus, not only the possible disease is displayed, but several top possibilities will also appear and are ranked accordingly to the confidence level.
4. Because accuracy for health-related diagnosis is of utmost importance, and all newly-generated predictions will have a "Pending confirmation" status, indicating it is waiting for confirmation of a licensed doctor. After that, the symptoms and the correct disease will be added to the dataset used in future training.

4 Results

4.1 Evaluation and Comparison with Prediction Performance of Bi-RNN

The proposed Bi-RNN architecture comprises the layers described in Sects. 3.2 and 3.3, stacked into a sequential model. Figure 5 displays the details of used Bi-RNN. There were a total of 304,777 trainable parameters. Training and testing losses were very close, and both decreased quickly over the first several epochs. In the beginning, the losses were about 2.2, but after 20 epochs, they went down to about 0.5,

```
Model: "sequential"
_________________________________________________________________
Layer (type)                 Output Shape              Param #
=================================================================
embedding (Embedding)        (None, 20, 128)           7552
_________________________________________________________________
bidirectional (Bidirectional (None, 256)               263168
_________________________________________________________________
dropout (Dropout)            (None, 256)               0
_________________________________________________________________
dense (Dense)                (None, 128)               32896
_________________________________________________________________
dense_1 (Dense)              (None, 9)                 1161
=================================================================
Total params: 304,777
Trainable params: 304,777
Non-trainable params: 0
_________________________________________________________________
```

Fig. 5 The architecture and the number of parameters of the Bidirectional model

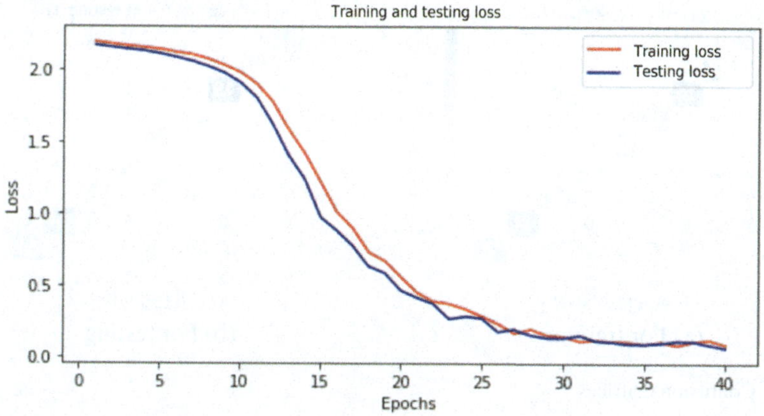

Fig. 6 Training and testing losses of the model

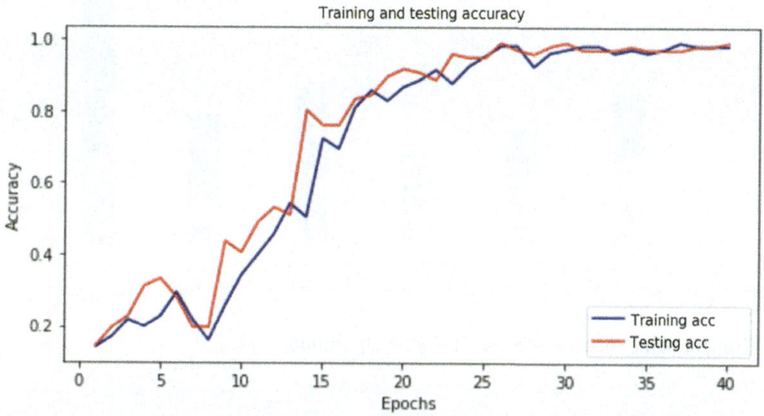

Fig. 7 Training and testing accuracies of the model

and we decided to stop at epoch 40 when the losses were 0.09 for training and 0.07 for testing (see Fig. 6).

Similarly, the accuracies of the model increased from about 14% at the beginning to about 87% at epoch 20. After epoch 30, they stabilized, and at the end of training (epoch 40), the figures are 98–99% for both training and testing (see Fig. 7). The confusion matrices for training and testing are also included (see Fig. 8), from which many other evaluative metrics like precision, recall, or F_1 score can be computed.

We also ran several other algorithms on the same dataset, namely Logistic Regression, Naive Bayes, Decision Tree, Random Forest, and SVM (using default parameters recommended by these algorithms). As revealed in Fig. 9, Bi-RNN achieves the highest results with 0.98, while other algorithms obtain performances that are lower than 0.9, ranging from 0.82 to 0.89.

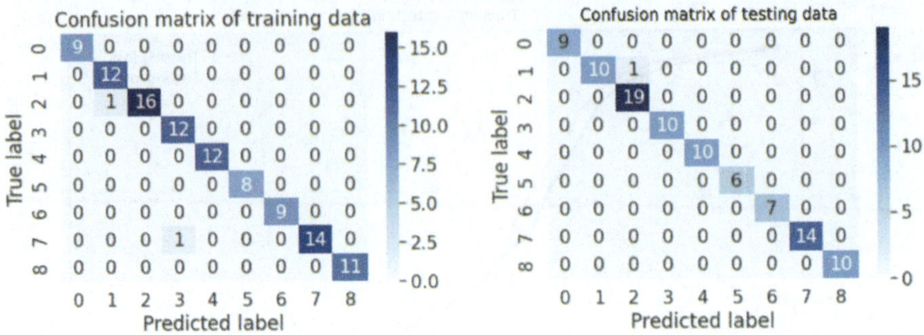

(a) For training (b) For testing

Fig. 8 Confusion matrices

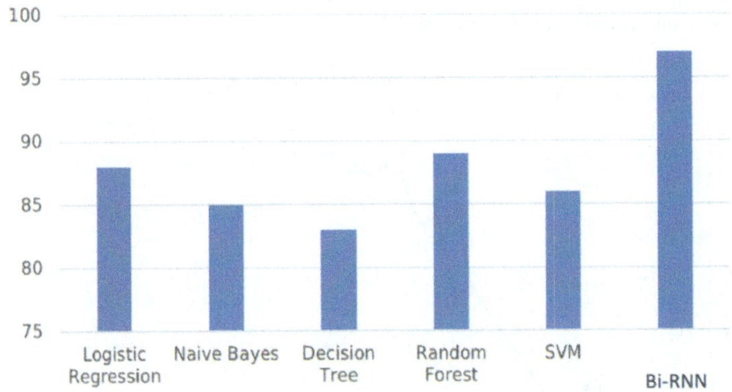

Fig. 9 Comparison of accuracy across different algorithms

4.2 Practical Testing

After deploying our model as part of an Android application, we acted as patients and gave different input to the program as tests. Specifically, six tests were given for each of the nine diseases, three of which included never-seen-before symptoms. The result was quite encouraging: the model only failed four tests, reaching a practical accuracy of 95 samples over 96 cases. Finally, we have built and deployed the proposed approach in an Android application as exhibited in Fig. 10.

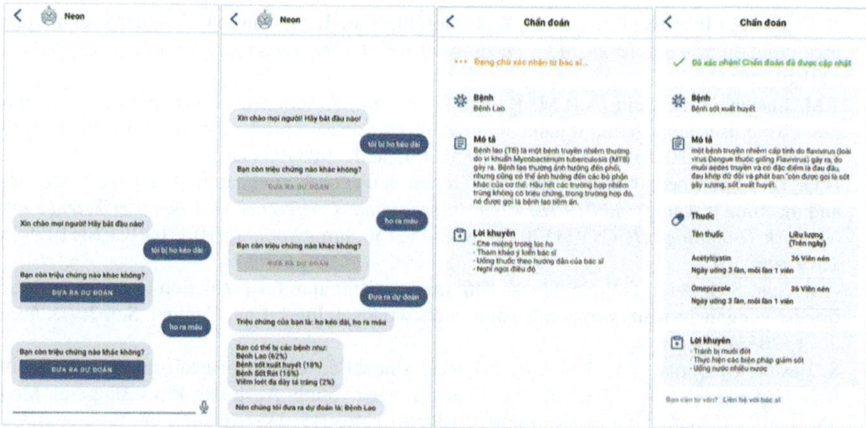

Fig. 10 An illustration of the Android application using chatbot implemented with the proposed method

5 Conclusion and Future Works

Throughout the research, we successfully created a private symptom-disease dataset and built a text-based classification model that gives predictions based on user-input symptoms with high accuracy. Although our work fulfilled our research objectives, there is still room for improvement: The dataset is still extremely limited in size, considering the wide range of symptoms and common diseases. The Vietnamese text processing module is still too rigid; it does not consider common spelling mistakes, abbreviations, or the non-diacritic way of typing Vietnamese some people usually use. Therefore, we intend to extend our dataset to include more symptoms and diseases in our future works. We would improve our natural language processing capability to allow for the non-standard written Vietnamese just mentioned. We aim to incorporate a speech-to-text service to allow patients to conveniently speak, rather than type, their symptoms into the application.

Acknowledgements To conduct this study, we would like to express our gratitude to Dr. Trinh Hoang Thanh in 121 Military Hospital in Can Tho, Vietnam, for providing valuable data for our work.

References

1. World Development Indicators, Physicians (per 1,000 people), The World Bank (2021). https://data.worldbank.org/indicator/SH.MED.PHYS.ZS
2. The Global Health Observatory Indicators, Medical doctors (per 10 000 population), World Health Organization (2021). https://www.who.int/data/gho/data/indicators/indicator-details/GHO/medical-doctors-(per-10-000-population)

3. N. Dragu, F. Elkhoury, T. Miyazaki, R.A. Morelli, N. di Tada, Ontology-based text mining for predicting disease outbreaks, in *Proceedings of Twenty-Third International FLAIR Conference* (2010)
4. T.M. Laskar, T. Hossain, A.R.M. Kamal, N. Rashid, Automated disease prediction system (ADPS): a user input-based reliable architecture for disease prediction. Int. J. Comput. Appl. **133**(15), 24–29 (2016). https://doi.org/10.5120/ijca2016908193 (Jan.)
5. H.Q. Yu, Experimental disease prediction research on combining natural language processing and machine learning, in *2019 IEEE 7th International Conference on Computer Science and Network Technology (ICCSNT)* (2019), pp. 145–150. https://doi.org/10.1109/ICCSNT47585.2019.8962507
6. A.K. Sen, S.B. Patel, D.P. Shukla, A data mining technique for prediction of coronary heart disease using neuro-fuzzy integrated approach two level. Int. J. Eng. Comput. Sci. **2**(9), 2663–2671 (2013). (Sep.)
7. S. Bashir, U. Qamar, F.H. Khan, BagMOOV: a novel ensemble for heart disease prediction bootstrap aggregation with multi-objective optimized voting. Australas. Phys. Eng. Sci. Med. **38**(2), 305–323 (2015). https://doi.org/10.1007/s13246-015-0337-6 (Jun.)
8. T.T. Tran, H.H. Pham, T.V. Nguyen, T.T. Le, H.T. Nguyen, H.Q. Nguyen, Learning to automatically diagnose multiple diseases in pediatric chest radiographs using deep convolutional neural networks, in *ICCV Workshop 2021* (2021). arXiv:2108.06486v1 [eess.IV]
9. H.H. Pham, T.T. Le, D.Q. Tran, D.T. Ngo, H.Q. Nguyen, Interpreting chest X-rays via CNNs that exploit hierarchical disease dependencies and uncertainty labels. Neurocomputing **437**, 186–194 (2021). https://doi.org/10.1016/j.neucom.2020.03.127 (May)
10. N.T.B. Diep, N.N. Lan, N.T.H. Ly, Chatbot technology and expert system for diagnostic automatic disease. TNU J. Sci. Technol. **200**(7), 193–198 (2019)
11. D.Q. Nguyen, D.Q. Nguyen, T. Vu, M. Dras, M. Johnson, A fast and accurate vietnamese word segmenter, in *Proceedings of the 11th International Conference on Language Resources and Evaluation (LREC 2018)* (2017). arXiv:1709.06307v2
12. P. Compton, R. Jansen, A philosophical basis for knowledge acquisition, in *Knowledge Acquisition* (1990). https://doi.org/10.1016/S1042-8143(05)80017-2
13. D.E. Rumelhart, G.E. Hinton, R.J. Williams, Learning representations by back-propagating errors. Nature **323**, 533–536 (1986). https://doi.org/10.1038/323533a0 (Oct.)
14. C. Olah, Understanding LSTM Networks. https://colah.github.io/posts/2015-08-Understanding-LSTMs/. Accessed 6 Sep 2021
15. VLSP 2013 datasets for Word Segmentation and POS tagging tasks, Association for Vietnamese Language and Speech Processing (2013). https://vlsp.org.vn/resources
16. VNESEcorpus, Vietnamese JNLP. http://viet.jnlp.org/download-du-lieu-tu-vung-corpus. Accessed 10 Aug 2021
17. M. Abadi et al., TensorFlow: large-scale machine learning on heterogeneous systems (2015). Software available from tensorflow.org
18. I. Goodfellow, Y. Bengio, A. Courville, *Deep Learning* (MIT Press, Cambridge, 2016)
19. P.T. de Boer, D.P. Kroese, S. Mannor, R.Y. Rubinstein , A tutorial on the cross-entropy method. Ann. Oper. Res. **134**, 19–67 (2005). https://doi.org/10.1007/s10479-005-5724-z
20. D.P. Kingma, J. Ba, Adam: a method for stochastic optimization (2017). arXiv:1412.6980v9 [cs.LG]

Chest X-Ray Image Analysis with ResNet50, SMOTE and SafeSMOTE

Nam Anh Dao and **Xuan Tho Dang**

Abstract Our study extends the analysis of biological information from chest X-ray images to explore the specific patterns, essential basis of classification. Following inference approach, we address the representation of features of medical images and class distribution as a comprehensive study of how symptoms of COVID-19 in chest X-ray images can be detected. The method extracts distinct features using a convolutional neural network that allows exploring hierarchical pattern and gather the patterns for images. The class distribution of images in this work is considered as a significant subject of learning process. We were able to discover symptoms of the disease with less positive samples than the negative samples by implementing the equalization technique, i.e., using a positive samples to generate similar samples. We show the efficiency of our method by contributing classification results on real medical images.

Keywords Chest X-ray image · COVID-19 · ResNet50 · SMOTE · SafeSMOTE · Class imbalance

1 Introduction

According to WHO [1], corona viruses are a group of viruses that affect the respiratory system of mammals, including humans. Corona virus species cause symptoms ranging from mild to severe such as the common cold, fatigue pneumonia, severe acute respiratory syndrome (SARS) and can also affect the gastrointestinal tract. Different strains of the human corona virus cause severe symptoms and can spread rapidly, such as MERS-CoV, which causes Middle East Respiratory Syndrome (MERS), or SARS-CoV, which causes severe acute respiratory syndrome (SARS).

N. A. Dao (✉)
Electric Power University, Hanoi, Vietnam
e-mail: anhdn@epu.edu.vn

X. T. Dang
Hanoi National University of Education, Hanoi, Vietnam

N. H. Phuong and V. Kreinovich (eds.), *Biomedical and Other Applications of Soft Computing*, Studies in Computational Intelligence 1045,
https://doi.org/10.1007/978-3-031-08580-2_18

In December 2019, a new strain of virus called COVID-19 that caused an outbreak was first detected in Wuhan City, Hubei Province, China and has since rapidly spread to other countries. As of September 2021, according to a WHO report [1], the whole world has more than 221 million people infected with COVID-19, including more than 4 million deaths. This is one of the dangerous pandemics with a rapid spread of the disease and the appearance of many deadly variants.

At this time, tests that accurately diagnose COVID-19 can only be conducted in medical facilities that are authorized to perform the test. Techniques to identify strains of COVID-19 can be divided into 3 groups including: (1) The PCR test is a molecular biology test (Realtime RT-PCR), which helps to determine the presence of the virus through the detection of the genetic material of the COVID-19 virus [2]. Samples are taken from the respiratory tract using a special sampler. This method requires special medical equipment and is not easily accessible to all medical facilities. Besides, this method needs at least 12 h before giving the test results. This will likely affect the isolation and monitoring of COVID-19 positive patients. In addition, the false negative rate is also one of the disadvantages of this method [3]. Combining the RT-PCR test with another method may give better results. (2) The second method for diagnosing COVID-19 is based on computed tomography (CT), which involves analyzing 3D X-ray images from different angles [4, 5]. This method takes about 15 min per patient and gives relatively good results based on some reports. However, it is also expensive equipment and is often not available at local medical hospitals in many countries around the world. (3) A third method is the use of chest X-ray images (CXRs) to aid in the diagnosis of COVID-19 [6–8]. This method is evaluated to be simpler and more effective than the previous two methods with some advantages such as less cumbersome equipment, can be easily transported and most of them are already equipped in local hospitals. Thus, the time for chest X-ray imaging test is only about 15 s for each patient, much faster.

However, besides the advantages of the chest X-ray method of COVID-19 diagnosis, there are still some disadvantages such as the quality of the X-ray image is not as sharp as the CT scan [9]. Some symptoms of COVID-19 are similar to viral pneumonia, which can sometimes lead to misdiagnosis of COVID-19 by doctors, especially when hospitals are overwhelmed and working non-stop. Inaccurate diagnoses can lead to patients with non-COVID-19 pneumonia being diagnosed with suspected COVID-19 infection, resulting in compromised, expensive, and dangerous treatment. Therefore, it is necessary to have a new method with the help of computers and algorithms to help doctors more accurately diagnose COVID-19 through chest X-ray images. Based on current research in brain tumor detection, breast cancer detection, etc., solutions based on artificial intelligence, deep learning techniques can extract many features, which are not obvious in original images can help doctors and improve diagnostic efficiency [10–12].

To address the limitations of chest X-ray-based COVID-19 diagnosis, we propose a deep learning model using an appropriate layer of the ResNet50 [13] to improve diagnostic efficiency. Besides, we also found that the chest X-ray image data for machine learning has imbalanced data distribution, and the data also has many different class labels such as Normal, Lung Opacity, Viral Pneumonia, and COVID-19. These problems also affect the effectiveness of COVID-19 diagnostics, and in this article we will try to address them.

2 Related Work

In recent months, there has been an increasing number of studies on the diagnosis of COVID-19 based on chest X-ray images based on the capabilities of deep learning models. The results are promising with a high degree of feasibility, however, there are still many areas for improvement both at the data level and the model design level. Some works related to this study presented here are described as follows.

F. Ucarr et al. introduced COVIDiagnosis-Net as a method of using chest X-rays to diagnose COVID-19 [14]. The authors have improved the SqueezeNet structure combined with optimal Bayesian algorithm to build a learning model that works effectively with the actual data set. COVIDiagnosis-Net was used to classify a dataset of chest X-ray images labeled in 3 groups: Normal, Pneumonia, and Covid. This method also incorporates a multi-scale augmentation process to overcome the problem of unbalanced data, thereby making the proposed network work more efficiently than existing networks and resulting in higher COVID-19 diagnostics.

A. Waheed et al. showed that there are abnormalities on chest X-ray images of COVID-19 patients, this is the basis for deep learning and reinforcement learning systems that can be applied to accurately diagnose COVID- 19 [15]. In this study, the authors developed a CovidGAN method based on the Auxiliary Classifier Generative Adversarial Network (ACGAN) and built on the pretrained VGG-16. This proposed method has improved the performance of CNN for COVID-19 detection, however, the experimental data set of the study is small, resulting in higher quality of the trained system if the data is supplemented and collected from many other factual data sources.

T. Ozturk et al. also suggested using chest X-ray images and deep learning models for early detection of COVID-19 cases [16]. In this study, the authors introduced a newly developed model of 17 complex layers and different filters on each layer. In addition, the authors also applied a new model for accurate diagnosis for binary classification (COVID vs. No-Findings) and multi-class classification (COVID vs. No-Findings vs. Pneumonia).

A. I. Khan et al. proposed a new method, CoroNet, a deep-converged neural model network for automatic detection of COVID-19 from chest X-ray images [17]. The CoroNet model has been upgraded based on the Xception architecture, which has achieved promising results. However, the experimental data set in the article was not very large with about 290 chest X-ray images of COVID-19 patients.

E. E. D. Hemdan has proposed a new deep learning framework, named COVIDX-Net, based on chest X-ray images to automatically diagnose COVID-19 [18]. This method uses seven different architectures of deep learning models such as VGG-19, Google MobileNet to analyze the normalized intensity of X-ray images to classify the patient's condition as negative or positive for COVID- 19. Data were used in this paper with over 50 Chest X-ray images with 25 confirmed positive COVID-19 cases.

S. Calderon-Ramirez et al. also showed that the method of applying deep learning to diagnose COVID-19 based on chest X-ray images has become a popular and useful diagnostic method [19]. However, one limitation when using deep learning is that the data has large labels and hence the challenging problem is the imbalance of data distribution. Especially in the case of a virus outbreak, this becomes even more of a challenge when the data with COVID-19 positive labels is small. In this study, the authors use MixMatch semi-supervised deep learning architecture and combines weight changes for class labels to improve predictive efficiency for binary classification (COVID-19 positive and normal cases), and for multi-class classification (COVID-19 positive, pneumonia and normal cases).

3 The Method

Given a set of chest X-ray images s and their label c, the goal of classification is to define specific features and clarify the relation of the features to the labels of the images. The feature needed to extracted from images with condensed patterns and convolutional neural networks are capable for this task. The ResNet-50 proposed by He et al. [13] was built with 48 convolutional layers, one MaxPool and one Average Pool layer. The Convolutional Neural Network (CNN) was pre-trained on a million images. By implementing the ResNet-50 we can get a set of deep features for learning. We can consider the relation of the features with the labels in the inference process, with support of the Bayes rule [20]:

$$p(c|s) = \frac{p(s|c)p(c)}{p(s)} \tag{1}$$

A set of classes is predefined using binary classification which labels image samples into positive c_1 and negative c_2 labels. Often this prognosis process is to estimate the probability of each label for a given image sample. The estimation of the positive label, proceeds as follows:

$$p(c = c_1|s) = \frac{p(s|c = c_1)p(c = c_1)}{p(s)} \tag{2}$$

To specify a specific label to an image sample, the relevancy of these labels are compared. This was received as a consequence of the formula (2):

$$\frac{p(c = c_1|s)}{p(c = c_2|s)} = \frac{p(s|c = c_1)}{p(s|c = c_2)} \times \frac{p(c = c_1)}{p(c = c_2)} \tag{3}$$

Usually, the prior models of labels are necessary for evaluating (3). However, this is estimable by ratio r, having the evidence of sample images in the training data set:

$$r := \frac{p(c = c_1)}{p(c = c_2)} = \frac{\sum_s p(c = c_1|s)p(s)}{\sum_s p(c = c_2|s)p(s)} \tag{4}$$

The first element of (3) is basically comprised of the likelihood $p(\hat{s}|c)$ of image sample $\hat{s}$ in the test set. Here, the likelihood can be represented with a set of likelihood $p(s|c)$ of training samples s:

$$\frac{p(\hat{s}|c = c_1)}{p(\hat{s}|c = c_2)} = \frac{\sum_s p(\hat{s}|s)p(s|c = c_1)}{\sum_s p(\hat{s}|s)p(s|c = c_2))} \tag{5}$$

For instance, the similarity $p(\hat{s}|s)$ of a test sample $\hat{s}$ and a training sample s is used for (5) and this can be derived by a function of inverse distance d:

$$p(\hat{s}|s) \approx d(\hat{s}, s)^{-1} \tag{6}$$

For the classification, a dominant prior should not happen: the ratio r from (4) should be around 1 by having the same number of positive samples and negative samples. When the ratio r is too small, meaning to have a small number of samples for label c_1 in training, a positive sample in test will be hard to be noticed by (3). For overcoming the class imbalanced problem, Chawla, et al. proposed the Synthetic Minority Oversampling TEchnique, or SMOTE in [21] to synthesize new samples from the minority class. The new samples s' have small distance to existing minority samples s which are positive samples in our case study.

$$s' : d(s', s)^{-1} < \epsilon, \ p(c_1|s) = 1 \tag{7}$$

As the samples of majority class are not considered in the SMOTE, synthesized samples could be placed near by samples of the class. The SafeSMOTE proposed by Dang et al. [22] extended the SMOTE by involving a condition related to the majority class:

$$s' : d(s', s)^{-1} < \epsilon, \ p(c_1|s) = 1; d(s', s)^{-1} > \kappa, \ p(c_2|s) = 1 \tag{8}$$

where ϵ, κ are the coefficients for controlling the distance of synthesized samples with the existing samples of class c_1, c_2. We can use different learning method including k-nearest neighbors (KNN) [23], Random Forest (RF) [24], Support vector machines (SVMs) [25] for the final learning task. The SVM learner realizes classification by searching border line that maximize the width of the gap between the two labels;

$$c(\hat{s}) = \begin{cases} c_1, \ if \ \frac{p(c=c_1|\hat{s})}{p(c=c_2)|\hat{s}} \geqslant \lambda \\ c_2, \quad otherwise \end{cases} \tag{9}$$

For estimating performance of the described method, $Recall$, $Specificity$, $Gmean$, $Accuracy$ and $Precision$ are used with support of True Positive (n_{TP}), False Positive (n_{FP}), True Negative (n_{TN}), and False Negative (n_{FN}):

$$Recall = \frac{n_{TP}}{n_{TP} + n_{FN}} \tag{10}$$

$$Specificity = \frac{n_{TN}}{n_{TN} + n_{FP}} \tag{11}$$

$$Gmean = \sqrt{Recall \times Specificity} \tag{12}$$

$$Accuracy = \frac{n_{TP} + n_{TN}}{n_{TP} + n_{FP} + n_{TP} + n_{FN}} \tag{13}$$

$$Precision = \frac{n_{TP}}{n_{TP} + n_{FP}} \tag{14}$$

The $Gmean$ groups samples by class and estimates mean for each group so it is not effected by the imbalanced class. The other metrics are sensitive with the imbalance. The following Fig. 1 illustrates our method by pseudo codes. Due to availability of labels for data set we can implement the method for two classes as mentioned above or for multi-classes as required.

ALGORITHM 1. Chest X-ray Image Analysis with ResNet50, SMOTE and SafeSMOTE
Input: the training images s, label c
the test images $\hat{s}$, label $\hat{c}$
Output: the prediction $\hat{c}_{pred}$; $performance$

$Feature\ extraction\ by\ ResNet - 50$
$Split\ Data\ to\ datasets\ with\ imbalanced\ classes$
for $each\ split$ **do**
    for $number\ of\ classes\ by\ 2\ or\ 4$ **do**
        for $each\ balancing\ method:\ SMOTE,\ SafeSMOTE$ **do**
            $Apply\ balancing\ for\ training\ set$ ▹ by (8)
            for $each\ learning\ method:\ KNN,\ RF,\ SVM$ **do**
                $Training\ by\ learning\ method$ ▹ by (9)
                $\hat{c}_{pred}$=$Prediction\ for\ test\ set\ by\ learning\ method$
                $Measure\ performance\ (\hat{c}_{pred}, \hat{c})$ ▹ by (10-14)
            end for
        end for
    end for
end for

Fig. 1 Chest X-ray Image Analysis with ResNet50, SMOTE and SafeSMOTE

Table 1 Data structure in experiments

Number of samples for a label	2 labels		4 labels	
	Test	Training	Test	Training
COVID	313	134	313	134
Lung opacity	941	404	941	404
Normal	941	404	941	404
Viral pneumonia	941	404	941	404
Ratio	1:9	1:9	1:3:3:3	1:3:3:3

4 Experiments

In order to validate the method expressed in this work we carried out a number of experiments. The COVID-19 Radiography Database [26] is used as major data for experiments. The database provides a thousand images labeled to $COVID$, $Lung\ Opacity$, $Normal$, and $Viral\ Pneumonia$. To perform a the imbalanced class problem, the data sets are selected from the database for two cases: classification with two labels and with four labels. In each case we specified number samples for each label so that the number of positive samples is less than the number of negative samples. The ratio is $1:9$ for binary classification and $1:3:3:3$ in case of four labels. The structure of data is shown in Table 1. For cross validation, we created five splits for training and test. The metrics described by (10)–(14) were used for each split and then their averaged values were computed for the final performance report.

Figure 2 represents scores of $Accuracy$ for all splits in case of two labels: each point corresponds to one split. The red points show results of KNN, RF and SVM without class equalization. The points in blue are reserved for implementation of SMOTE before learning, and the green color is the mark for SafeSMOTE. By finding color points which have their location in the figure we can see that the red points are usually lower than other, meaning that there is improvement by implementing balancing techniques. The best average accuracy of 96.12 for the case of two labels was achieved by SVM with SafeSMOTE, see Table 2. The Fig. 3 is used to demonstrate $Gmean$ results for the case showing that SVM with SafeSMOTE by green points is also the best with average score of 96.34.

The following two figures show results for the case of four labels. The $Accuracy$ is reported by Fig. 4. Here, SafeSMOTE with SVM has 94.64 and SMOTE with SVM has 94.67. The scores are mostly the same level. Figure 5 demonstrates the $Gmean$ for the case. Thus, the SMOTE with SVM is the best by 92.83, while SafeSMOTE with SVM is in the second place by 92.78.

Experimental results show that the best accuracy for classification with two labels is come from SafeSMOTE with SVM. In case of four labels, the SMOTE with SVM is the best. The described method combines advantages of ResNet-50, SMOTE,

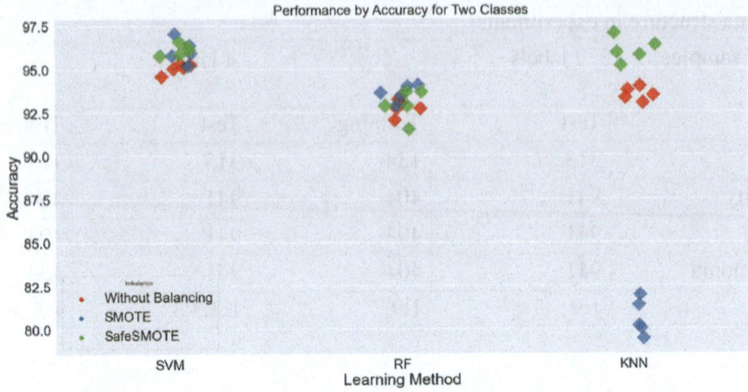

Fig. 2 Accuracy of KNN, RF, SVM using SMOTE, SafeSMOTE for the case of two labels

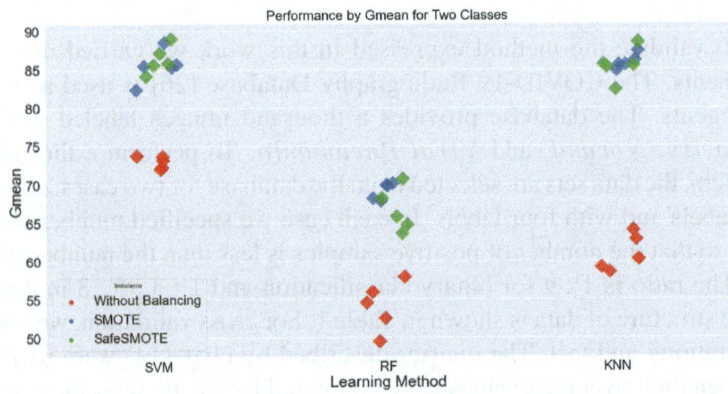

Fig. 3 Gmean of KNN, RF, SVM using SMOTE, SafeSMOTE for the case of two labels

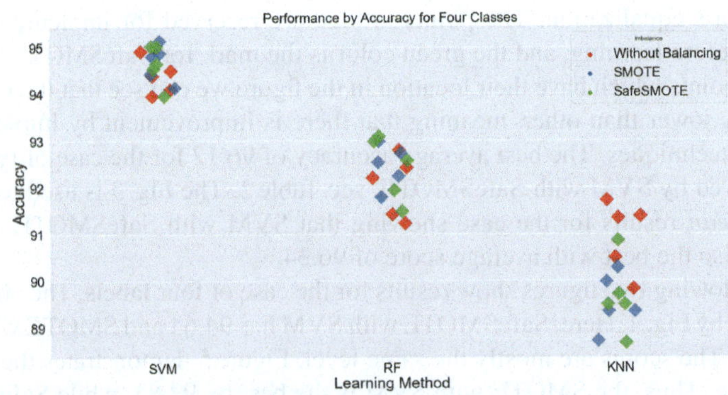

Fig. 4 Accuracy of KNN, RF, SVM using SMOTE, SafeSMOTE for the case of four labels

Table 2 Performance Report KNN, RF, SVM using SMOTE, SafeSMOTE

Imbalance	Number of labels	Learning method	Specificity	Precision	Recall	Accuracy	Gmean
No balancing	2	KNN	99.72	93.87	37.31	93.51	60.96
		RF	99.75	92.82	29.40	92.75	54.08
		SVM	99.65	94.61	53.43	95.05	72.97
	4	KNN	93.97	81.90	81.90	90.95	87.72
		RF	94.68	84.03	84.03	92.01	89.19
		SVM	96.24	88.74	88.74	94.37	92.41
SMOTE	2	KNN	79.19	33.14	93.13	80.58	85.86
		RF	98.58	79.76	48.51	93.60	69.14
		SVM	98.46	84.48	74.18	96.05	85.44
	4	KNN	93.01	79.05	79.05	89.53	85.75
		RF	94.97	84.90	84.90	92.45	89.80
		SVM	96.45	89.35	89.35	94.67	**92.83**
SafeSMOTE	2	KNN	98.46	84.48	74.18	96.05	85.44
		RF	98.20	74.11	45.22	92.93	66.60
		SVM	98.36	84.01	75.82	**96.12**	86.34
	4	KNN	93.10	79.28	79.28	89.64	85.91
		RF	94.94	84.83	84.83	92.41	89.74
		SVM	96.42	89.27	89.27	94.64	92.78

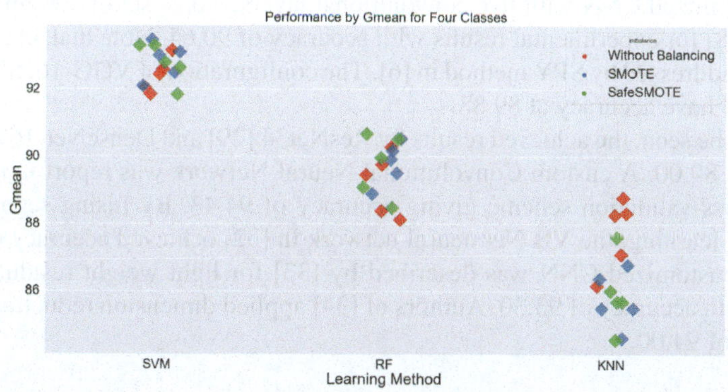

Fig. 5 Gmean of KNN, RF, SVM using SMOTE, SafeSMOTE for the case of four labels

SafeSMOTE and SVM for realizing the classification problem. The solution can be implemented to deal with other medical images.

To highlight the distinction in our method from those of other studies, the performance of recent published articles is displayed in the Table 3 along with their methods and data sets used in experiments. Features were generated by Keras ImageDataGenerator in [27] for image augmentation before classification. The work used the

Table 3 Performance Report by Other Works

Method	Database	Accuracy
Influence of batch size and image size [27]	COVID-19 Radiography Database [26]	87.00
Five convolutional layers CNN [28]	2 databases	90.64
VGG-16, SPY, SVM class imbalance [6]	COVID-19 image data collection	89.88
Dual sampling attention network ResNet34 [29]	4982 images	87.90
DenseNet-169 [30]	COVID-CT with masks	89.00
Custom CNN model [31]	BIMCV-COVID19+ data set	94.43
The VB Net [32]	300 CT images for validation. 249 CT images for training.	91.60
The COVID-Net with a lightweight residual design pattern [33]	COVIDx data set of 16,756 images	93.30
Dimensionality reduction [34]	420 2-D X-ray images	94.00
SMOTE with SVM for 4 classes (our)	COVID-19 Radiography Database [26]	94.64
SafeSMOTE with SVM for 2 classes (our)	COVID-19 Radiography Database [26]	96.12

COVID-19 Radiography Database [26] and accuracy of 87.00 is reported. By proposing a customized CNN with five convolutional layers, a data set of 700 images was used in [28] for experimental results with accuracy of 90.64. Note that, class imbalance was addressed by SPY method in [6]. The configuration of VGG-16, SPY, SVM allowed to have accuracy at 89.88.

As can be seen, the achieved results by ResNet34 [29] and DenseNet-169 [30] are 87.90 and 89.00. A custom Convolutional Neural Network was report in [31] with 5-fold cross-validation scheme giving accuracy of 94.43. By fusing segmentation with deep learning, the VB Net neural network in [32] achieved accuracy of 91.60. Another customized CNN was described by [33] for light weight residual design pattern with accuracy of 93.30. Authors of [34] applied dimension reduction to have accuracy of 94.00.

5 Conclusions

Class balancing is a method that is very appropriate for unbalanced class distribution in machine learning. Our study case addressed chest X-ray images with features extracted by pre-trained CNN for diagnosis of COVID-19. The imbalance was shown in data set with ratio $1:9$ and $1:(3+3+3)$ for classification with two classes and four classes. By implementing SMOTE and SafeSMOTE we have derived a CNN based feature and class balancing based method which accesses to both feature and class before conducting machine learning. Whereas the approaches covering KNN, RF and SVM, being learning algorithm, the SVM is proved to be robust for classification for the unbalanced data. Additionally, unbalance techniques, such as SafeSMOTE, can be applied to detect positive label for other medical images. Studies are at present going on the extension of CNN based features being able to deal with multi-labels image classification.

Acknowledgements This research was supported by the Vietnam Ministry of Education and Training, project B2022-SPH-04.

References

1. Rolling updates on coronavirus disease (COVID-19). World Health Organization (2021). https://www.who.int/emergencies/diseases/novel-coronavirus-2019
2. A. Tahamtan, A. Ardebili, Real-time RT-PCR in COVID-19 detection: issues affecting the results. Expert Rev. Mol. Diagn. **20**(5), 453–454 (2020)
3. A.T. Xiao, Y.X. Tong, S. Zhang, *False-negative of RT-PCR and prolonged nucleic acid conversion in COVID-19: rather than recurrence J* (Med, Virol, 2020)
4. K. Nakajima, H. Kato, T. Yamashiro, T. Izumi et al., COVID-19 pneumonia: infection control protocol inside computed tomography suites. Jpn. J. Radiol. **38**(5), 391–393 (2020)
5. D. Caruso, T. Polidori, G. Guido, Nicolai, et al., Typical and atypical COVID-19 computed tomography findings. World J. Clin. Cases **8**(15), 3177 (2020)
6. X.T. Dang, N.A. Dao, *Imbalance in Learning Chest X-Ray Images for COVID-19 Detection Soft Computing: Biomedical and Related Applications* (Springer, Berlin, 2021)
7. A. Abbas, M.M. Abdelsamea, M.M. Gaber, Classification of COVID-19 in chest X-ray images using DeTraC deep convolutional neural network. Appl. Intell. **51**(2), 854–864 (2021)
8. X.T. Dang, & N.A. Dao, Deep Learning-Based Imbalanced Data Classification for Chest X-Ray Image Analysis, in International Conference on Intelligent Systems & Networks (Springer, 2021), pp. 109–115
9. S. Tabik, A. Gomez-Rios, J.L. Martin-Rodriguez et al., COVIDGR dataset and COVID-SDNet methodology for predicting COVID-19 based on Chest X-Ray images. IEEE J. Biomed. Health Inform. **24**(12), 3595–3605 (2020)
10. S. Asif, Y. Wenhui, H. Jin, S. Jinhai, Classification of COVID-19 from chest X-ray images using deep convolutional neural network, in *IEEE 6th International Conference on Computer and Communications*
11. M. Togacar, B. Ergen, Z. Comert, BrainMRNet: brain tumor detection using magnetic resonance images with a novel convolutional neural network model. Med. Hypotheses **134**, 109531 (2020)
12. J. Zheng, D. Lin, Z. Gao, S. Wang, M. He, J. Fan, Deep learning assisted efficient AdaBoost algorithm for breast cancer detection and early diagnosis. IEEE Access **8**, 96946–96954 (2020)
13. K. He, X. Zhang, S. Ren, J. Sun, Deep residual learning for image recognition, in *IEEE Conference on Computer Vision and Pattern Recognition* (2016), pp. 770–778
14. F. Ucar, D. Korkmaz, COVIDiagnosis-Net: deep Bayes-squeezeNet Based diagnosis of the coronavirus disease 2019 (COVID-19) from X-ray images. Med. Hypotheses **140**, 109761 (2020)
15. A. Waheed, M. Goyal, D. Gupta, A. Khanna et al., Covidgan: data augmentation using auxiliary classifier gan for improved covid-19 detection. IEEE Access **8**, 91916–91923 (2020)
16. T. Ozturk, M. Talo, E.A. Yildirim et al., Automated detection of COVID-19 cases using deep neural networks with X-ray images. Comput. Biol. Med. **121**, 103792 (2020)

17. A.I. Khan, J.L. Shah, M.M. Bhat, CoroNet: a deep neural network for detection and diagnosis of COVID-19 from chest x-ray images. Comput. Methods Prog. Biomed. **196**, 105581 (2020)
18. E.E.D. Hemdan, M.A. Shouman, M.E. Karar, Covidx-net: a framework of deep learning classifiers to diagnose covid-19 in x-ray images (2020). arXiv:2003.11055
19. S. Calderon-Ramirez, S. Yang, Moemeni et al., Correcting data imbalance for semi-supervised COVID-19 detection using X-ray chest images. Appl. Soft Comput. **111**, 107692 (2021)
20. D. Barber, *Bayesian Reasoning and Machine Learning* (Cambridge University Press, Cambridge, 2012)
21. N.V. Chawla, K.W. Bowyer, L.O. Hall, W.P. Kegelmeyer, SMOTE: synthetic minority oversampling technique. J. Artif. Intell. Res. **16**, 321–357 (2002)
22. X.T. Dang, O. Hirose, D.H. Bui, T. Saethang, V.A. Tran, L.A.T. Nguyen, K. Satou, A novel over-sampling method and its application to cancer classification from gene expression data. Chem.-Bio Inf. J. **13**, 19–29 (2013)
23. N.S. Altman, An introduction to kernel and nearest-neighbor nonparametric regression. Amer. Stat. **46**(3), 175–185 (1992)
24. T.K. Ho, Random decision forests, in *International Conference on Document Analysis and Recognition, Montreal, QC, 14–16 August 1995* (1995), pp. 278–282
25. C. Cortes, V.N. Vapnik, Support-vector networks (PDF). Mach. Learn. **20**(3), 273–297 (1995). CiteSeerX 10.1.1.15.9362
26. T. Rahman, A. Khandakar, Y. Qiblawey, A. Tahir, Kiranyaz, et al., Exploring the effect of image enhancement techniques on COVID-19 detection using chest X-ray images (2020). arXiv:2012.02238, COVID-19 Radiography Database
27. A.B. Hasmadi, Interactive automation Of COVID-19 classification through X-Ray images using machine learning. J. Indep. Stud. Res. - Comput. (2020). https://doi.org/10.31645/10
28. F. Ahmed, S.A.C. Bukhari, F. Keshtkar, A deep learning approach for COVID-19 & viral pneumonia screening with X-ray images. Digit. Gov.: Res. Pract. **2**, 2 (2020)
29. O. Xi, H. Jiayu, X. Liming, et al., Dual-sampling attention network for diagnosis of COVID-19 from community acquired pneumonia (2020). arXiv:2005.02690 [cs.CV]
30. X. Yang, X. He, J. Zhao, Y. Zhang, S. Zhang, P. Xie, COVID-CT-dataset: a CT scan dataset about COVID-19 (2020). arXiv:2003.13865, https://github.com/UCSD-AI4H/COVID-CT
31. L. Duran-Lopez, J.P. Dominguez-Morales, J. Corral-Jaime, S. Vicente-Diaz, A. Linares-Barranco, COVID-XNet: a custom deep learning system to diagnose and locate COVID-19 in chest X-ray images. Appl. Sci. **10**(16), 5683 (2020)
32. F. Shan, Y. Gao, J. Wang et al., Lung infection quantification of covid-19 in ct images with deep learning (2020). arXiv:2003.04655
33. L. Wang, A. Wong, COVID-Net: a tailored deep convolutional neural network design for detection of COVID-19 cases from chest radiography images (2020)
34. A. Zargari Khuzani, M. Heidari, S.A. Shariati, COVID-Classifier: an automated machine learning model to assist in the diagnosis of COVID-19 infection in chest X-ray images. Sci. Rep. **11**, 9887 (2021)

Weakly Supervised Localization of the Abnormal Regions in Breast Cancer X-Ray Images Using Patches Classification

Nguyen Hoang Phuong, Ha Manh Toan, Le Tuan Linh, Nguyen Ngoc Cuong, and Bui My Hanh

Abstract We present an approach of a Weakly Supervised Localization of the Abnormal Regions in Breast Cancer X-Ray images using patches classification. We use the breast cancer X-Ray image dataset provided by radiologists of the Hanoi Medical University Hospital to generating patches with Normal, Benign and Malignant categories. After that we train the ResNet 34 model on the breast cancer X-Ray patches dataset for classifying Normal, Benign and Malignant categories. Our network achieves a macAUC of 0.817139. Finally, we apply this model to calculate heatmaps for visualization of Benign and Malignant areas in the Breast Cancer X-Ray images.

Keywords Heatmap · Weakly supervised localization · Breast cancer · X-Ray images

---

N. H. Phuong
Thang Long University, Nghiem Xuan Yem Road, Hoang Mai District, Hanoi, Vietnam
e-mail: nhphuong2008@gmail.com

H. M. Toan
Institute of Information Technology, Vietnam Academy of Science and Technology, Hanoi, Vietnam
e-mail: hmtoan@ioit.ac.vn

L. T. Linh · N. N. Cuong · B. M. Hanh (✉)
Hanoi Medical University Hospital, 1 Ton That Tung Road, Hanoi, Vietnam
e-mail: buimyhanh@hmu.edu.vn

L. T. Linh
e-mail: linhdhyhn2017@gmail.com

N. N. Cuong
e-mail: cuongcdha@hmuh.vn

N. H. Phuong and V. Kreinovich (eds.), *Biomedical and Other Applications of Soft Computing*, Studies in Computational Intelligence 1045,
https://doi.org/10.1007/978-3-031-08580-2_19

1 Introduction

Breast cancer is one of the top cancers for women of developed and developing countries in the world. In Vietnam, the rate of breast cancer was calculated as 23 per 100,000 women in 2012, and it has continuously increased [1]. According to GLOBOCAN, there were approximately 2.26 million new cases and nearly 685,000 deaths from breast cancer in 2020 [2]. In order to help radiologists to diagnose breast cancer with X-Ray images, we have developed a Deep Neural Networks Based Mammograms Classification System to classify the X-Ray mammograms into BI-RADS 045, BI-RADS 1, BI-RADS 23 [3]. To visualize the abnormal regions as benign and malignant regions in breast cancer X-Ray images of the Mammograms Classification System, we apply a method of heatmap calculation using patches classification. A weakly supervised localization problem is defined as the learning process when some partial information regarding the task of object location on a small subset of the data points is at our disposal [4]. In this paper we aim to develop a system for visualization of the abnormal regions in breast cancer X-Ray images using a heatmap calculation method.

To summarize, the main contributions of this paper are:

- We develop a patch image dataset of Normal, Benign and Malignant categories.
- We train a CNN with backbone of ResNet 34 on this patch image dataset for classifying the normal, benign and malignant categories.
- We apply a heatmap calculation method using the above trained model for visualizing a hurt area in breast cancer X-Ray image.

The rest of the paper is organized as follow: Sect. 2 reviews some related works. Section 3 present a method of weakly supervised localization of the cancer regions in breast cancer X-Ray images. Section 4 gives some conclusions.

2 Related Works

With regards to the general weakly-supervised object localization, there have been a number of recent works exploring weakly—supervised object localization using CNNs [3–8, 10–24]. Bergamo et al. [6] propose a technique for self-taught object localization involving masking out image regions to identify the regions causing the maximal activations in order to localize objects. Cinbis et al. [7] combines multiple-instance learning with CNN features to localize objects. Oquab et al. [8] propose a method for transferring mid-level image representations and show that some object localization can be achieved by evaluating the output of CNNs on multiple overlapping patches. For medical imaging applications, medical image analysis is one area where the weakly supervised object localization and detection methods are of critical importance as only few annotations of target objects by trained experts in bio-image. In other work, Hwang and Kim [25] develop a two-stream DNN model to localize the tuberculosis regions from the chest X-ray images. This work proposes a weakly supervised learning scheme without requiring any pre-trained network parameters. The proposed network contains a fully connected layer-based classification branch and a Class Activation Map (CAM)—based localization branch with shared convolutional layers for feature extraction. Both of the two branches are supervised by image label annotation, where a weighting parameter is introduced to dynamically control the relative importance between them to gradually switch the focus of the learning process from the classification branch to the localization branch. In application of object localization methods for breast cancer X-Ray images, in [26], the authors developed the heatmap model applied to the full resolution mammograms in a sliding window fashion to create two heatmaps for each image, one containing an estimated probability of a malignant finding for each pixel, and the other containing an estimated probability of a benign finding.

3 A Method of Weakly-Supervised Localization of the Cancer Regions in Breast Cancer X-Ray Images

3.1 Data Pre-processing

The patches dataset in Table 1 is developed by selecting the images of BI-RADS 1 category (called Normal), of BI-RADS 23 category (called Benign) and of BI-RADS 45 category (called Malignant) from the image dataset of The Hanoi Medical University Hospital (HMU) [3], in which the radiologists of the hospital performed the localization of lesions and identified BI-RADS 2, BI-RADS 3, BI-RADS 4, BI-RADS 5 by using ICY software [27]. The total number of the images of Normal, Benign and of Malignant categories is 4746 images which consists of 2195 Normal images, 1177 Benign images and 774 Malignant images. These images of Normal, Benign and of Malignant categories are converted into a PNG 16-bit format. Subsequently, these images are realized by pre-processing process. Data pre-processing process is described in Fig. 1. We normalized the images in the following way. Firstly, we flip horizontally the images of R-CC and R-MLO views so that the breast is always on the same side of the image. Secondly, the breast region is divided by image processing technique: we crop image by probing along the edge of the breast. After that this image is resized into 512 × 512 pixels (see Fig. 1).

Table 1 Patch dataset divided in training set, validation set and test set

Category	Training set	Validation set	Test set	Total
Normal	1 303 190	325 932	184 484	1 813 606
Benign	33 053	8 143	4 561	45 757
Malignant	14 637	3 645	2 735	21 017

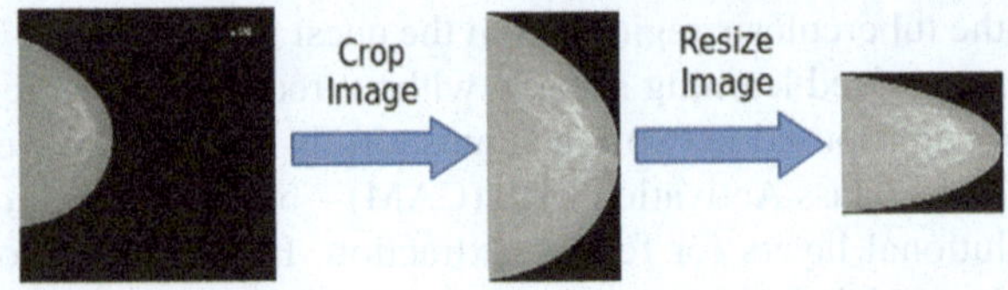

Fig. 1 Example of cropping and resizing an original image

Finally, each image is normalized by Eq. 1.

$$X_{i,j}norm = (X_{i,j} - \text{min\_}X)/(\text{max\_}X - \text{min\_}X) \quad (1)$$

In the Eq. 1, $X_{i,j}$ is a pixel in an input image, where min_X is the minimum of all the values $\{X_{i,j}\}$, and max_X is the maximum of all the values $\{X_{i,j}\}$. After using this normalization operation, each pixel $X_{i,j}norm$ of the normalized image will be a value in the range [0, 1].

3.2 Generating Patches

The process of generating patches is to realize over all images of Normal, Benign and Malignant categories provided by the breast cancer X-Ray radiologists using a sliding window fashion with the size of sliding window of 64 × 64 pixels and the window sliding step is 16 pixels. A window region is selected if in this window region there is a minimum of 410 breast image pixels which occupies about 10% of the area of the sliding window.

A selected window region is called Benign if it has a minimum of 41image pixels which are Benign. In similar way, a selected window region is called Malignant if it has a minimum of 41image pixels which are Malignant. If a selected window region satisfies the condition of Benign region and the condition of Malignant region, it will be labeled as Malignant category. If a selected window region not satisfies the condition of Benign region and not satisfies the condition of Malignant region, it will be labeled as Normal category.

The generated patch dataset is divided into three datasets such as training set, validation set and test set described in Table 1.

In the Table 1, we observe that the number of Normal patch images is higher than the number of Benign patch images and Malignant patch images.

3.3 Patch Augmentation

Before the training process, the patches will be augmented by the following transform techniques:

- Flipping: randomly flipping from left to right with the probability of 0.5.
- Scaling and cropping: randomly scaling images from 100% to 110% size then randomly cropping inside.
- Lighting: random linear changing light: x - > x*a + b, where a is a random value in [0.8, 1.25] and b is a random value in [−0.176, 0.176].

  (Here, we use the formula x -> a*x + b for all pixels in an image. This is a linear transform for adjusting the contrast and brightness of an image [28, 29]).
- Rotating: rotate by −10 to +10°. This means the images are rotated randomly from 0 to 10° (negative value: rotating to left; positive value: rotating to right).

In the above transform techniques, for random values, we use uniform distribution on each interval.

All transformation steps are realized on input images for creating new images, then these images are used for the training process.

3.4 *Patch Training*

After augmentation process, the patches will be put into the training model. Because of the number of Normal patch images is larger than number of images of Benign patches and of Malignant patches, we apply a balanced data method as the loss function of Weighted cross-entropy in [26]. The CNN model architecture used for patch classification with backbone ResNet 34 is described in Fig. 2. The ResNet 34 adds more the Pooling layer and the Flatten layer. The Pooling layer is used

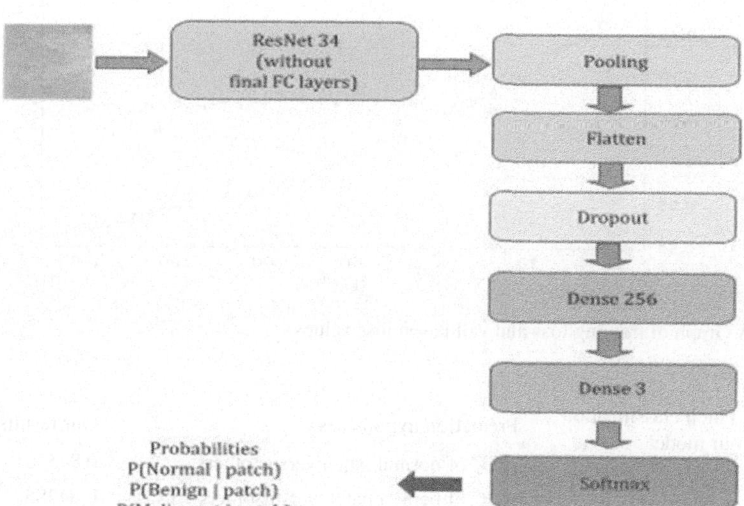

Fig. 2 The CNN model architecture for patch classification

to reduce the dimension of the feature map. The Flatten layer is the function that converts the feature map of the pooling layer to a single column that is passed to the fully connected layer. The Dropout layer generates the features having a form of 1 dimensional vector. After that the calculation process for 2 full connected layers are realized. The Dense layer 256 generates a 256-dimensional vector and the Dense layer 3 generates a 3 dimensional vector according to the output labels. Before putting to the first fully connected Dropout layer, the features are normalized by BatchNorm and removed randomly by Dropout technique. After this step, the features are putted to the activation function ReLU. In the similar way, the features are normalized by BatchNorm before they are put into the next layer. The output of the neural network is calculated by SoftMax function for classifying patches. The output is 3-dimensional vector represented as probabilities of Normal, Benign and Malignant patches.

A graph of the training loss and validation loss values is described in Fig. 3.

From the Graph of training loss and validation loss values, we choose an optimal model at the epoch 103, because after this epoch, the validation loss value is changed a little and the training rate value is quite small, it takes about 5×10^{-7} and continues to decrease.

The result of the patch classification based on ResNet 34 trained by the patch dataset in Table 1 is described in Table 2 as follows.

Table 3 presents a Confusion Matrix of Test dataset with the model based on ResNet 34 backbone. The result of Table 3 shows.

Among 184 484 patch images of Normal, the system has correctly recognized 130 682 patch images which is equivalent to 70.84%, and 27 041 patch images were

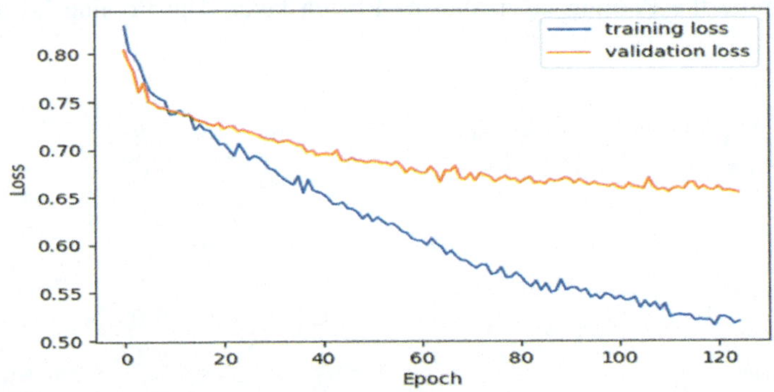

Fig. 3 A Graph of training loss and validation loss values

Table 2 Patch classification result of our model

Prediction hypotheses	Our results
AUC of normal patch versus others	0.875581
AUC of benign patch versus others	0.842852
AUC of malignant patch versus others	0.732985
macAUC	0.817139

Table 3 Confusion matrix of test dataset with the model based on backbone ResNet 34

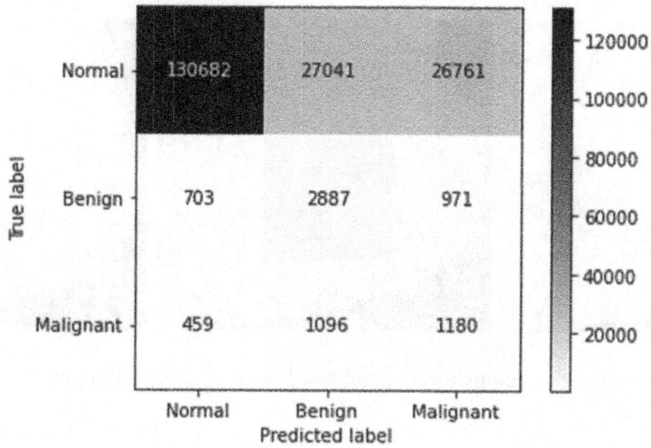

misrecognized as patch images of Benign which is equivalent to 14.66%, and 26 761 patch images were misrecognized as patch images of Malignant which is equivalent to 14.51%.

Among 4 561 patch images of Benign, the system has correctly recognized 2 887 patch images which is equivalent to 63.3%, and 703 patch images were misrecognized as patch images of Normal which is equivalent to 15.41%, and 971 patch images were misrecognized as patch images of Malignant which is equivalent to 21.29%.

Among 2 735 patch images of Malignant, the system has recognized correctly. 1 180 patch images which is equivalent to 43.14%, and 459 patch images were misrecognized as patch images of Normal which is equivalent to 16.78%, and 1 096 patch images were misrecognized as patch images of Benign which is equivalent to 40.07%.

3.5 Heatmaps Calculation

To use the trained model to generate heatmap, the sub-region of images created by applying the sliding window fashion in the input breast cancer X-Ray images.

With the extracted sub-regions of images, we apply the model to predict Benign and Malignant areas. It is an analogue as the patch generating process is used for training heatmap. The input images are resized to 512 × 512 pixels and the patch generating process is realized by the sliding window fashion with the window size of 64 × 64 pixels and the window sliding step of 16 pixels. In every position of the selected patch, the data will be put into the prediction model. The prediction result will be summarized to heatmap.

Given an input image I, the process of heatmap generation is the following:

1. Initiate to assign heatmapB (heatmap with Benign mass) and heatmapM (heatmap with Malignant mass) by value of 0.

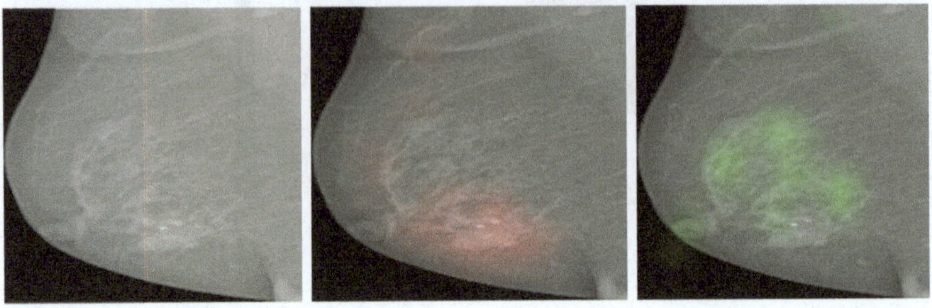

Fig. 4 An example of generated heatmap patches of Benign and Malignant

2. For every patch_roi:

   2.1 Calculating the patch prediction as I(patch_roi) - > realizing a patch classification of I(patch_roi)
   2.2 If I(patch_roi) is Benign then heatmapB(patch_roi) + = 1.
   2.3 If I(patch_roi) is Malignant then heatmapM(patch_roi) + = 1.

3. Calculating an everage for heatmapB and heatmapM.

An example of generated heatmap patches of Benign and Malignant from the input image is presented in Fig. 4.

Figure 4 shows that from the input image (**a**), the system results the Malignant heatmap over the image (**b**), in which the Malignant area gets a red color and the Benign heatmap over the image (**c**), in which the Benign area gets a green color.

4 Conclusions

This paper presents a method of a heatmap calculation based weakly supervised localization of the abnormal regions in breast cancer X-Ray images using patches classification. The radiologists of the Hanoi Medical University Hospital have developed the patch image dataset of Normal, Benign and Malignant categories. We have trained a CNN with backbone of ResNet 34 on this patch image dataset for classifying the normal, benign and malignant categories. We also applied a heatmap calculation method using the above trained model for visualizing the Benign and Malignant regions in breast cancer X-Ray images. In the future work, we will add this heatmap model to the system for BI-RADS classification of breast cancer X-Ray images [3] to visualize the Benign and Malignant heatmaps areas.

Acknowledgements This work is supported in part by foundation of the research and development contract between Thang Long University and Hanoi Medical University Hospital, Vietnam dated on 27 November, 2018 on "Developing a support system for diagnosis of breast cancer based on

X-Ray using Artificial Intelligence". We thank to Prof. Anh Nguyen from Auburn University, AL, USA for his advises of this work.

References

1. P.D.Y. Trieu, C. Mello-Thoms, P.C. Brennan, Female breast cancer in Vietnam: a comparison across Asian specific regions. Cancer Biol. Med. J. **12**(3), 238–245 (2015)
2. H. Sung, J. Ferlay, R.L. Siegel, M. Laversanne, I. Soerjomataram, A. Jemal, F. Bray, Global cancer statistics 2020: GLOBOCAN estimates of incidence and mortality worldwide for 36 cancers in 185 countries. CA: Cancer J. Clin. **71** 209–249 (2021)
3. L.T. Linh, B.M. Hanh, N.N. Cuong, H.M. Toan, V. Kreinovich, H.-P. Nguyen, Approach to classify deep neural networks based mammograms for breast cancer screening, in *Proceedings of the International Workshop on Soft Computing and Advances in Intelligent Systems* (Mexico City, Mexico, October 25 to 30, 2021)
4. D. Zhang, J. Han, G. Cheng, M.-H. Yang, Weakly supervised object localization and detection: a survey. arXiv:210407918V1. [cs.CV] 16 Apr 2021
5. B. Zhou, A. Khosla, A. Lapedriza, A. Oliva, A. Torralba, Learning deep features for discriminative localization, in *2016 IEEE Conference on Computer Vision and Pattern Recognition (CVPR)* 2921–2929 (2016)
6. A. Bergamo, L. Bazzani, D. Anguelov, L. Torresani, Self-taught object localization with deep networks (2014). arXiv:1409.3964
7. R.G. Cinbis, J. Verbeek, C. Schmid, Weakly supervised object localization with multi-fold multiple instance learning. IEEE Trans. Pattern Anal. Mach. Intell. **39**(1) (2015). https://doi.org/10.1109/TPAMI.2016.2535231
8. M. Oquab, L. Bottou, I. Laptev, J. Sivic, Learning and transferring mid-level image representations using convolutional neural networks, in *Proceedings of the 2014 IEEE Conference on Computer Vision and Pattern Recognition* (Columbus, OH, USA, 23–28 June 2014). https://doi.org/10.1109/CVPR.2014.222
9. M. Oquab, L. Bottou, I. Laptev, J. Sivic, Is object localization for free? weakly-supervised learning with convolutional neural networks, in *Proceedings of the 2015 IEEE Conference on Computer Vision and Pattern Recognition (CVPR)* (2015). https://doi.org/10.1109/CVPR.2015.7298668
10. P.O. Pinheiro, R. Collobert, From image-level to pixel-level labeling with convolutional networks, in *Proceedings of the 2015 IEEE Conference on Computer Vision and Pattern Recognition (CVPR)* (Boston, MA, USA, 7–12 June 2015). https://doi.org/10.1109/CVPR.2015.7298780
11. D.J. Crandall, D.P. Huttenlocher, Weakly supervised learning of part-based spatial models for visual object recognition, in *Computer Vision—ECCV 2006: 9th European Conference on Computer Vision, Graz, Austria, May 7–13, 2006. Proceedings, Part I* (Springer, 2006), pp. 16–29
12. N. Gonthier, Y. Gousseau, S. Ladjal, O. Bonfait, Weakly supervised object detection in artworks, in *European Conference on Computer Vision* (Springer, 2018), pp. 692–709
13. N. Inoue, R. Furuta, T. Yamasaki, K. Aizawa, Cross-domain weakly-supervised object detection through progressive domain adaptation, in *Proceedings of the IEEE Conference on Computer Vision and Pattern Recognition*, pp. 5001–5009 (2018)
14. R. Wang, B. Chen, D. Meng, L. Wang, Weakly-supervised lesion detection from fundus images. IEEE Trans. Med. Imaging **38**(6), 1501–1512 (2019). https://doi.org/10.1109/TMI.2018.2885376
15. W.M. Gondal, J.M. Köhler, R. Grzeszick, G.A. Fink, M. Hirsch, Weakly-supervised localization of diabetic retinopathy lesions in retinal fundus images, in *2017 IEEE International Conference on Image Processing (ICIP)* (IEEE, 2017), pp. 2069–2073

16. C. Li, K. Shirahama, M. Grzegorzek, Environmental microorganism classification using sparse coding and weakly supervised learning, in *Proceedings of the 2nd International Workshop on Environmental Multimedia Retrieval* (ACM, 2015), pp. 9–14
17. F. Dubost, H. Adams, P. Yilmaz, G. Bortsova, G. van Tulder, M.A. Ikram, W. Niessen, M. Vernooij, M. de Bruijne, Weakly supervised object detection with 2D and 3D regression neural networks (2019). arXiv:1906.01891
18. M. Liu, J. Zhang, C. Lian, D. Shen, Weakly supervised deep learning for brain disease prognosis using MRI and incomplete clinical scores. IEEE Trans. Cybern. **50**(7), 3381–3392 (2019)
19. Z. Ji, Y. Shen, C. Ma, M. Gao, Scribble-based hierarchical weakly supervised learning for brain tumor segmentation, in *Proceedings of the International Conference on Medical Image Computing and Computer-Assisted Intervention* (Springer, 2019), pp. 175–183
20. K. Wu, B. Du, M. Luo, H. Wen, Y. Shen, J. Feng. Weakly supervised brain lesion segmentation via attentional representation learning, in *Proceedings of the International Conference on Medical Image Computing and Computer-Assisted Intervention* (Springer, 2019), pp. 211–219
21. D. Bontempi, S. Benini, A. Signoroni, M. Svanera, L. Muckli, CEREBRUM: a fast and fully-volumetric convolutional encoder-decoder for weakly-supervised segmentation of brain structures from out-of-the-scanner MRI. Med. Image Anal. **62**, 101688
22. J. Zhou, L.-Y. Luo, Q. Dou, H. Chen, C. Chen, G.-J. Li, P.-A. Heng, Weakly supervised 3D deep learning for breast cancer classification and localization of the lesions in MR images. J. Magn. Reson. Imaging **50**, 1144–1151 (2019). https://doi.org/10.1002/jmri.26721
23. W.E. Fathy, A.S. Ghoneim, A deep learning approach for breast cancer mass detection. (IJACSA) Int. J. Adv. Comput. Sci. Appl. **10**(1) (2019)
24. R.R. Selvaraju, M. Cogswell, A. Das, R. Vedantam, D. Parikh, D. Batra, Grad-CAM: visual explanations from deep networks via gradient-based localization. arXiv:1610.02391v4, [cs.CV], 3 Dec 2019
25. S. Hwang, H.-E. Kim, Self-transfer learning for weakly supervised lesion localization. arXiv: 1602.01625v1, [cs.CV], 14 Feb 2016
26. N. Wu, J. Phang, J. Park, Y. Shen, Z. Huang, M. Zorin, S. Jastrzebski, T. Fevry, J. Katsnelson, E. Kim, S. Wolfson, U. Parikh, S. Gaddam, L.L.Y. Lin, K. Ho, J.D. Weinstein, B. Reig, Y. Gao, H. Toth, K. Pysarenko, A. Lewin, J. Lee, K. Airola, E. Mema, S. Chung, E. Hwang, N. Samreen, S.G. Kim, L. Heacock, L. Moy, K. Cho, K.J. Geras, Deep neural networks improve radiologists' performance in breast cancer screening. IEEE Trans. Med. Imaging **39**(4), 1184–1194 (2020)
27. http://icy.bioimageanalysis.org/ (ICY: an open community platform for Bioimage Informatics)
28. https://docs.opencv.org/3.4/d3/dc1/tutorial_basic_linear_transform.html
29. https://pytorch.org/vision/master/generated/torchvision.transforms.Normalize.html#torchvision.transforms.Normalize

Effects Evaluation of Data Augmentation Techniques on Common Seafood Types Classification Tasks

Hai Thanh Nguyen, Ngan Kim Thi Nguyen, Chi Le Hoang Tran, and Huong Hoang Luong

Abstract The applications which aim to classify seafood on the ships or markets are essential. However, when fishers have caught a large amount of seafood on board, they have to manually classify each seafood type for a long time, affecting seafood quality after catching. Moreover, When classifying seafood by manual method, fishers who have prolonged exposure to seafood are prone to skin diseases such as skin infections, fungal infections, cracked fingers. In addition, they also often suffer from spinal diseases due to having to stand in one position and move continuously during the classification process. Therefore, this study focuses on developing a method to effectively detect and classify some popular species of saltwater seafood using MobileNetV2 based on selected hyper-parameters via experiments on 9000 images of nine seafood types. Experimental results have reached 0.9956 in accuracy metric on the test set with the selected hyper-parameters combining considered data augmentation techniques. Also, the work evaluates the effects of eliminating each data augmentation technique to compare their influence on image classification tasks improvement. We can consider removing augmentation techniques that have revealed less influence in the image classification performance of seafood types.

H. T. Nguyen (✉)
College of Information and Communication Technology, Can Tho University, Can Tho 900100, Vietnam
e-mail: nthai.cit@ctu.edu.vn

N. K. T. Nguyen
Faculty of Information Technology, Kien Giang College, Kien Giang, Vietnam
e-mail: ntkngan@kgc.edu.vn

C. L. H. Tran
Department of Information Technology, FPT Polytechnic, Can Tho 900100, Vietnam
e-mail: lechi98.ct@gmail.com

H. H. Luong
FPT University, Hanoi, Vietnam

N. H. Phuong and V. Kreinovich (eds.), *Biomedical and Other Applications of Soft Computing*, Studies in Computational Intelligence 1045,
https://doi.org/10.1007/978-3-031-08580-2_20

1 Introduction

Fishing and exploitation of seafood are some of the strengths of many countries globally. However, fishing efficiency and seafood quality after exploitation are still a concern. Because seafood after catching should be washed by fishermen with seawater and required to be classified for preservation in different forms, each seafood type can have a different form of preservation, but now fishers still classify seafood by manual method based on long experience. In addition, some Seafood types have similar shapes, leading to confusion between precious (allowed to exploit) seafood of high economic value and those of low economic value, causing significant economic loss. Moreover, the long time to classify seafood affects the quality, human health and causes economic losses due to spoiled seafood. According to fishers, currently, fishing boats do not have enough modern equipment to prolong the time of each trip. Besides, scientific research on species composition and geographical distribution of marine organisms has tremendous significance for the study of marine resources and the protection of rare and precious marine species [1]. Therefore, conserving marine ecosystems and preserving marine species is an essential aspect. In order to do this, it is necessary to identify Seafood type right at the ship after catching, but finding experts in the field of ichthyology is very difficult [2]. Previously, Seafood types were identified manually by observation, so it was necessary to memorize or study different characteristics to identify them. Therefore, it can lead to misclassification among Seafood types.

In this study, we have deployed methods using the MobileNetV2 with chosen hyper-parameters on the large scale Fish dataset in.[1] In addition, the work has performed data pre-processing techniques and normalization to improve the model's accuracy. Besides, we have fine-tuned the hyper-parameters, including initial learning Rate, Learning Decay, Epochs, and we compare the efficiency of existing augmentations, including techniques related to rotation, shifting, shearing, flipping, etc.

2 Related Work

The manual method of classifying seafood takes a lot of time and effort and affects the quality after catching. Therefore, it is necessary to apply modern technology to classify seafood right at the ship. Classifying seafood by images has been studied and gives results with high accuracy.

In the research [3], CNN is the appropriate choice when it comes to solving the problem of Seafood types and has proven to be successful in the areas of image recognition and taxonomy. Besides, the authors have also emphasized the importance of selecting a CNN architecture and hyperparameter combination to achieve a high value of performance measure. However, the classification performance is generally poor due to the low image quality and the small number of images collected.

[1] https://www.kaggle.com/aungpyaeap/fish-market.

The authors in [1] proposed a fish classification algorithm based on Inception-V3. Besides, transfer learning is also applied to improve prediction accuracy. Experimental results show that the proposed method can effectively improve the classification accuracy, reaching about 89% for fish species. Another CNN model, the VGG network, is also widely used. The research of Rum et al. [4] proposed a predictive model of the FishDeTec is developed using the VGG16 to detect freshwater fish species found in Malaysia through an image processing approach. Besides, Hridayami et al. [5] proposed using VGG16 pre-trained on ImageNet via transfer learning method to improve the recognition rate of seafood taxonomy of previous studies. The fish dataset in their study included 50 species and trained the model on four different types of datasets. The results showed that blending image mixed with RGB image trained model exhibited the best actual acceptance rate (GAR) value of 96.4%. According to research by Iqbal et al. [6], authors used a reduced version of the AlexNet model and achieved higher accuracy than the original Alexnet model. The other achievement of Montalbo et al. [7] proposed using VGG16 retraining, fine-tuning, and optimization to provide better accuracies in classifying specific Verde Island fish species. Their study also concluded that a pre-trained model like VGG16 could improve by refining, optimizing, and enhancing the data to classify specific fish species. In [8], K. Knausg et al. proposed to identify and classify fish in temperate regions by implementing an in-depth deep learning-based approach. First, they used You Only Look Once (YOLO) object detection to detect individual fish in an image, independent of species and sex. Next, they applied a CNN with Squeeze-and-Excitation (SE) architecture and transferred learning to classify each fish in the image without pre-filtering. The experimental results have shown that the YOLOv3 technique can successfully detect an individual fish in different complex environmental conditions. In addition, the CNN-SENet architecture achieves the highest accuracy on the Fish4Knowledge dataset without any data augmentation or image pre-processing. Another study by Raza et al. [9] presented research on fish identification using YOLOv3 and Transfer learning technique. They used the ImageNet dataset and 1000 trained classes for feature extraction. In the fine-tuning process, they adjusted several parameters, including input image size, learning rate, batch size, to improve performance and percentage. The fish identification results showed that the improved version performed better than the original YOLOv3 model. The authors in [10] used YOLOv5 as the target development network model, trained using the underwater marine product dataset. Simultaneously, it combined six underwater image enhancement recovery algorithms to enhance and restore images before they were detected.

In the study [11], the authors have used six CNN-based architectures with transfer learning to classify seven types of Bangladesh freshwater fish. Performance metrics have been calculated to check the effectiveness of the applied models. The results showed that the InceptionResnetV2 and Xception approaches achieved the highest accuracy. InceptionV3, MobilenetV1, and MobilenetV2 models were used to compare and evaluate the performance of the methods [12]. The results showed that the MobilenetV2 model with transfer learning could be the best choice for real-time marine animal image classification than the other two models.

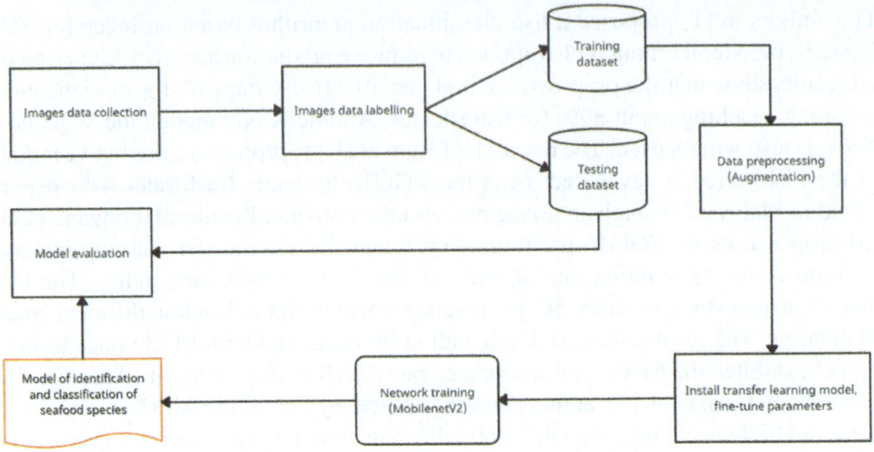

Fig. 1 General proposed workflow for seafood types classification tasks

3 Method

The general scheme for the Seafood types classification tasks is shown in Fig. 1. First, we collected the dataset from Kaggle[2] and divided it into a training set consisting of 90% and 10% for the test set. The training set is then divided into folds with 5-fold-cross validation for evaluating and selecting hyper-parameters. Finally, the best set of hyper-parameters is trained on the training set and evaluated on the test set. Also, we compare the performance with each augmentation technique when we remove it from the list of original augmentation techniques. Details will be presented in the following sections.

3.1 Data Collection

The data is collected from [13]. We have collected 9000 images with nine types of seafood: Striped Red Mullet, Sea Bass, Trout, Red Sea Bream, Red Mullet, Horse Mackerel. Each Seafood type has 1000 color images with 256 × 256 pixels or more (Fig. 2).

[2] https://www.kaggle.com/aungpyaeap/fish-market.

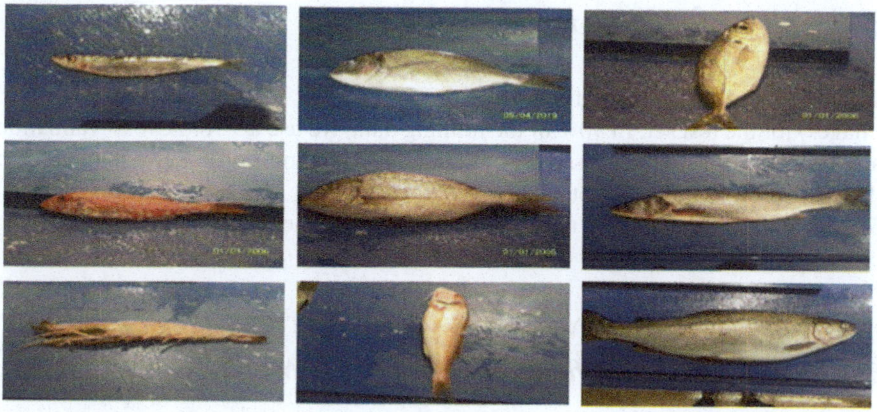

Fig. 2 An illustration of several types of saltwater seafood

Table 1 The considered augmentation techniques and their values in the experiments

Enhancement methods	Range
rotation_range	30
width_shift_range	0.2
height_shift_range	0.2
shear_range	0.1
zoom_range	0.2
horizontal_flip	True
vertical_flip	True
fill_mode	Nearest

3.2 Data Preprocessing and Augmentation Techniques

Before performing model training, we normalize the input data according to the format of the MobileNet network (the pre-processing technique of mobilenet with the function mobilenet_v2.preprocess_input) as the input pixel values are scaled [-1, 1].[3] Besides applying Data augmentation technique, which is a very effective technique to increase the size of the dataset by applying several image editing measures such as rotation, width shift, height shift, shear, zoom, horizontal flip, vertical flip, the image, etc. are described in detail in Table 1. The number of images used for training is 8100. After performing image enhancement, the number of generated images is 162,400.[4]

[3] https://www.tensorflow.org/api_docs/python/tf/keras/applications/mobilenet_v2/.

[4] https://www.researchgate.net/post/How-many-images-does-Imagedatagenerator-generate-in-deep-learning.

3.3 Split Data for the Evaluation

The original data is divided into 90% (8100 images) for training and 10% (900 images) for the test set. In 90% of the training data, we use a 5-fold Cross-Validation approach to split the data used for training and validation of hyper-parameters on the validation set. On the other hand, the test data is kept separate and reserved for the final evaluation step to test the "reaction" of the model when encountering completely new data.[5]

3.4 Building a Classification Model Using MobileNetV2

As shown in Fig. 3, the MobileNetV2 is conducted from depth-separated convolutions, which gives it higher accuracy, fewer parameters, and fewer computations (about nine times). The model has a 30-layer architecture. The first layer is a convolutional layer with a stride of 2. The second layer is a depth-based convolution layer with a stride of 1. The third layer is a point convolution layer with a 1×1 filter. The fourth layer is a depth convolution layer with a stride of 2 (different from layer 2 with a stride of 1), etc., different layers such as AveragePooling2D with 7×7, Flatten, Dropout, and Dense. The depth-based convolutional layer used to filter the final input is a 1×1 convolutional layer that helps to reduce the number of channels or the bottleneck. It brings data with many dimensions (channels) into a scale with a much lower number of dimensions. In the Dense final layer, use the softmax function to output a vector representing the probability of each class. After each convolution layer, MobileNetV2 will use Batch Normalization (BN) and non-linear ReLU, except for the fully connected layer without non-linearity [14]. The last layer is the Softmax layer, used for classification.

The MobileNetV2 can provide a high accuracy while the number of parameters is only 3.5 million (about 1/40 of the parameters of VGG16). Therefore, MobileNet can operate smoothly even on low-profile hardware. This model is very suitable for embedded systems or applications that need to run on mobile. We use a Pretrained model trained on the imageNet dataset, a very popular dataset in CV Deep Learning with more than 14 million images and more than 20,000 different classes, each layer has at least a few hundred images, so the accuracy is relatively high. This famous dataset was used in the vision competition with the name ILSVRC to build a model with high accuracy for the image classification problem [15]. ImageNet is a standard dataset for benchmarking newly created image classification algorithms. This dataset has become a de facto standard for classification algorithms in CVs. Therefore, almost every classification problem has a label that appears in the imageNet and can reuse the pre-trained—model.

[5] https://medium.com/the-owl/k-fold-cross-validation-in-keras-3ec4a3a00538.

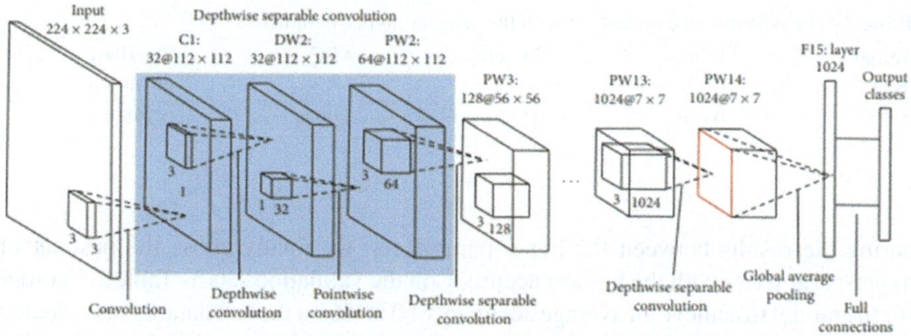

Fig. 3 MobileNetV2 Network Architecture (https://www.pluralsight.com/guides/transfer-learning-in-deep-learning-using-tensorflow-2.0)

4 Experimental Results

4.1 Hyper-parameters Fine-Tuning

In this section, we have performed hyper-parameter tuning tasks such as Initial Learning Rate, Decay, Optimizer, Batch Size, Epochs to help determine the optimal combination of hyper-parameter of the model.

After performing training on multiple parameter sets (Table 2) and simultaneously evaluating the model on the validation dataset. The obtained results are compared to consider the influence of the combination of different hyper-parameter on the model's performance as a basis for selecting the best model for the problem. As shown in Table 2, the set of hyper-parameter with the highest accuracy is 0.993456 include a learning rate initialized at 0.02, a decay of 0.005, and running to 15 epochs. In addition, the input pre-processing with the function mobilenet_v2.preprocess can also improve the performance to 0.4467, which is used in all experiments in our study.

We used the K-Fold Cross-Validation with $K = 5$ and simultaneously performed training and evaluation on various sets of hyper-parameters (Table 2). After com-

Table 2 The average results on validation sets with various hyper-parameter

Initial learning Rate	Decay	Epochs	ACC	Standard deviation
0.02	0.0004	10	0.9806	0.0117
0.05	0.0004	10	0.5304	0.2381
0.02	0.005	15	0.9934	0.0029
0.03	0.004	15	0.9607	0.0032
0.03	0.005	10	0.9571	0.0269

Table 3 The performance on test set with the selected hyper-parameters

Initial learning Rate	Decay	Epochs	ACC	Standard deviation
0.02	0.005	15	0.9934	0.2962

paring the results between the hyper-parameters, we finally chose the best set of hyper-parameters with the highest accuracy on the validation sets as Table 2. In order for the model to achieve an average accuracy of 0.9956 on the test data set (as revealed in Table 3), we performed the classification with selected hyper-parameters. Finally, we evaluated the effectiveness of the classification model through measures on the test set. After training with the MobileNetV2 network, the model showed an average accuracy (ACC) of 0.9956.

4.2 Effect of Data Augmentation Techniques on Seafood Types Classification

Table 4 describes the increased accuracy when one of data augmentation techniques in Table 1 is eliminated to compare their effects. As observed from Table 4, some techniques such as rotation, height shift, and zoom can affect the performance greater than other techniques. Without them, the performance can decrease to 0.0023 for each case. Some techniques such as shear and vertical flip only cause minor effects on the performance with decreases of 0.001, 0.0009, respectively.

The effectiveness of the seafood classification task on each class and all classes with various metrics is exhibited in Tables 5 and 6, respectively. We also evaluate results by the confusion matrix on the MobileNetV2 network model for each seafood type (Fig. 4). The horizontal axis of the confusion matrix represents the actual label, the vertical axis the predicted label. It can be concluded that the classes Black Sea

Table 4 The effectiveness of preprocessing techniques

Eliminated data augmentation technique	Decreased (−)/Increased(+) accuracy
rotation_range=30	−0.0023
width_shift_range=0.2	−0.0019
height_shift_range=0.2	−0.0023
shear_range=0.1	−0.0010
zoom_range=0.2	−0.0023
horizontal_flip=True	−0.0012
vertical_flip=True	−0.0009
fill_mode='Nearest'	−0.0012

Table 5 Results in accuracy and recall for each Seafood types

	Black sea sprat	Gilt-Head bream	Horse Mackerel	Red mullet	Red sea bream	Sea bass	Shrimp	Striped red mullet	Trout
Precision	1.0000	1.0000	1.0000	0.9900	0.9800	1.0000	1.0000	0.9900	1.0000
Recall	1.0000	0.9800	1.0000	1.0000	1.0000	1.0000	0.9900	0.9900	1.0000

Table 6 Efficiency of the model

Model	Size	ACC	AUC	MCC	Precision	Recall
MobileNetV2	224x224	0.9956	0.9999	0.9950	0.9954	0.9957

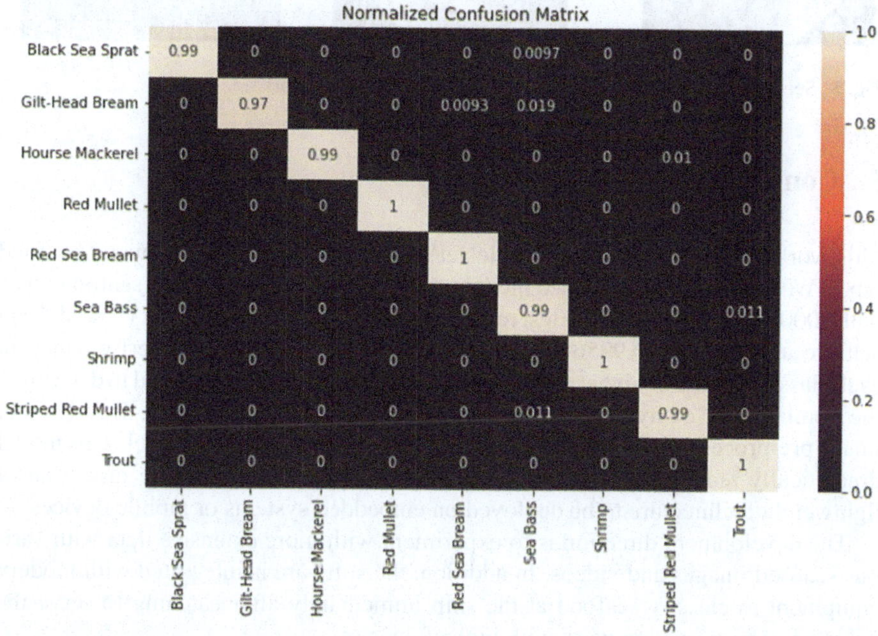

Fig. 4 Confusion matrix results with MobileNetV2

Sprat, Horse Mackerel, Red Mullet, Red Sea Bream, Sea Bass, Trout correctly classify 1.0000 on the test dataset, the remaining classes correctly classify 0.9800 or more. The results show that the correct classification model for each class is relatively high. Some seafood types are classified using the MobileNetV2 network model illustrated in Fig. 5.

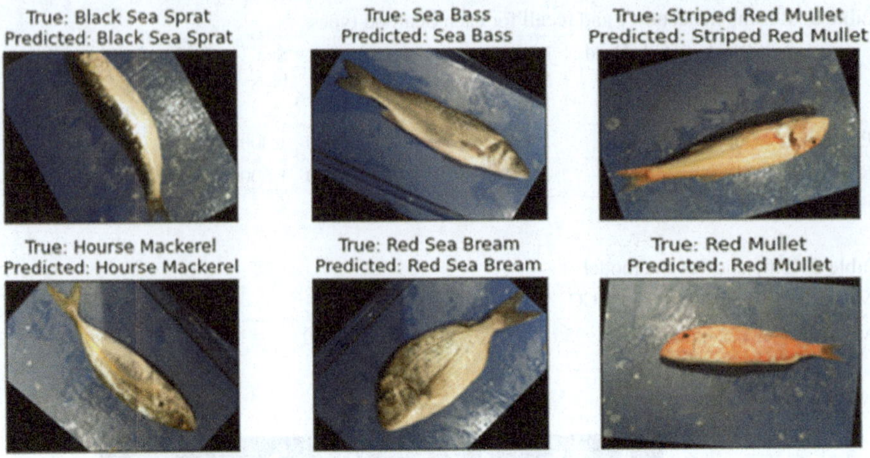

Fig. 5 Seafood is identified and classified by MobileNetV2 network

5 Conclusion

This work proposed using the MobileNetV2 network to classify saltwater seafood types. We evaluated the proposed model on a dataset consisting of nine seafood types with 9000 images. Through the test results, we found that the MobileNetV2 model can achieve an accuracy of 0.9956 and simultaneously has a classification performance on each class of 0.9800 or higher. In order to get high accuracy, it is essential to determine the combination of hyper-parameters and augmentation techniques. Besides, the image pre-processing techniques also improve the accuracy of the model. This model dramatically reduces the time for manual processing. Moreover, the model has a lightweight architecture to be deployed on embedded systems or mobile devices.

The development direction is to experiment with more extensive data with various seafood images and videos. In addition, the software is integrated with modern equipment to classify seafood at the ship immediately after catching to serve the fishing and conservation work with high efficiency.

References

1. X. Lan, J. Bai, M. Li, J. Li, Fish image classification using deep convolutional neural network, in *Proceedings of the 2020 International Conference on Computers, Information Processing and Advanced Education*, CIPAE 2020 (Association for Computing Machinery, 2020), pp. 18–22. http://dx.doi.org/10.1145/3419635.3419643
2. D. Rathi, S. Jain, S. Indu, Underwater fish species classification using convolutional neural network and deep learning, in *2017 Ninth International Conference on Advances in Pattern Recognition (ICAPR)* (IEEE Publishing, 2017), pp. 1–6. https://doi.org/10.1109/icapr.2017.8593044

3. D. Štifanić, Z. Car, Use of convolutional neural network for fish species classification. J. Maritime & Trans. Sci. **59**(1), 131–142 (2020). https://doi.org/10.18048/2020.59.08
4. S.N.M. Rum, F. Az, FishDeTec: a fish identification application using image recognition approach. Int. J. Adv. Comput. Sci. Appl. **12**(3) (2021). https://doi.org/10.14569/ijacsa.2021.0120312
5. P. Hridayami, I. Putra, K.S. Wibawa, Fish species recognition using vgg16 deep convolutional neural network. J. Comput. Sci. Eng. **13**(3), 124–130 (2019). http://dx.doi.org/10.5626/JCSE.2019.13.3.124
6. M.A. Iqbal, Z. Wang, Z.A. Ali, S. Riaz, Automatic fish species classification using deep convolutional neural networks. Wireless Personal Commun. **116**(2), 1043–1053 (2019). https://doi.org/10.1007/s11277-019-06634-1
7. F.J.P. Montalbo, A.A. Hernandez, Classification of fish species with augmented data using deep convolutional neural network, in *2019 IEEE 9th International Conference on System Engineering and Technology (ICSET)* (IEEE Publishing, 2019), pp. 396–401. https://doi.org/10.1109/ICSEngT.2019.8906433
8. K.M. Knausgård, A. Wiklund, T.K. Sørdalen, K.T. Halvorsen, A.R. Kleiven, L. Jiao, M. Goodwin, Temperate fish detection and classification: a deep learning based approach. Appl. Intell. (2021). https://doi.org/10.1007/s10489-020-02154-9
9. K. Raza, S. Hong, Fast and accurate fish detection design with improved YOLO-v3 model and transfer learning. Int. J. Adv. Comput. Sci. Appl. **11**(2) (2020). https://doi.org/10.14569/ijacsa.2020.0110202
10. G. Huang, A comparative study of underwater marine products detection based on yolov5 and underwater image enhancement. Int. Core J. Eng. **7**(5), 213–221 (2021). https://doi.org/10.6919/ICJE.202105_7(5).0028
11. A. Majumder, A. Rajbongshi, M.M. Rahman, A.A. Biswas, Local freshwater fish recognition using different CNN architectures with transfer learning. Int. J. Adv. Sci. Eng. Inf. Technol. **11**(3), 1078 (2021). https://doi.org/10.18517/ijaseit.11.3.14134
12. X. Liu, Z. Jia, X. Hou, M. Fu, L. Ma, Q. Sun, Real-time marine animal images classification by embedded system based on mobilenet and transfer learning, in *OCEANS 2019 - Marseille* (IEEE Publishing, 2019), pp. 1–5. https://doi.org/10.1109/oceanse.2019.8867190
13. O. Ulucan, D. Karakaya, M. Turkan, A large-scale dataset for fish segmentation and classification, in *2020 Innovations in Intelligent Systems and Applications Conference (ASYU)* (IEEE publishing, 2020), pp. 1–5. https://doi.org/10.1109/asyu50717.2020.9259867
14. A.G. Howard, M. Zhu, B. Chen, D. Kalenichenko, W. Wang, T. Weyand, M. Andreetto, H. Adam, Mobilenets: efficient convolutional neural networks for mobile vision applications, vol. abs/1704.04861 (2017). http://arxiv.org/abs/1704.04861
15. M.M. krishna, M. Neelima, M. Harshali, M.V.G. Rao, Image classification using deep learning. Int. J. Eng. & Technol. **7**(2.7), 614–617 (2018). https://doi.org/10.14419/ijet.v7i2.7.10892

Image Caption Generator with a Combination Between Convolutional Neural Network and Long Short-Term Memory

Duy Thuy Thi Nguyen and Hai Thanh Nguyen

Abstract Automatically describing the content of an image holds an essential role in numerous applications. Some practical applications include providing more accurate and detailed images or videos in scenarios like information search by image or video surveillance systems. The technique generates picture captions that are generally semantically informative and grammatically accurate by acquiring image and caption pairings. Humans use natural languages to describe scenes because they are short and compact. On the other hand, machine vision systems characterize the scene by capturing an image that is a two-dimensional array. The concept is to combine the image and captions into one area and then map from the image to the sentences. This study proposes a merge model to combine the image vector and the partial caption. It can be implemented through three steps: processing the sequence from the text, extracting the feature vector from the image, and decoding the output by concatenating the above two layers. Besides that, to evaluate model performance, we generate multiple sentences using Beam Search and BLEU. The experiments show that the method can generate captions with relatively accurate content and less training memory. We use the Flickr8k dataset consisting of 8000 images paired with five different captions, which provide precise descriptions of the salient entities and events. For training, we use 6000 images, 1000 for test and 1000 for development, while Flickr8k text includes text files describing train set, test set. The best results were obtained when testing the model with BLEU-1, Greedy, and Beam Search with k = 5 or k = 7 all over 60 in BLEU scores.

1 Introduction

Automatically generating image descriptions is gaining popularity as a challenging and essential topic of artificial intelligence. Image captioning's purpose is to con-

D. T. T. Nguyen · H. T. Nguyen (✉)
College of Information and Communication Technology, Can Tho University, Can Tho 900100, Vietnam
e-mail: nthai.cit@ctu.edu.vn

N. H. Phuong and V. Kreinovich (eds.), *Biomedical and Other Applications of Soft Computing*, Studies in Computational Intelligence 1045,
https://doi.org/10.1007/978-3-031-08580-2_21

struct linguistically reasonable statements that are semantically faithful to the image's content. As a result, there are two main parts to image description: visual comprehension and language processing [1]. CV and NLP approaches should be utilized to deal with the challenges caused by the complementary modalities and appropriately integrate them to ensure that the generated sentences are grammatically and semantically correct. Retrieval-based approaches [2] and template-based methods [3] are two traditional ways of image captioning. Farhadi et al. [1] used description templates to infer the final text after detecting objects to construct a triple of scene elements. Through tests, Devlin et al. [2] highlighted and summarized the aspects of retrieval-based approaches as well as the important difficulties that are required to be addressed in image description tasks. They identified the important components contributing to this method's good descriptive effect. On the other hand, these two approaches rely on the training corpora, and the generated text descriptions lack diversity.

Early machine learning-based image processing methods employ image processing operators to extract image features, then use Support Vector Machine (SVM) [4] as a classifier to retrieve the image's target, and then use the acquired target and its attributions as the foundation for a produced sentence. The extraction of image features and the criteria for producing sentences rely excessively on this technique. The complexity of training big datasets limits captioning. Most approaches use pre-trained Convolutional Neural Network models to extract visual information in the form of picture features, followed by modification of the visual information using a Caption Generator module to generate output phrases [5]. Another technique combines current standard deep neural network methods, such as the NIC (Neural Image Caption) model, with semi-supervised deep learning ideas like pseudo labels and N-gram. This technique uses the N-gram Search algorithm to produce pseudo labels and increases the model's effect by using previous knowledge of the N-gram table and people's descriptive habits [6].

Some neural network optimization algorithms have recently been linked to image captioning. Based on machine learning algorithms, especially CNN and RNN, many researchers have developed deep learning models to create descriptions for images. Images caption generator can be defined as automatically generating the textual descriptions for an image. This process includes detecting objects in the image, the relevance, properties of the objects. Manipulating deep learning networks based on the feature that they can learn from already trained instances and providing predictions based on what they have learned is creating possibilities for the development of this field. The level of abstraction will gradually increase through each layer of the network until the output layer, thereby giving the prediction results and labels at the output layer. The main contributions of this work include as follows.

- Using CNN and LSTM, we develop a novel encoder-decoder framework to improve the accuracy of predicted sentences for image captioning.
- Through comprehensive experiments, we validate the effectiveness of our model on Flickr8k dataset with different parameters, including different epochs and batch sizes, to choose the best value for the training model.

- Moreover, we combine the use of Greedy, Beam Search, and BLEU score algorithms to evaluate the model's predictive performance. As a result, our technique produces state-of-the-art outcomes in image captioning, as measured by the evaluation metrics used.

2 Related Work

The challenge of describing static images with natural words has recently piqued people's curiosity. We can drive natural language generation systems by using recent advancements in object identification, characteristics, and locations, albeit these systems are restricted in their expressivity. For example, the work in [1] proposed methods to create a triplet of scene elements. Then, they are then translated to text using templates. Li et al. [7] did the same thing, starting with detection and piecing together a final description from phrases including identified items and relationships.

Using natural language processing, the technique presented in [8] produces an image description automatically. Image processing and natural language processing are also involved. The many models for picture captioning are discussed in this study. For image captioning, they primarily concentrate on object identification and machine translation. They looked at how these challenges aided in improving the image captioning job. It identifies the top-n matching photos and their corresponding captions, and the system outputs these captions. The solution presented in [9] is based on the picture captioning task's hardware requirements. The system can run on low-end hardware used in hand-held devices. It employs an encoder-decoder implementation that has undergone substantial changes and optimizations, allowing it to be used on hand-held devices. Image retrieval for image captioning [10] employs visual similarity to get a set of captioned pictures for a query image to decrease the effects of noisy image estimates in existing techniques. First, it uses computer vision algorithms to interpret and represent images using triplets (objects, activities, and spatial connections). Then, based on the visual recognition findings, they define picture description as a tree-generating process. Then, for rating sentences for a query image, combine sentence and image fragments in a shared space. As sentence fragments, they employ dependency tree relations from a sentence, and as picture fragments, they use detection results from the Region Convolutional Neural Network technique.

The work in [11] suggested a face action detection system. It is based on the analysis of facial expressions and the changing behavior of facial muscles. It increases the detection rate by 7.8%. Authors in [1] detection are used to infer a triplet of scene elements, which are then translated to text using templates.

In the study [12], a face detection technique was employed; in this method, a set of rules is defined as a knowledge-based method, and it has rules on an input picture. First, scanning is applied to the face, and all possible faces are discovered by scanning. Then, multiple criteria are created for face identification from images,

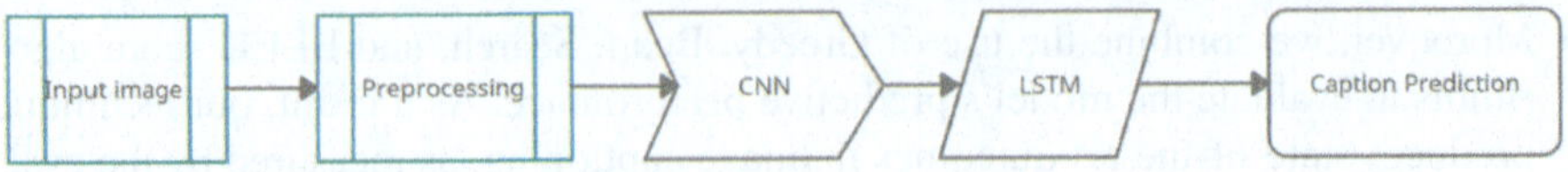

Fig. 1 Image captioning process diagram

and different rules are used, such as defining a human face at the highest level and describing facial features at the lowest level. Finally, a multilayer of picture hierarchy is formed by averaging and subsampling.

3 Methodology

The general idea of creating the description for an image can be explained as follows. Initially, the input images are converted to grayscale images (or RGB images) and processed by CNN into objects. Then these objects will be converted to linear feature vectors and fed to the LSTM model to generate a textual description (a string of English words) using LSTM. The process of this work is shown in Fig. 1 and details of components will be presented in the following sections.

3.1 Preprocessing

3.1.1 Word Embedding

Word embeddings are a form of word representation that bridges the human understanding of language to a machine. They have learned text representations in an n-dimensional space where words with the same meaning have an equal representation. Meaning that two similar words are represented by almost similar vectors that are very closely placed in a vector space. Word2Vec [13] and GloVe popular technique to learn word embeddings. In this study, we have deployed GloVe [14] for mapping words to a vector space, where similar words can group while different words can be separated. GloVe is more benefits compared to Word2Vec. GloVe relies on the local context of words and incorporates global word co-occurrence to obtain word vectors.

3.1.2 Preprocessing Image

A CNN can provide a rich representation of the input image by embedding it to a fixed-length vector representation to caption the images. Various CNN can be deployed, including VGG, InceptionV3, ResNet. In this study, we use the InceptionV3[1] model

[1] https://www.pluralsight.com/guides/introduction-to-lstm-units-in-rnn/.

created by Google Research as an encoder. This model was pre-trained on the ImageNet dataset. We have removed the last layer of the model used for classification and pre-processed images with the InceptionV3 model and have extracted features.

3.2 *Training with Deep Learning Approaches*

3.2.1 Convolutional Neural Network (CNN)

This study applies a convolutional neural network (CNN) that maps an RGB image to a visual feature vector. CNN will consist of three layers: convolutional pooled and fully connected. In addition, the Rectified Linear Unit (ReLU) is used as a non-linear activation function. ReLU is used faster than Tanh(x) or traditional activation functions. Finally, the dropout layer is used to prevent overfitting. The dropout sets the output of each hidden neuron to zero with a certain probability. Thus, these neurons do not affect the forward pass and do not join in the backpropagation process.

3.2.2 Long Short-Term Memory (LSTM)

LSTM is an extended version of Recurrent Neural Network (RNN) designed to solve long-term dependencies. RNN is a neural network containing loops. This network is capable of storing information transmitted from one layer to another. The output of the hidden layer depends on the information of the layers at all times. RNNs have been widely used in natural language processing or problems with sequential data. However, because the architecture of RNN is quite simple, the ability to link layers over long distances is not good. It is essentially incapable of remembering information from long-distance data, and therefore the first element in the input sequence usually do not have much effect on the result of element prediction for the output sequence in the following steps. It is because the RNN is affected by the descending of the gradient in the learning process—the vanishing gradient. The LSTM network is designed to overcome this problem. The mechanism of operation of LSTM is to remember only relevant information necessary for prediction, and non-relevant information will be ignored. The sigmoid function is a type of non-linear activation function. The gate keeps it contained. Sigmoid, unlike Tanh, keeps the values between 0 and 1. It aids the network's ability to update or forget data. If the multiplication yields a zero, the data is considered lost. If the value is 1, the information remains. It can aid the network in determining which data can be ignored and which data must be kept. Tanh is a non-linear activation function. It controls the values that go over the network, keeping them between -1 and 1. A function with a long-lasting second derivative is required to avoid information fading. Some values may grow gigantic, leading other values to become inconsequential.

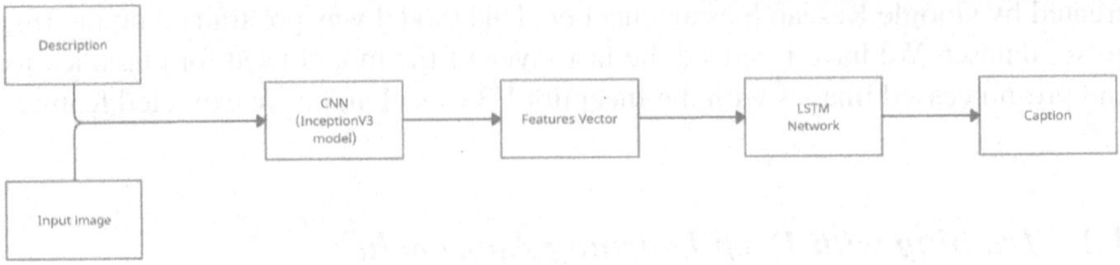

Fig. 2 Processing flow diagram

3.2.3 The Architecture of the Proposed Combination

We have based on the architecture in [15], the CNN-LSTM architecture in "Learning CNN-LSTM Architectures for Image Caption Generation"[2] and further developed our CNN-LSTM architecture. Instead of using VGG16 as in [15], we use InceptionV3 for image pre-processing. Moreover, we also use GloVe vector [14] to pre-process word vector, mapping words to a vector space, where similar words are grouped, and different words are separated.

The critical reason to use LSTM instead of other architectures based on traditional RNN is that the range of contextual information available to RNN is limited, and Back-Propagation through time does not work well. It is visible in the network's vanishing or exploding outputs. This topic is known as the vanishing gradient problem or the bursting gradient problem in the literature. Because of the vanishing gradient problem, while the network is learning to bridge significant time gaps, it takes a long time or does not operate at all. Furthermore, the exploding gradient causes fluctuating weights, lowering the network's quality. In reality, this means that when a recurrent network learns to retain information across extended time intervals, it takes a long time since there is not enough decaying error backflow. Hochreiter and Schmidhuber provide their unique, efficient, gradient-based approach termed "Long Short-Term Memory" in [16] to solve these issues (LSTM) specifically. The LSTM addresses erroneous backflow difficulties through carousels in their specific units accomplished with computational complexity, and the LSTM improves the RNN's capacity to bridge temporal intervals. Therefore, we have used LSTM in the proposed model architecture in this study. We present a neural and probabilistic framework for generating descriptions from images in this research. To generate picture captioning in this study, We used an encoder-decoder architecture that included a visual attention mechanism (see Fig. 2). The encoder is based on CNN, whereas the decoder is based on the visual attention module. The encode-decode model is embedded into the model. For the encoder, a CNN may create a rich representation of the input image by embedding it to a fixed-length vector representation in the encoder-decoder architecture for image captioning. Many other CNNs, including VGG, Inception V3, and ResNet, can be employed. This article employs Google Research's Inception V3 model as an encoder. This model was pre-trained on the ImageNet dataset, where it

[2] https://cs224d.stanford.edu/reports/msoh.pdf.

Fig. 3 Model Architecture

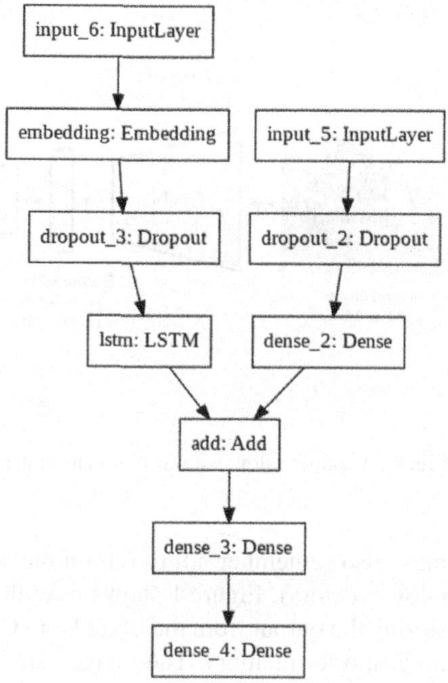

placed second in the ILSVRC 2015 image classification competition. It is important to note that we do not need to classify the images here. All we need to do is extract an image vector. As a result, the softmax layer is removed from the inceptionV3 model. We used the Inception V3 model to pre-process pictures and extract features. On the other hand, a decoder converts the image into ordinary words given the image representations. In addition, a decoder is a kind of RNN usually built using LSTM or GRU. As a decoder, we utilized LSTM in this case. In addition, unlike RNN, LSTM does not have the problem of vanishing gradients. The caption generator is a straightforward decoder with a Dense 512 layer activated by ReLU. Concatenation combines the outputs of the picture feature encoder and the word sequence encoder, which are then sent into the dense layer. The dense layer creates a softmax prediction for each word in the lexicon to be the following word in the sequence and then chooses the word with the highest likelihood. This procedure is repeated until an end token is produced.

The proposed architecture is illustrated in the following Fig. 3. The partial caption Input3 is sent into the embedding layer, and it has a maximum length of 34 characters. The 200-d Glove embedding is mapped to the words here to prevent overfitting. After that, the sequence is put into the LSTM to be processed. Our InceptionV3 network extracted an image vector, which we call Input2. It has then fed into a Fully Connected layer, which has a dropout of 0.5 to avoid overfitting. After that, the Image and Language models are combined and put into another Fully Connected layer. Our 1660-word vocabulary has a softmax layer that offers probability. We

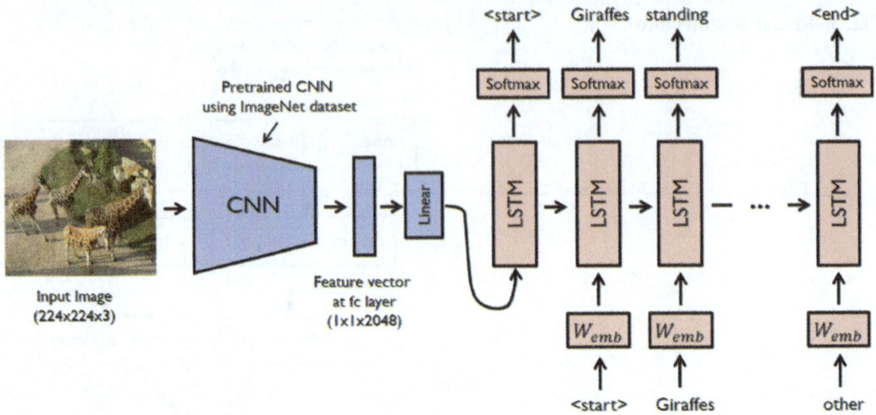

Fig. 4 A training flow with a combination between CNN and LSTM [17]

must also remember not to retrain the weights in our embedding layer (pre-trained Glove vectors). Figure 4 shows how the model is trained using some images and storing the output from the LSTM and CNN as the convolutional neural network and long, shorter memory. The image data is processed and stored in vector by LSTM, which repeats until the final meaningful captions are obtained, and the entire image is processed.

The method greedily picks the word with the highest probability to get the next word prediction because the model generates a 1660 long vector with a probability distribution across all the words in the vocabulary. Greedy Search is the name of this strategy. Beam Search aims to take top k predictions, feed them again into the model and then arrange them using the probabilities returned by the model. So, the list will always contain the top k predictions. In the end, we take the one with the highest probability and go through it till we encounter the end of the sequence or reach the maximum caption length. Then, for each word in the vocabulary, the result of the caption generator is converted into a probability score. The greedy algorithm selects the word with the best probability for each step. This system may not consistently deliver the best potential caption because any word's prediction is based on all previously anticipated words. As a result, choosing the sequence with the best overall score from a list of candidates is more efficient. So, we employ the beam search technique with a beam size of k. The initial decode stage considers the top k candidate words. We can change the value of k and test them in the model to get the best result.

4 Experimental Results

4.1 Dataset Description

Several datasets are used to train, test, and evaluate the image captioning methods. The datasets differ in various perspectives, including the number of images, captions per image, the format of the captions, and image size. Three datasets consist of Flickr8k, Flickr30k, and MS COCO popularly used. In Flickr8k dataset, [11] each image is related to five captions that describe the entities and events. Flickr8k is a good starting dataset as it is small in size and can be trained quickly on low-end laptops/desktops using a CPU. It includes Flickr8k Dataset and Flickr8k text. Flickr8k dataset contains 8092 images in JPEG format with different shapes and sizes, of which 6000 are used for training, 1000 for the test, and 1000 for development, while Flickr8k text includes text files describing train set, test set. In addition, flickr8k.token.txt contains five captions for each image, a total of 40460 captions.

4.2 Performance Comparison

BLEU (Bilingual Evaluation Understudy) [18] is a quality score metric for evaluating prediction results. This evaluation technique counts the number of n-grams that the candidate translation and the reference text have in common, where n refers to the number of ordered words in the comparison (1-g compares each word and 2-g compares each word pair). Based on the accuracy analysis, it calculates the n-gram correlation between prediction descriptions and reference sentences. The closer the prediction and reference descriptions are, the higher the BLEU score is. In BLEU@N (N = 1, 2, 3, 4), N represents the length of the n-gram. The longer the n-gram, the more difficult it is to match. If only BLEU for N = 1, it is easy to obtain a relatively high evaluation on a description of low quality. Thus, we calculated many levels of BLEU scores in the actual comparison. BLEU score is calculated according to the following Formula (1).

$$BLEU = \min\left(1, \frac{output - length}{reference - length}\right)\left(\prod_{i=1}^{4} precision_i\right)^{\frac{1}{4}} \tag{1}$$

where : $output - length$ is the length of sentence output, $reference - length$ is the length of description sentence input and precision is the precision of n-gram of size 1 to 4, for example, BLEU-1 denotes BLEU with a n-gram of 1.

This section examines the model's performance quantitatively and qualitatively. While the evaluation metrics provide a numerical indication of the captions' validity, any result can be misinterpreted. As shown in Table 1, we trained the model ten times with different parameters. Besides, we have performed an accuracy comparison,

Table 1 Performance comparison of hyper-parameters

Times	Train set	Test set	Epoch	Batch size	Training time	Accuracy
1	6000 images	1000 images	20	4	11009 s/epoch	0.583
2	6000 images	1000 images	20	3	11120 s/epoch	0.647
3	6000 images	1000 images	25	4	11088 s/epoch	0.672
4	6000 images	1000 images	25	3	11292 s/epoch	0.724
5	6000 images	1000 images	30	4	11120 s/epoch	0.791
6	6000 images	1000 images	30	3	11090 s/epoch	0.825
7	6000 image	1000 image	35	4	12352 s/epoch	0.764
8	6000 image	1000 image	35	3	12511 s/epoch	0.749
9	6000 image	1000 image	40	4	12788 s/epoch	0.653
10	6000 image	1000 image	40	3	12698 s/epoch	0.618

which is the best result is saved through each training to select optimal parameters for our model training. As can be seen in Table 1, the model with an epoch number of 30 and a batch size of 3 produces the best results.

The slight differences in the generated captions compared to the original human language description can be evaluated via qualitative analysis. The models' predicted captions were evaluated using the existing evaluation metrics BLEU. The most commonly used metric in the image description literature has been the BLEU score, precision of word n-grams between generated and reference sentences. Even though this metric has some obvious drawbacks, it has been shown to correlate well with human evaluations [19].

We can see the comparison of the quality of the captions generated by the models in Table 2 and Fig. 5. The model achieves the best performance with BLEU-1, as can be observed. In addition, Greedy and Beam Search with k = 5 and k = seven can improve the model's evaluation outcomes. We also compare our model to state-of-the-art approaches. Table 3 shows the experimental results when using other models for image captioning. We can observe from the experiment results in Table 3 that our model outperforms the other image captioning algorithms on numerous assessment metrics. For example, our model's BLEU-1, BLEU-2, BLEU-4 on the other evaluation metrics improved 0.011–0.031 when compared to the PoS model [20], while the scores in BLEU-3 of our model is slightly lower than PoS model by 0.015,

Table 2 Comparison in average BLEU scores on test set of our model

Method	BLEU-1	BLEU-2	BLEU-3	BLEU-4
Greedy	0.633	0.492	0.257	0.227
Beam Search k = 3	0.628	0.451	0.244	0.201
Beam Search k = 5	0.669	0.463	0.292	0.232
Beam Search k = 7	0.652	0.411	0.281	0.212

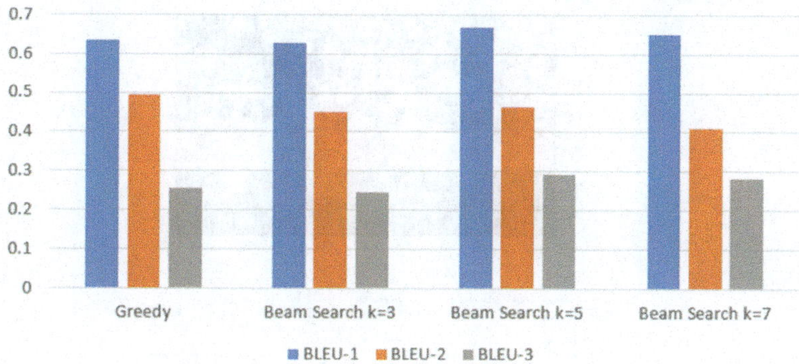

Fig. 5 The chart shows BLEU score of model evaluation

Table 3 Comparison in BLEU scores of our model and other model

Methods	BLEU-1	BLEU-2	BLEU-3	BLEU-4
PoS [20]	0.638	0.446	0.307	0.221
NIC [21]	0.630	–	–	–
Our method	0.669	0.463	0.292	0.232

respectively. Meanwhile, the results on BLEU-1 of our model have a higher value than the NIC model [21]. The previous discussion and comparative results suggest that by using CNN combined with LSTM to get the features of images and generate a caption for them, our proposed model can improve the model-generation caption's performance.

We provided some qualitative examples in Fig. 6, for a better understanding of our model. The images are randomly taken from the test dataset. The generated descriptions show that using Beam Search with a suitable k value, the image captioning system performs better than the Greedy Search. In addition, object recognition was influenced by many factors in complicated scenes and images that are difficult to describe, which needs further research.

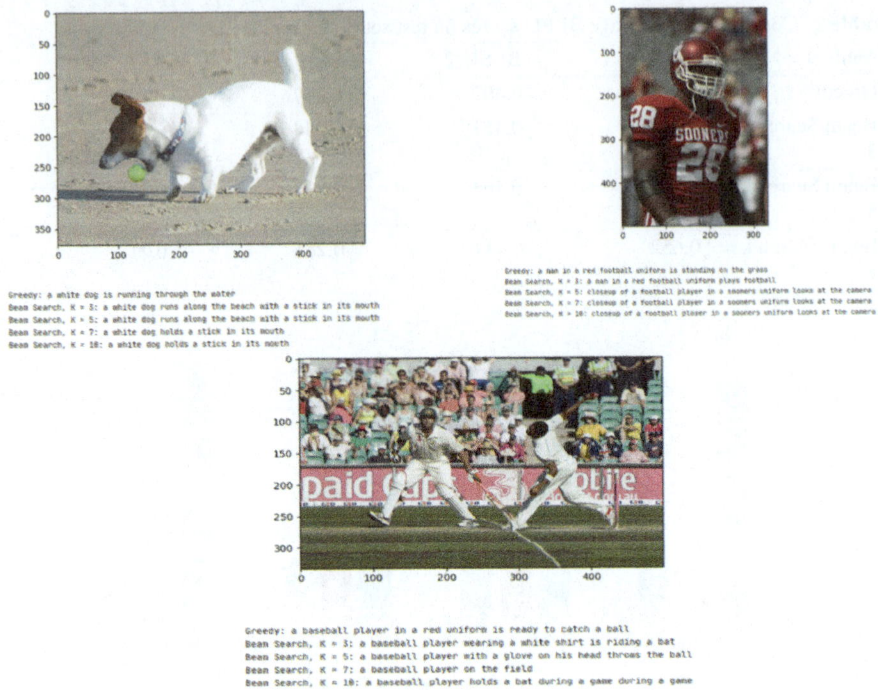

Fig. 6 The test images and generation sentence results

4.3 Conclusion

Instead of using a typical sequence-to-sequence model, this study provides an image caption generation system based on a CNN-LSTM merge encoder-decoder model. On the Flickr8k dataset, a thorough test is carried out with good results, validating the efficacy of our suggested model. Furthermore, the experimental results suggest that the proposed model, which combines CNN and LSTM, produces excellent performance in automatically generating image descriptions. However, because the data set is biased towards human subjects, the proposed model has trouble admitting non-human things. Therefore, we want to create a dataset with even more detailed and diversified captions.

References

1. A. Farhadi, M. Hejrati, M.A. Sadeghi, P. Young, C. Rashtchian, J. Hockenmaier, D. Forsyth, Every picture tells a story: generating sentences from images, in *Computer Vision – ECCV 2010* (Springer, Berlin, Heidelberg, 2010), pp. 15–29. https://doi.org/10.1007/978-3-642-15561-1_2
2. J. Devlin, H. Cheng, H. Fang, S. Gupta, L. Deng, X. He, G. Zweig, M. Mitchell, Language models for image captioning: the quirks and what works, in *Proceedings of the 53rd Annual Meeting of the Association for Computational Linguistics and the 7th International Joint Conference on Natural Language Processing (Volume 2: Short Papers)* (Association for Computational Linguistics, 2015). https://doi.org/10.3115/v1/p15-2017
3. Y. Choi, A.C. Berg, V. Premraj, S. Li, G. Kulkarni, S. Dhar, T.L. Berg, Baby talk: understanding and generating simple image descriptions, in *2013 IEEE Conference on Computer Vision and Pattern Recognition* (IEEE Computer Society, Los Alamitos, CA, USA, 2011), pp. 1601–1608. https://doi.ieeecomputersociety.org/10.1109/CVPR.2011.5995466
4. I. Kotsia, I. Pitas, Facial expression recognition in image sequences using geometric deformation features and support vector machines. IEEE Trans. Image Proc. **16**(1), 172–187 (2007). https://doi.org/10.1109/tip.2006.884954
5. S. Katiyar, S. Kumar, Comparative evaluation of CNN architectures for image caption generation. Int. J. Adv. Comput. Sci. Appl. **11**(12) (2020). https://doi.org/10.14569/ijacsa.2020.0111291
6. C. Cheng, C. Li, Y. Han, Y. Zhu, A semi-supervised deep learning image caption model based on pseudo label and n-gram. Int. J. Appr. Reas. **131**, 93–107 (2021). https://doi.org/10.1016/j.ijar.2020.12.016
7. S. Li, G. Kulkarni, T.L. Berg, A.C. Berg, Y. Choi, Composing simple image descriptions using web-scale n-grams, in *Proceedings of the Fifteenth Conference on Computational Natural Language Learning* (Association for Computational Linguistics, Portland, Oregon, USA, 2011), pp. 220–228. https://aclanthology.org/W11-0326
8. K. Fu, J. Jin, R. Cui, F. Sha, C. Zhang, Aligning where to see and what to tell: image captioning with region-based attention and scene-specific contexts. IEEE Trans. Pattern Anal. Mach. Intell. **39**(12) (2017)
9. R. Socher, A. Karpathy, Q.V. Le, C.D. Manning, A.Y. Ng, Grounded compositional semantics for finding and describing images with sentences. Trans. Assoc. Comput. Ling. **2** (2014). https://aclanthology.org/Q14-1017
10. P. Young, A. Lai, M. Hodosh, J. Hockenmaier, From image descriptions to visual denotations: New similarity metrics for semantic inference over event descriptions. Trans. Assoc. Comput. Ling. **2**, 67–78 (2014). https://aclanthology.org/Q14-1006
11. M. Hodosh, P. Young, J. Hockenmaier, Framing image description as a ranking task: data, models and evaluation metrics. J. Artif. Intell. Rese. **47**, 853–899 (2013). https://doi.org/10.1613/jair.3994
12. W. Ge, C. Han, W. Quan, Face detection in complex background based on adaboost algorithm and YCbCr skin color model, in *MIPPR 2015: Pattern Recognition and Computer Vision*, ed. by T. Zhang, J. Liu (SPIE, 2015). https://doi.org/10.1117/12.2209428
13. T. Mikolov, K. Chen, G. Corrado, J. Dean, Efficient estimation of word representations in vector space, in *1st International Conference on Learning Representations, ICLR 2013, Scottsdale, Arizona, USA, May 2–4, 2013, Workshop Track Proceedings*, ed. by Y. Bengio, Y. LeCun (2013). arXiv:abs/1301.3781
14. J. Pennington, R. Socher, C. Manning, Glove: global vectors for word representation. Association for Computational Linguistics (2014). https://doi.org/10.3115/v1/d14-1162
15. A. Mohamed, Image caption using CNN and LSTM (2020). https://www.researchgate.net/publication/342407897_Image_Caption_using_CNN_LSTM
16. S. Hochreiter, J. Schmidhuber, Long short-term memory. Neural Comput. **9**(8), 1735–1780 (1997). https://doi.org/10.1162/neco.1997.9.8.1735

17. G. Geetha, T. Kirthigadevi, G. Ponsam, T. Karthik, M. Safa, Image captioning using deep convolutional neural networks (CNNs). J. Phys.: Conf. Ser. **1712**(1), 012015 (2020). https://doi.org/10.1088/1742-6596/1712/1/012015
18. L. Abisha Anto Ignatious., S. Jeevitha., M. Madhurambigai., M. Hemalatha, A semantic driven cnn - lstm architecture for personalised image caption generation, in *2019 11th International Conference on Advanced Computing (ICoAC)* (2019), pp. 356–362
19. K. Papineni, S. Roukos, T. Ward, W.-J. Zhu, BLEU, in *Proceedings of the 40th Annual Meeting on Association for Computational Linguistics - ACL '02* (Association for Computational Linguistics, 2001). https://doi.org/10.3115/1073083.1073135
20. X. He, B. Shi, X. Bai, G.-S. Xia, Z. Zhang, W. Dong, Image caption generation with part of speech guidance. Pattern Recogn. Lett. **119** (2019). https://doi.org/10.1016/j.patrec.2017.10.018
21. O. Vinyals, A. Toshev, S. Bengio, D. Erhan, Show and tell: a neural image caption generator, in *2015 IEEE Conference on Computer Vision and Pattern Recognition (CVPR)* (IEEE, 2015). https://doi.org/10.1109/cvpr.2015.7298935

Clothing Classification Using Shallow Convolutional Neural Networks

Mai Truc Lam Nguyen and Hai Thanh Nguyen

Abstract The online fashion industry is continuously growing with more requirements from applications such as 3D printing, digital clothing, and the Internet of Things to provide outstanding support for buyers. Many customers expect an algorithm capable of recognizing clothes may help clothing merchants better understand the profile of potential customers, focus sales targeting specific niches, design campaigns based on consumer preferences, and improve user experience. Therefore, artificial intelligence methods that detect and categorize human clothes are necessary to increase sales or better understand customers. Recent advancements in deep learning have sparked a slew of computer vision-based commercial applications. Object identification is used in various industries to speed up business processes where deep learning tools and techniques are leveraged in a variety of ways to improve service quality. One of them is clothing classification in the clothing business. The trained deep learning model can predict the name of any clothing by displaying an image of it, and this process can be repeated at a much quicker pace to tag thousands of items in a short amount of time with high accuracy. This study has examined shallow convolutional neural network architectures conducted from VGGNet architecture but in more compact forms with some stacks of layers, including convolutional, max pooling, and dropout layers. The proposed architecture has reached 0.9359 in classification accuracy on dataset Fashion-MNIST with 70000 samples, including images of the shirt, bag, and boot. The shallow convolutional neural networks architecture can outperform the classic machine learning method and better performance than VGGNet in classifying small-size images.

M. T. L. Nguyen · H. T. Nguyen (✉)
College of Information and Communication Technology, Can Tho University, Can Tho 900100, Vietnam
e-mail: nthai.cit@ctu.edu.vn

N. H. Phuong and V. Kreinovich (eds.), *Biomedical and Other Applications of Soft Computing*, Studies in Computational Intelligence 1045,
https://doi.org/10.1007/978-3-031-08580-2_22

1 Introduction

The fashion industry is constantly changing, and keeping up with the current trends is essential. People all across the world, regardless of their socio-economic circumstances, are eager to spend money to stay fashionable. In addition, customers may now shop on the go thanks to the internet and mobile gadgets. While individuals follow fashion trends, machine learning influences the fashion industry. All oversized and undersized brands use machine learning approaches to enhance their growth by enticing clients and staying ahead of the trend. Every day, systems keep track of every sale and upcoming trend, providing companies with a wealth of information about users' interests. For example, we may like a particular sort of clothing while watching a movie or television show. When this happens, people want to know where they can get a similar piece.

Many cultures use clothing to express their social position, age, gender, and lifestyle. Given the importance of clothing in culture, "clothing recognition" can be used in a variety of ways [1–3]. Predicting the specifics of a garment in an image, for example, can aid in the discovery of comparable apparel in an e-commerce database. Similarly, clothing identification based on the user's preferences can make recommendations. In video surveillance, real-time apparel identification can be valuable when information about a person's garments can be utilized to identify criminal suspects. Apparel classification also allows for the automatic tagging of photographs with clothing-related tags or descriptions, allowing for better information retrieval in situations like a user's photo from social media. The primary difficulty to be solved will vary depending on the application of garment classification. Our main focus is on improving clothing recognition and retrieval of related clothes for a query image item. Convolutional neural networks (CNN) are the most current innovation in the field of image retrieval and classification [4]. CNN, a type of Deep Neural Network (DNN), can be used to turn research data incorporated and processed in industrial and commercial applications into useful information [5]. It is utilized for image processing, human action recognition, and fashion image classification[6–8].

This study has leveraged well-known deep learning architectures as Convolutional neural network (CNN) to label the dataset without the need for too much training or processing on activation. Furthermore, we do not need to preprocess the Fashion-MINST dataset because it has already been processed. As a result, we found that the Convolutional neural network (CNN) has revealed the best performance in clothing classification from the obtained results.

2 Related Work

The Fashion-MNIST datasets for image classification with 70,000 annotated real-life images have been published recently. This section shall briefly review the Works on the Fashion-MNIST dataset.

Shobhit Bhatnagar et al. [9] proposed a state-of-the-art model for the classification of fashion article images which trained convolutional neural network-based deep learning architectures to classify images in the Fashion-MNIST dataset. They have also suggested three different convolutional neural network architectures and used batch normalization and residual skip connections to accelerate learning. The results showed that their proposed model reports improved accuracy of around 2% over the existing state-of-the-art systems in the literature.

In 2017, Xintong Han, Zuxuan Wu, Yu-Gang Jiang, and Larry S. Davis [10] suggested the Long Short-Term Memory Networks (LSTMs) build a model that uses the Fashion-MNIST dataset for image classification.

The multiclass SVM was also used in [11]. Selecting appropriate techniques for feature extraction and choosing the best classifier algorithm remains a challenging task for attaining good classification accuracy. However, the experimental results show impressive results on this new benchmarking dataset Fashion-MNIST.

3 Dataset

The original Modified National Institute of Standards and Technology (MNIST) dataset contained much handwriting, and the current Fashion-MNist [12] (Fig. 1) dataset is a natural alternative to the original dataset. MNIST is a set of handwritten digits containing 70000 grayscale 28×28 images of 10 labels. All images of the Fashion MNIST are of 28×28 matrices, with 28 pixels in their height and 28 pixels in their widths Fashion-MNIST dataset has the same structure as the MNIST dataset, but the images are fashion products instead of numbers.

The Fashion-MNIST dataset has been separated into two 785-column CSV files. Each CSV row represents a picture, with 784 columns describing the 28×28 pixel image with pixel luminosity values ranging from 0 to 255 and a label column (numbered from 0 to 9). It corresponds to the existing data definition, and a typical assignment would be as follows 0: 'T-shirt/top', 1: 'Trouser', 2: 'Pullover', 3: 'Dress', 4: 'Coat', 5: 'Sandal', 6: 'Shirt', 7: 'Sneaker', 8: 'Bag', 9: 'Ankle boot'. Every row represents a distinct picture, each column has a class name, and each pixel value represents the pixel's level of darkness. For example, the first column represents the clothing category. The pixel values for the image based on the lightness or darkness properties set with the 0–255 integer value will be in the remaining columns. Because these are the only pictures utilized, the CNNs tested and trained with them will not immediately transfer to real-time fashion data sets on the market.

Label	Description	Examples
0	T-Shirt/Top	
1	Trouser	
2	Pullover	
3	Dress	
4	Coat	
5	Sandals	
6	Shirt	
7	Sneaker	
8	Bag	
9	Ankle boots	

Fig. 1 Fashion-MNIST samples [13]

4 Method

4.1 Overview

The general proposed workflow for clothing types classification tasks is illustrated in Fig. 2. We evaluate the proposed methods on the Fashion-MNIST dataset from Zalando's study, consisting of a training set of 60,000 examples and a test set of

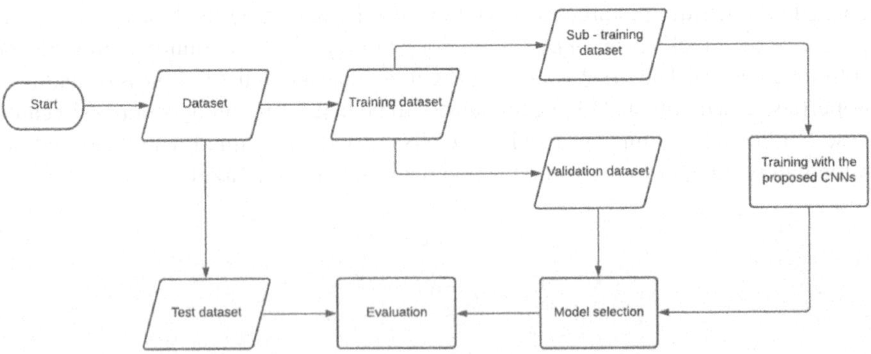

Fig. 2 General proposed workflow for clothing types classification tasks

10,000 examples. Then for training images, around 48,000 images are allocated. It represents almost 80% of the total number of images in the training set, and the remaining 20% of the images are 12,000 images, are set aside for validation. Our architecture is expected to label the dataset without too much training or processing on activation. As a result, developers may utilize it to create real-time applications like online marketplaces and search engines. The next section will provide more information.

4.2 *CNN-1*

A convolutional neural network is different from a standard artificial neural network and may involve convolutional, pooling, fully connected, and softmax layers. This model uses consecutive blocks containing: a convolution, a max pooling, and finally, a dropout. In the architecture of CNN, as exhibited in Fig. 3 proposed in [14], the architecture receives input images with the size of 28 × 28. The first convolution layer with 32 kernels of 3 × 3 (stride 1). The second convolution layer has 64 kernels of 3 × 3. The next convolution layer has filters 128, with each filter being 3 × 3. In this architecture, we use dropout equals 0.25 is the percentage of hidden units we want to train each time. Moreover, the convolution layer is followed by max-pooling with filter size 2 × 2 (stride 2). In addition, when passing through the convolution layer and max pooling. The next layer is flattened, and it connects with the Fully-connected layer. After that, we have 512 nodes, and we use a dropout layer with a rate of 0.25 again. The last layer is a Fully-connected layer that has ten nodes. This model has 163,850 trainable parameters (weights).

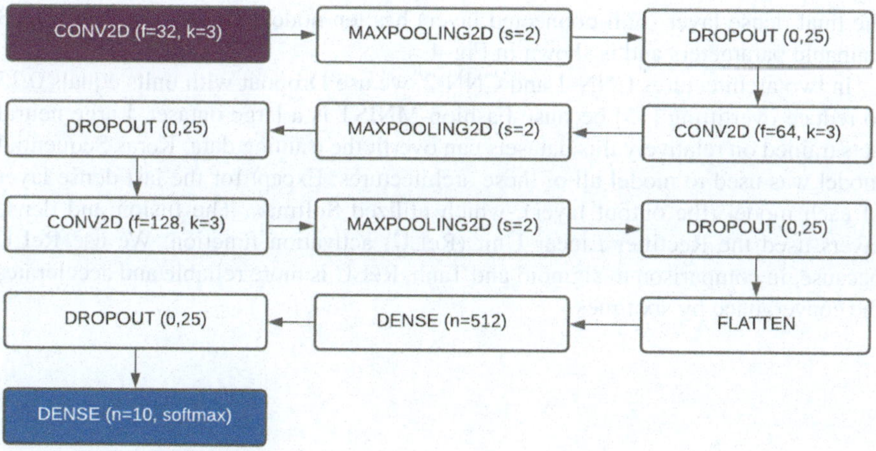

Fig. 3 The architecture of CNN-1

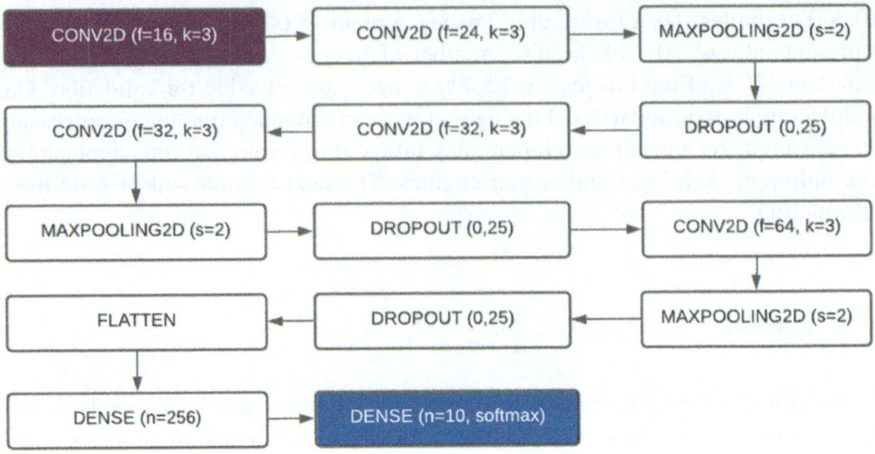

Fig. 4 The architecture of CNN-2

4.3 CNN-2

The CNN-2 architecture is quite similar to the CNN-1 architecture. Input images have the size of 28 × 28 for all datasets. Therefore, before each max pooling, it has two layers of convolutions. In CNN-2, the first convolution layer has 16 kernels of 3 × 3, and the second convolution layer has 24 filters with 3 × 3 for each filter. Two layers follow it by max-pooling with 2 × 2 (stride 2). Two next convolution layers include 32 kernels of 3 × 3. The next convolution has 64 kernels of 3 × 3. This layer follows it by max-pooling with 2 × 2 (stride 2). In this architecture, we use a dropout layer with a rate of 0.25, like CNN-1. The Full connected layer has 256 nodes, and the final dense layer (Full-connected layer) has ten nodes. This model has 57,538 trainable parameters and is shown in Fig. 4.

In two architectures CNN-1 and CNN-2, we use Dropout with units equals 0.25 to reduce overfitting [15] because Fashion-MNIST is a large dataset, Large neural nets trained on relatively this datasets can overfit the training data. Keras Sequential model was used to model all of these architectures. Except for the last dense layer of each model (the output layer), which utilized Softmax. The fusion and dense layers used the Rectifier Linear Unit (ReLU) activation function. We use ReLU because, in comparison to sigmoid and Tanh, ReLU is more reliable and accelerates the convergence by six times.

5 Experimental Results

The experiments were designed and implemented in Python using Keras and TensorFlow libraries. We have deployed the experiments in Google Colaboratory (Colab) notebook using a Tesla K80 GPU with 12GB of GDDR5 VRAM. The internet can be required not to interrupt the training to train on Colab stably.

On the training dataset, the most accurate model was CNN-1, reaching an average accuracy of 92.23% while CNN-2 obtained an accuracy of 93.94% as shown in Fig. 5a. Validation accuracy of the two models had the same results (Fig. 5b) as Fig. 5a, with CNN-2 having the best results with 92.89% and CNN-1 are lower with 91.22%.

In the loss of two architectures, CNN-2 gets the best results, followed by CNN-1. The result is show in Fig. 6a. But in validation loss Fig. 6b, CNN-1 have results better than CNN-2. We also evaluate results by the confusion matrix on the CNN-1 and CNN-2 network model for each cloth type Figs. 7 and 8. The horizontal axis of the confusion matrix represents the true label, and the vertical axis the predicted label. In CNN-1, it can be concluded that the classes Trousers, Dress, Sandal, Sneaker,

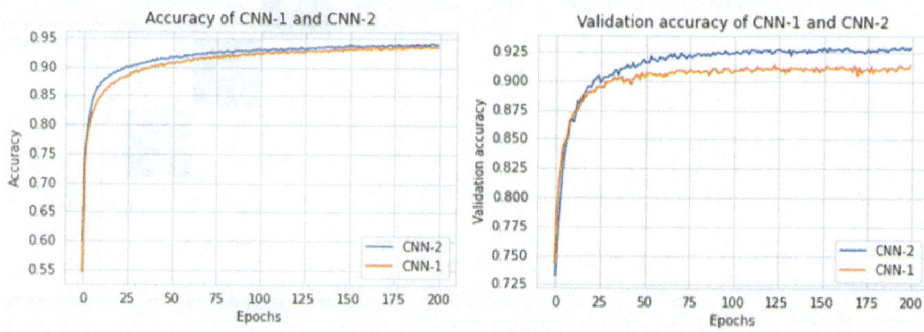

Fig. 5 Performance in Accuracy of CNN-1 and CNN-2 on a sub-training set (left chart) and a validation set (right chart)

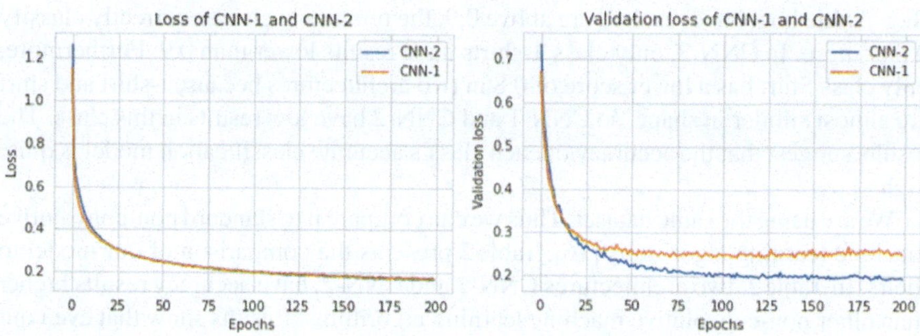

Fig. 6 Performance in Loss of CNN-1 and CNN-2 on a sub-training set (left chart) and a validation set (right chart)

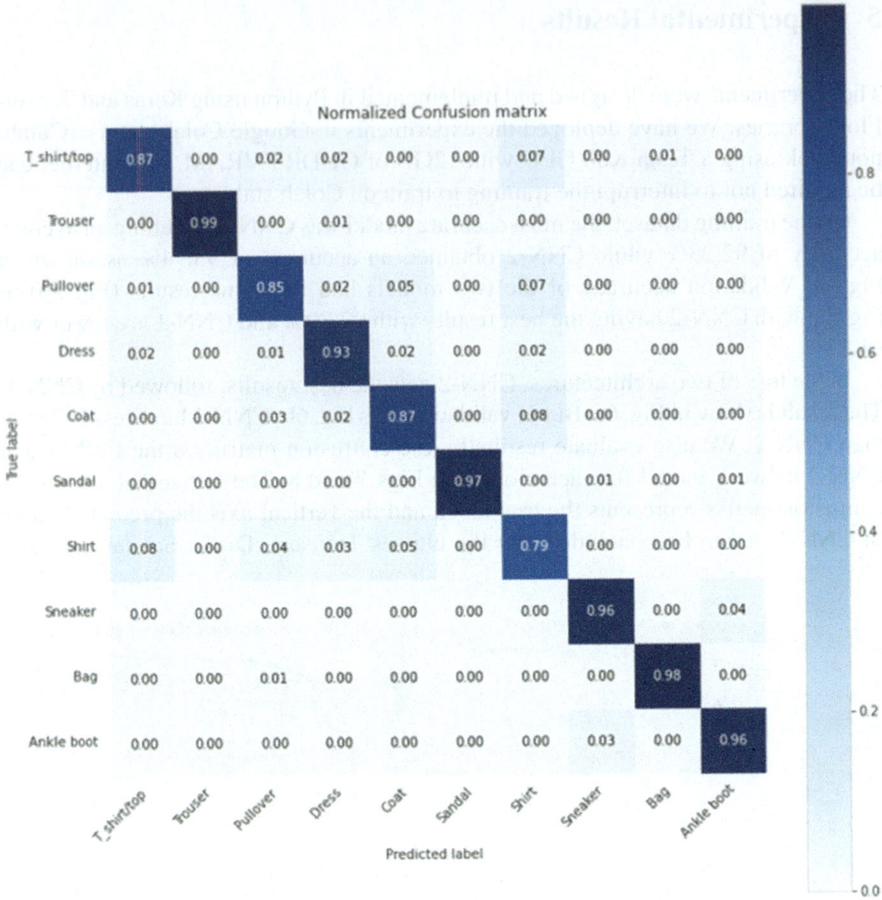

Fig. 7 Confusion matrix on a validation set with CNN-1

Bag, Ankle boot have resulted are above 0.9, the remaining classes correctly classify 0.8 or more. In CNN-2, only class T-shirts have results lower than 0.9. Furthermore, only class Shirt has a lower score of 0.8 in two architectures because t-shirt and shirt are almost similar in shape, so CNN-1 and CNN-2 have low results in this class. The results suggest that the accuracy of each class's accurate classification model is quite high.

We are using the same dataset. Thus we can compare it to standard non-convolutive machine learning algorithms [16]. Table 2 presents the comparison of our model to theirs. In Table 2, two architectures CNN-1 and CNN-2, have accuracy results higher than other non-convolutive machine learning algorithms. Results show that even our worst results (CNN-1) got better results than the best result in [8, 16].

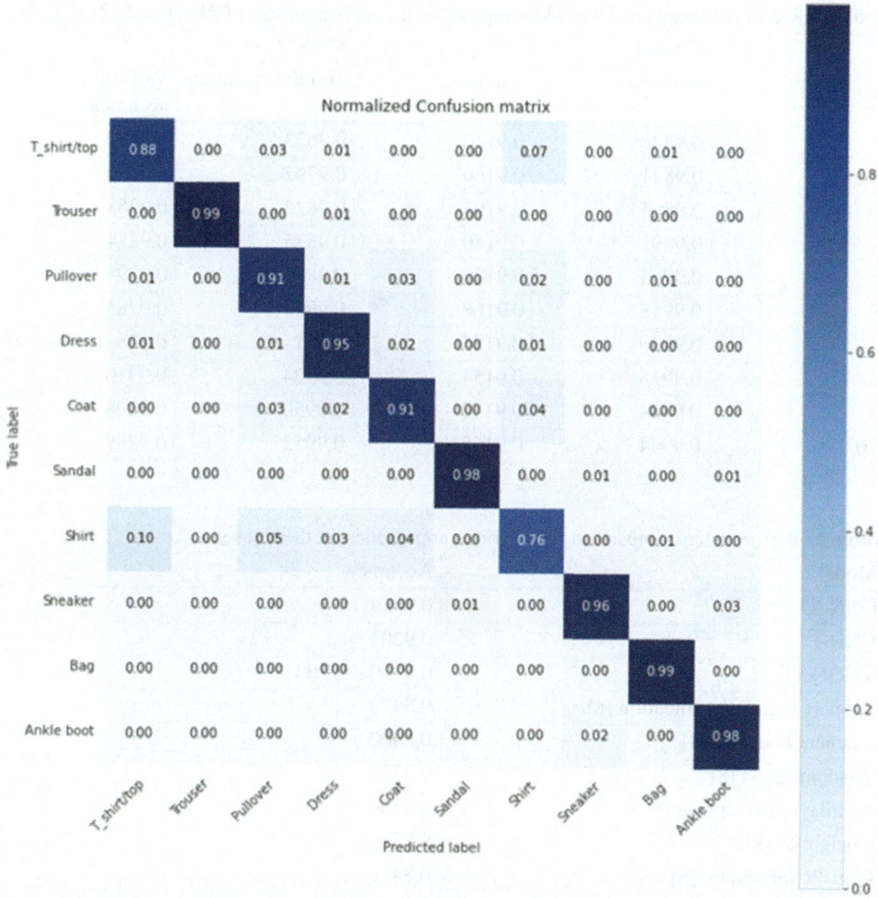

Fig. 8 Confusion matrix on a validation set with CNN-2

6 Discussion

In general, the proposed use of CNN to cloth recognize gives a better performance than non-convolutive machine learning algorithms in the same dataset (Table 1). Besides the two architectures we proposed, CNN has other architectures [17]. Table 2 contains our main results about it and can be used in future works to assist researchers and developers in finding the best classification technique.

Table 1 Results of details of 10-holdout runs using the architectures of CNN-1 and CNN-2

	CNN-1		CNN-2	
	Training accuracy	Validation accuracy	Training accuracy	Validation accuracy
1	0.9748	0.9130	0.9623	0.9160
2	0.9841	0.9160	0.9792	0.9250
3	0.9864	0.9103	0.9822	0.9258
4	0.9891	0.9140	0.9835	0.9284
5	0.9905	0.9120	0.9895	0.9298
6	0.9918	0.9168	0.9902	0.9285
7	0.9930	0.9155	0.9914	0.9289
8	0.9935	0.9154	0.9924	0.9310
9	0.9944	0.9150	0.9950	0.9298
10	0.9944	0.9165	0.9952	0.9299

Table 2 Performance comparison with various approaches on the test set

Model	Accuracy
CNN-2	0.9359
CNN-1	0.9203
VGGNet [17]	0.9100–0.9200
Support vector classification [8]	0.8970
Gradient boosting [8]	0.8800
Random forest [8]	0.8730
Multilayer perceptron [8]	0.8700
k neighbor [8]s	0.8540
Logistic regression [8]	0.8420
Linear support vector classification [8]	0.8360
Stochastic gradient descent [8]	0.8190
Decision tree [8]	0.7980
Perceptron [8]	0.7820
Passive aggressive classifier [8]	0.7760
Extra tree [8]	0.7750
Gaussian naive bayes [8]	0.5110

7 Conclusion

This study introduced a clothing classification method using the convolutional neural networks (CNN) conducted using VGG architectures with CNN-1 and CNN-2. We also analyze different architectures of CNN to identify which of the CNN presents better accuracy in classification and identification. As the results after using the CNN model to train on the Fashion-MNIST dataset, it can be shown that the CNN model works quite well compared to the remaining algorithms. Furthermore, the results reveal that shallow convolutional neural networks architecture can approximate the performance of intense architecture such as VGGNet in classifying images with small sizes. Other comparisons show that shallow CNN outperforms classic robust machine learning such as support vector machines, random forest, and Gradient Boosting.

References

1. A. Hodecker, A.M.R. Fernandes, A. Steffens, P. Crocker, V.R.Q. Leithardt, Clothing classification using convolutional neural networks, in *2020 15th Iberian Conference on Information Systems and Technologies (CISTI)* (2020), pp. 1–6
2. D. Beohar, A. Rasool, Handwritten digit recognition of mnist dataset using deep learning state-of-the-art artificial neural network (ann) and convolutional neural network (cnn), in *International Conference on Emerging Smart Computing and Informatics (ESCI)* (2021), pp. 542–548
3. S. Shubathra, P. Kalaivaani, S. Santhoshkumar, Clothing image recognition based on multiple features using deep neural networks, in *International Conference on Electronics and Sustainable Communication Systems (ICESC)* (2020), pp. 166–172
4. A. Krizhevsky, I. Sutskever, G.E. Hinton, Imagenet classification with deep convolutional neural networks. Commun. ACM **60**, 84–90 (2012)
5. A. Karpathy, G. Toderici, S. Shetty, T. Leung, R. Sukthankar, L. Fei-Fei, Large-scale video classification with convolutional neural networks, in *IEEE Conference on Computer Vision and Pattern Recognition* (2014), pp. 1725–1732
6. S. Ji, W. Xu, M. Yang, K. Yu, 3d convolutional neural networks for human action recognition. IEEE Trans. Pattern Anal. Mach. Intell. **35**(1), 221–231 (2013)
7. A.R. Pope, International journal of computer vision. Int. J. Comput. Vis. **40**(2), 149–167 (2000). https://doi.org/10.1023/a:1026502202780
8. R.L.S.D. Alisson Steffens Henrique, A.M. da Rocha Fernandes, R. Lyra, V. Reis Quietinho Leithardt, S.D. Correia, P. Crocker, Classifying garments from fashion-MNIST dataset through CNNs. Adv. Sci. Technol. Eng. Syst. J. **6**(1), 989–994 (2021)
9. S. Bhatnagar, D. Ghosal, M.H. Kolekar, Classification of fashion article images using convolutional neural networks, in *Fourth International Conference on Image Information Processing (ICIIP)* (2017), pp. 1–6
10. X. Han, Z. Wu, Y.-G. Jiang, L.S. Davis, Learning fashion compatibility with bidirectional LSTMs, in *Proceedings of the 25th ACM International Conference on Multimedia* (ACM, 2017). https://doi.org/10.1145/3123266.3123394
11. Y. Shin, I. Balasingham, Comparison of hand-craft feature based svm and cnn based deep learning framework for automatic polyp classification, in *2017 39th Annual International Conference of the IEEE Engineering in Medicine and Biology Society (EMBC)* (2017), pp. 3277–3280
12. H. Xiao, K. Rasul, R. Vollgraf, Fashion-mnist: a novel image dataset for benchmarking machine learning algorithms (2017). arXiv:abs/1708.07747

13. J. Zhao, L. Jiao, Sparse deep tensor extreme learning machine for pattern classification. IEEE Access **7**, 119 181–119 191 (2019)
14. (2020) Fashion apparel recognition using convolutional neural network. https://analyticsindiamag.com/fashion-apparel-recognition-using-convolutional-neural-network/
15. N. Srivastava, G.E. Hinton, A. Krizhevsky, I. Sutskever, R. Salakhutdinov, Dropout: a simple way to prevent neural networks from overfitting. J. Mach. Learn. Res. **15**, 1929–1958 (2014)
16. M. Kayed, A. Anter, H. Mohamed, Classification of garments from fashion mnist dataset using cnn lenet-5 architecture, in *International Conference on Innovative Trends in Communication and Computer Engineering (ITCE)* (2020), pp. 238–243
17. Y. Zhang, Evaluation of cnn models with fashion mnist data (2019). https://lib.dr.iastate.edu/creativecomponents/364/

Similar Vietnamese Document Detection in Online Assignment Submission System

Hai Thanh Nguyen, Trinh Kieu Nguyen, Minh Tri Pham, Chi Le Hoang Tran, Tran Thanh Dien, and Nguyen Thai-Nghe

Abstract Plagiarism is considered a serious problem in academic and educational contexts. Tools for detecting plagiarism are concerned by editors, writers, teachers, students, etc., to ensure that the produced papers are original and avoid plagiarism. Therefore, the algorithms of checking document similarity with high accuracy are essential. In this study, we present a method to check the similarity of students' assignments compared to the archived assignments. This method is used to compute the similarity between each pair of sentences, integrated with semantic checking. In addition, we collect and build a dataset of synonyms, antonyms, and auxiliary verbs with negative meanings for semantic checking. Moreover, we manually generate a dataset to evaluate the similarity computation and seek similar assignments. Experimental results on 200 documents reveal an average precision of 0.9666 and a recall of 0.9100 in determining similar assignments with various levels of revisions.

H. Thanh Nguyen (✉) · T. Kieu Nguyen · M. Tri Pham · T. Thanh Dien · N. Thai-Nghe
College of Information and Communication Technology, Can Tho University, Can Tho 900100, Vietnam
e-mail: nthai.cit@ctu.edu.vn

T. Kieu Nguyen
e-mail: trinhnguyenk43@gmail.com

M. Tri Pham
e-mail: haiminh2231999@gmail.com

T. Thanh Dien
e-mail: thanhdien@ctu.edu.vn

N. Thai-Nghe
e-mail: ntnghe@ctu.edu.vn

C. Le Hoang Tran
Department of Information Technology, FPT Polytechnic, Can Tho 900100, Vietnam
e-mail: lechi98.ct@gmail.com

N. H. Phuong and V. Kreinovich (eds.), *Biomedical and Other Applications of Soft Computing*, Studies in Computational Intelligence 1045,
https://doi.org/10.1007/978-3-031-08580-2_23

1 Introduction

According to the definition in [1], plagiarism is "the unauthorized use of language and thoughts of another author and representation of them as one's own". In developed countries, copying text has long been recognized as a severe offense in academia and is highly regulated. However, in Vietnam, we have only begun to make more proactive decisions on this issue in the past few years. Plagiarizing other people's ideas will limit creativity, prevent the development of reading and writing skills, and above all, misrepresent students' abilities. Therefore, each university has its policy on certain levels of plagiarism to prevent students from copying academic papers and dissertations. However, plagiarism still exists and tends to increase in the school environment in Vietnam. From the above extremely urgent reasons, the introduction of checking the similarity between student submissions into the academic environment is indisputable. In addition, in the past few years, many universities have applied the checking of student submissions into the assignment submission system to improve the quality of assignments, essays, projects and improve academic performance.

The Cosine method combined with Term Frequency-Inverse Document Frequency (TF-IDF) used in the text-similarity comparison problem obtained many positive results. By sliding the pairs of sentences together and checking, the method solved the problem of verbatim copying and changing the sentence order. Above all, it can be used and improved to match semantically between pairs of sentences. It is essential to consider pairs of synonyms, antonyms, and auxiliary verbs with negative forward meaning to get more accurate results. The problem uses data from word-similarity[1] with 12,780 data lines, including pairs of synonyms and antonyms. In addition, a set of stop words is used—which are words that have no meaning and, when omitted, do not affect the semantics of a sentence consisting of 1,376 data lines taken from FramgiaBlog.[2] The experimental process surveyed a dataset of 50 articles, including copy and paste, copy and paste with minor changes (10–15%), copy and paste with significant changes (30–45%). It incorporates using the similarity marker method to get a broader view of the results, making reviewing them more accessible and more objective. The similarity test is applied to the student submission system to make it easier to check the submissions, to ensure the quality of the students' submitted assignments.

2 Related Work

Many studies have proposed various methods to detect whether a paragraph of one document is contained in another source document. However, most of them are pretty expensive, use English, software, and applications that support the Vietnamese language are limited.

---

[1] https://github.com/NLP-Projects/Word-Similarity.

[2] https://github.com/quangph-1686/FramgiaBlog/blob/master/Blog01_Word_embedding/word2vec_wiki/stopwords.csv.

The study [2] was conducted by researchers from seven countries to present a collaborative test of 15 text-matching websites to check for plagiarism in eight different languages. The systems mentioned in the article are Plagiarism Software, Turnitin, PlagScan, Unicheck, etc. In addition, plagiarism checker software[3] is a handy plagiarism detection tool. It is used to test a specific paragraph, and the user only needs to paste the text (no more than 250 words) into the test; the returned results include the link to the related content.

Turnitin plagiarism[4] is famous software in recent years. It is a tool to help detect similar work used by many universities today with large data warehouses (including 24 billion web pages, 300 million student submissions, and many other publications) integrated with cloud services such as Google Drive. The study in [3], supported Vietnamese language analysis based on search and string matching. However, this method is only effective if the copying is "verbally", i.e., copying the whole paragraph from the source document used for testing without any changes. It cannot detect a minor change in sentence structure, nor can we check if we change the order in which the sentences appear. In addition, with the richness and diversity of the Vietnamese language, checking synonyms and antonyms has posed a great challenge for studying similarities between the two texts. Therefore, comparing the similarity between two Vietnamese texts will be more complicated than those in foreign languages. Checking the semantic similarity of the Vietnamese language requires finding a large word dataset and a method that meets the above factors. In,[5] the author has presented in many cases the determination of similarity between two documents is based on the matching of single words and the number of their occurrences in the text. Therefore, this method cannot determine semantic similarity, and the obtained results will be inaccurate when the frequency of occurrence of similar words is too high. In addition, it also has difficulty in identifying sentences when there are synonyms or negative auxiliary verbs.

Potthast et al. [4] depicted the data processing and steps to check text similarity. First, the test area is located, the model will filter out the set of candidate data corresponding to the test data to shorten the access time. Next, they conduct grouping between the test documents and the candidate documents; after the alignment, compare and identify the similarity. Finally, process data and return sentences with high similarity based on previously learned knowledge. In [5], the authors concluded that although it is time-consuming to process, the Cosine similarity method is more suitable than the Jaccard similarity method in detecting the plagiarism level of text documents. Besides, the authors in [6] proposed the TF-IDF method to ensemble Cosine similarity and word-order similarity of documents. Experimental results show that the methods proposed by the author are feasible in detecting plagiarism, especially with Vietnamese. In a related study [7], the author used the Vector Space Model and Cosine Similarity method to measure the degree of plagiarism of the thesis. At

[3] https://plagiarismdetector.net/.

[4] https://www.turnitin.com/.

[5] http://www.khcn.vimaru.edu.vn/sites/khcn.vimaru.edu.vn/files/nguyen_kim_anh_-_thuyet_minh_nckh.pdf.

the same time, TF and TF-IDF techniques were used to calculate the similarity of the two documents. The results showed that the TF-IDF technique has a smaller value than the TF technique. The authors [8] presented the use of IF-IDF Weighting in data preprocessing. Besides, they also emphasize that Vector Space Model (VSM) with Cosine Similarity and Conditional Probability can be implemented in the plagiarism detection system. Paul Donner [9] recommended combining the Cosine similarity method to collect and identify related articles in Cumulative dissertations. Furthermore, Singh et al. [10] have researched and created a bookmarked title and link extraction tool that has been created to extract top news from Google's news feed. The research has used Cosine similarity, Jaccard similarity, Euclidean distance measure to calculate news similarity score. The results showed that it is possible to efficiently identify top articles and measure similarities between news reports.

3 The Proposed Method

The input (test document) as exhibited in Fig. 1 is a Vietnamese document to be checked the similarity with documents in a repository. The scheme is extended from [3] with adding semantic similarity technique using synonym words and cosine distance to compute the similarity between documents. The processing process is carried out through steps described in detail. In the scheme, we propose to perform the experiments with x = 95% and y = 40% (in cosine similarity) in decisions in the scheme. Various values of x, y can be considered in future works.

3.1 Data Preprocessing and Cosine Similarity Computation

Let say Document 1 is the test document to compare the similarity to documents in the repository while Document i ($i = 1..n$, where n is the number of documents in the repository) is the document taken from the repository for the similarity comparison. First, we perform some preprocessing steps for these documents.

- We remove the list of references from Document 1 (in the case we do not want to compare the list of references between documents) and special characters. Moreover, meaningless sentences less than four words in length are also eliminated. Next, we convert the document into a list of sentences (in this step, we need to check the "," which occurs to be used to link compound sentences or to be used to list) and counter the list of sentences.
- Each sentence in Document 1 is compared each sentence in Document i one by one to calculate the cosine similarity combined with TF-IDF. Some cases can be considered depending on the similarity scores between sentences.

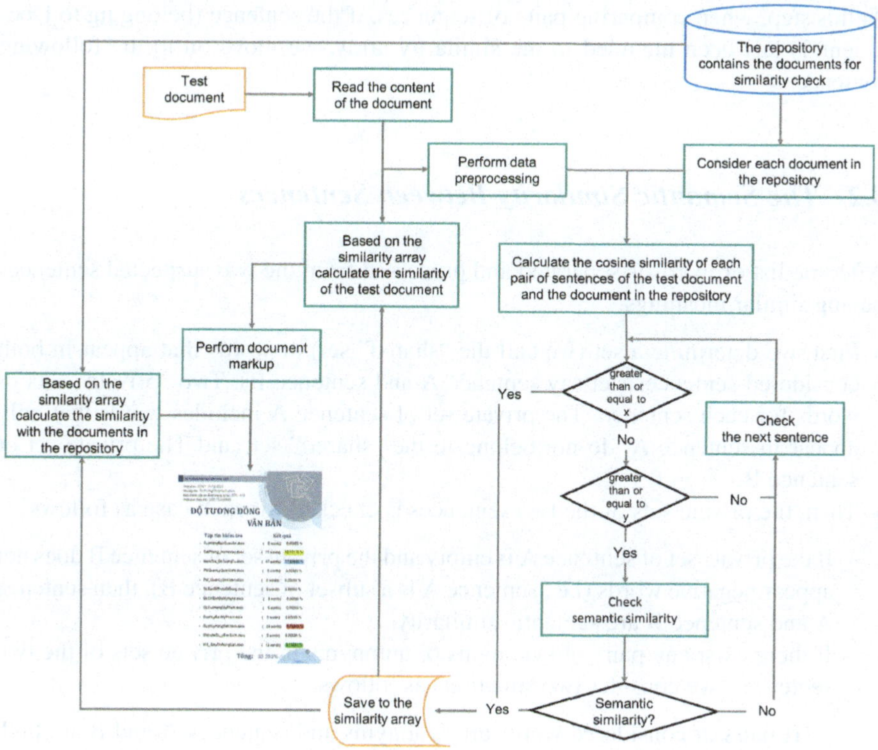

Fig. 1 The proposed scheme for the similarity checker

- Similarity score >= 95%: at this level, the sentences are copied and pasted, without editing and any changes compared to the original sentence, so it can be considered a similar sentence and put in the similarity array.
- Similarity score < 40%: when checking between two sentences, there is no sign of interpreting the same topic, so it is considered dissimilar.
- Similarity score ranges from < 95% and >= 40%: this is a doubtful result and should be checked for semantic similarity. Furthermore, the similarity between sentences that fall into this range has been copied and pasted from other sources with various ratios. Therefore, it needs to be further processed by determining the semantic similarity of the pair of sentences as follows.
  - The words in sentences are changed to lowercase. Then each sentence is converted into a list of words.
  - Stop words in the list of words in each sentence are removed. The negative word can be deleted if the negative word appears in both sentences simultaneously. For example, Given Sentence 1: "tôi không đồng ý với ý kiến này" (I do not agree with this idea) and Sentence 2: "tôi không nhất trí với ý kiến này" (I am not unanimous on this idea), we can eliminate "không" (not) in both sentences.

At this step, when comparing pairs of sentences, if the sentence (belonging to Document 1) has been included in the similarity array, we move on to the following sentence.

3.2 The Semantic Similarity Between Sentences

After the list of words is separated and processed from the two suspected sentences having similar meanings.

- First, we determine a set (we call the "shared" set) of words that appear in both considered sentences (let say sentence A and sentence B). Two "private" sets of words for each sentence. The private set of sentence A includes words that only appear in sentence A, do not belong to the "shared" set, and The private set of sentence B.
- Then, the private sets of the two sentences is checked for each case as follows:
  - If the private set of sentence A is empty and the private set of sentence B does not appear negative words (i.e., sentence A is a subset of sentence B), then sentence A and sentence B are semantic similarity.
  - If there exist any pairs of synonyms or antonyms in the private sets of the two sentences, we consider two situations as follows.

    If pairs of considered words are synonyms and sentences A and B are both affirmative or negative; the considered sentences are semantic similarity.
    If pairs of considered words are antonyms and only one of two sentences is negative; the considered sentences are semantic similarity.

After completing this step, we have a return result of "True" or "False" to determine the semantic similarity between the two sentences, from which it can be applied to all sentences of the document. The final received result is an array of sentences considered similar sentences.

3.3 Calculate the Similarity Between the Documents

Based on the similarity array checked in step Sect. 3.2, we calculate the overall similarity score in percentage entire the document. In this step, each sentence in the similarity array can be split separated single words. Moreover, the overall similarity calculation is based on the number of words counted in the array divided by the number of words in the whole document (the sentences that occur in the similarity array are multiplied by the values counter made in step Sect. 3.1 to determine the number of occurrences of the sentence). Finally, the similarity score is rounded to the fourth decimal. This calculation can be deployed if we compare the similarity between one document and a set of many documents. The similarity array also stores and adds separate sentences and performs the calculation based on the percentage of the total.

4 Experimental Results

To evaluate the plagiarism detection program proposed above, we implement a plagiarism detection program as shown in Fig. 1. We use a computer with CPU Intel ® 64 Cores, 128GB memory, Ubuntu version 20.4 LTS, and Visual Studio Code version 1.60.2.0 software to implement the implementation steps of the program. The operation of the program is shown in Fig. 1 and the Sect. 3. We performed two tests: one to measure the accuracy and the other to measure the computational time of the system.

4.1 Data Description

To evaluate the efficiency of the proposed method, we have sampled some documents from the source documents, which are original and dissimilar sets of documents. The first three columns in Table 1 show the number of documents and illustrate some information on how to generate and the number of documents to be examined. For copy-and-paste documents with minor changes, we copy at 10–20% of the original post and make corrections, replacing synonyms and antonyms. To copy and paste documents with significant changes, we copy at a rate of 30–45% of the original post and make corrections, replace synonyms and antonyms, add negatives, and make corrections and prepositions of words in the sentence. Finally, we copy 80–100% from the original document to copy and paste documents without change.

4.2 Performance Evaluation

The test results on the dataset are shown in Table 1. We measure two evaluation factors, including precision and recall [11] calculated in two cases. The first case is to consider the results returned within the test threshold. In the second case is considering the results returned within the check threshold and slightly exceeded the check threshold, and it can be understood that this is the result when detecting the copied document. Our program generally gives quite good results in the above two evaluation factors (Tables 1 and 2).

Table 1 The results of finding similar documents in Precision and Recall scores

Dataset	Ratio of copy to source material	Number of suspicious documents	Case 1		Case 2	
			Precision	Recall	Precision	Recall
Documents modified with Minor Similarity	10–20%	200	0.9412	0.9500	0.9423	0.9800
Documents modified with Moderate Similarity	30–45%	200	0.9535	0.8200	0.9574	0.9000
Documents modified with Significant Similarity	80–100%	200	1.0000	0.8400	1.0000	0.8400
Average			0.9649	0.8700	0.9666	0.9100

Table 2 Statistical deviation of the result from the test threshold

Copy ratio (%)	Min	Max	Mean
10–20	5.1000	20.2000	15.1000
30–45	11.7000	63.4000	35.9000
80–100	17.4000	99.9000	84.7000

Precision (Formula 1) is based on the documents retrieved and the Recall (Formula 2) is based on the relevant documents in collection [12].

$$Precision = \frac{|\text{RelRetrieved}|}{|\text{Retrieved}|} \tag{1}$$

$$Recall = \frac{|\text{RelRetrieved}|}{|\text{Rel in Collection}|} \tag{2}$$

However, the results are affected by several factors: some documents have formatting errors, so the program cannot retrieve them. In addition, in the data preprocessing process, removing invalid sentences (too short sentences are usually titles, indexes, etc.) and direct quotes also affect the returned results.

Table 3 shows the time to check one file with 1, 2, 10, 50, 100, 150 files, respectively (we consider the average of 50 checks).

Table 3 Execution time of the number of files

The number of files	1	2	10	50	100	150
Execution time (hours, minutes, seconds)	0:00:40	0:02:48	0:44:22	2:09:20	3:19:51	4:29:05

4.3 Similarity Between Sentences

Checking for the similarity between pairs of sentences will increase the accuracy of the problem; splitting to check causes a runtime disadvantage. However, the runtime issue will be resolved and developed later to ensure the correct element. Using the algorithm described above to check to help detect two sentences that are similar in terms of semantics, here we consider two cases in terms of semantics (Table 4).

"True" is a synonym, and "False" is antonyms. We present examples in detail as follows.

- With two synonym sentences:
  - Sentences with pure synonyms
    - vector1 = "Tôi bị cuốn hút bởi những lập luận trên" (I am fascinated by the above arguments)
    - vector2 = "Tôi bị lôi cuốn bởi những lập luận trên" (I am intrigued by the above arguments)
      - Set of common words: {"trên", "tôi", "bởi", "lập luận"}(on, I, by, argument)
      - Set of own words vector1: ["cuốn hút"] (fascinate)
      - Set of own words vector1: ["lôi cuốn"] (intrigue)
    - Pair of synonyms: "cuốn hút" - "lôi cuốn" (fascinate - intrigue)
    - Semantic test result: True
    - Cosine test result: 0.8019
  - Sentences with a negative auxiliary verb

Table 4 Describe the results of checking the similarity between pairs of sentences

Sentence type	Negative word	Result
Synonymous	No	True
	Yes	False
Antonyms	No	False
	Yes	True

- vector1 = "Tôi không bị cuốn hút bởi những lập luận trên" (I am not fascinated by the above arguments)
- vector2 = "Tôi bị lôi cuốn bởi những lập luận trên" (I am intrigued by the above arguments)

  - Set of common words: {"trên", "tôi", "bởi", "lập luận"}(on, I, by, argument)
  - Set of own words vector1: ["cuốn hút", "không"] (fascinate, not/no)
  - Set of own words vector1: ["lôi cuốn"] (intrigue)

- Pair of synonyms can be "cuốn hút" - "lôi cuốn" (fascinate - intrigue), Semantic test result is False while Cosine test result exhibits a value of 0.7327

- With two antonym sentences:
  - Sentences with pure antonyms
    - vector1 = "Ý kiến trên được hội đồng chấp thuận" (The above opinion was accepted by the council)
    - vector2 = "Hội đồng bác bỏ ý kiến được đề xuất trên" (The council rejected the proposed opinion above)

      - Set of common words: {"hội đồng", "ý kiến", "được", "trên"}(the council, opinion, above)
      - Set of own words vector1: ["chấp thuận"] (accept)
      - Set of own words vector1: ["bác bỏ", "đề xuất"] (reject, propose)

    - Pair of synonyms includes "chấp thuận" - "bác bỏ" (accept - reject). Semantic test result gives a value of False while Cosine value is 0.7327.
  - Sentences with a negative auxiliary verb
    - vector1 = "Ý kiến trên không được hội đồng chấp thuận" (The above opinion was not approved by the council)
    - vector2 = "Hội đồng bác bỏ ý kiến được đề xuất trên" (The council rejected the proposed opinion above)

      - Set of common words: {"hội đồng", "ý kiến", "được", "trên"}(the council, opinion, above)
      - Set of own words vector1: ["chấp thuận", "không"] (accept, not/no)
      - Set of own words vector1: ["bác bỏ", "đề xuất"] (reject, proposee)

    - Pair of synonyms is "chấp thuận" - "bác bỏ" (accept - reject). Semantic test result is True and Cosine score is 0.4211.

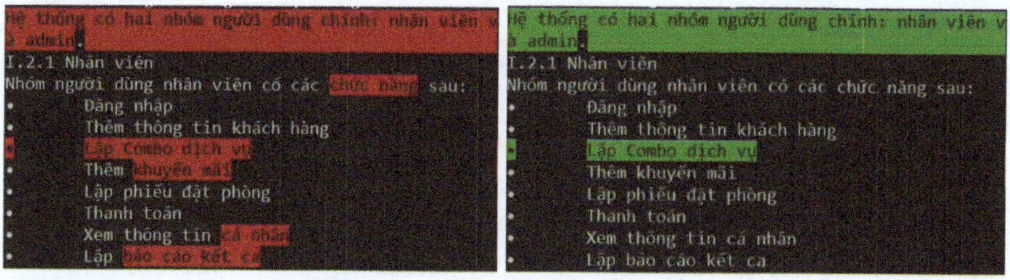

Fig. 2 An illustration of word and sentence matchings

We proceed with some matchings of words and sentences as examples in Fig. 2.

When upgrading the above algorithm, the matching based on the similarity of sentences, we found that the similarity results were significantly reduced (Fig. 2). However, this matching cannot determine similarity when there is a minor change in the sentence (adding stop words, using synonyms, changing sentence structure, etc.). Therefore, we need to check the semantics between pairs of sentences through the above two experiments to get the expected results. The above experiment can easily recognize whether it is a synonym or antonym; the cosine result still has no noticeable change and cannot be recognized at that threshold as a synonym or antonym. The semantic test results are in line with the original expectations, from which it is possible to determine whether the sentence in question is copied from another sentence or not.

4.4 Similarity Between Documents

Based on the results of the similarity test on each sentence, we get the similarity array (from considering the cosine similarity and semantic similarity), we proceed to calculate the similarity on the whole document by Formula (3).

$$Similarity = \frac{\sum(sws \times n)}{swd} \times 100\% \tag{3}$$

where *sws* denotes the total number of words in each sentence of the similarity array while *swd* reveals the total number of words in the document (spaces and special characters removed), and *n* exhibits the number of occurrences of the sentence in the similarity array. From here, we will get the overall similarity of the document. However, the results will be difficult to appear in the range of 80–100% because, in the data preprocessing process, too short sentences (usually title names, indexes, etc.) and direct quotations have been removed. Therefore, perform similarity check between the file "CTUD_QLDatPhong.docx" with five given data files by three methods: Cosine combines TF-IDF on the whole document, cosine combines TF-IDF on each sentence, cosine combined TF-IDF and semantic check for each

Table 5 Results of three methods of checking the similarity between the file "CTUD_QLDatPhong.docx" and the given five data files

Dataset name	Cosine & TF-IDF on the whole document	Cosine & TF-IDF on each sentence	Cosine & TF-IDF and semantic check by sentence
QuanCafe.docx	58.0813%	23.71542%	27.2162%
KiemTraDoTuongTu.docx	37.9842%	10.7284%	12.3658%
DPHomestay.docx	75.0947%	42.4618%	46.5838%
MoiGioiBatDongSan.docx	35.0718%	9.3167%	11.5189%
BanDiThoaiDiDong.docx	32.9554%	6.4935%	7.3969%
Total execution time	0:00:01.018276	0:04:03.281798	0:03:23.477995

sentence, we obtain the following results in Table 5. The results show that the data obtained when testing cosine combined with TF-IDF on the whole document gives relatively high results because the test is based on word matching without regard to its semantics, so the results are not accurate. Besides, cosine measure combined with TF-IDF on each sentence, better results are obtained but cannot be detected when the sentence is slightly modified (adding or removing words) or interpreted differently (using synonyms, antonyms). However, with Cosine combined TF-IDF and semantic check by sentence, we get more significant results than the sentence matching method because excluding entirely similar sentences. In addition, it can also check semantically similar sentences (as tested in Sect. 4.3).

4.5 The Application to Assignment Submission System

The method is applied to the student assignment submission system to check the similarity between documents and obtain positive results. After students submit the assignment, the program processes and output the resulting file to obtain the similarity between the test set and the data warehouses (files with more than 15% similarity will be highlighted), the results are aggregated to be able to calculate the overall similarity and output the corresponding execution time (Fig. 3). If the similarity results are too large, the system can send warnings to students' emails.

5 Conclusion

The results show that the system can check the sentence similarity, whether the sentence in question is copied "verbally" from another document. At the same time, it also can check the semantics of sentences, including synonyms, antonyms, and aux-

Fig. 3 Results of students' document similarity test

iliary verbs with negative meanings combined to give sentence similarity results. In addition, similar sentences are highlighted for the most comprehensive view of the test assignments. In the end, the aggregate is obtained after calculating the similarity of the test document with the dataset.

Although the method has met the proposed requirements, the test is inevitably flawed due to the limited dataset. Besides, the execution time is relatively slow, leading to ambiguity in the testing process. Nevertheless, experimental results can be promising, but the problem needs to be studied and improved in the processing of Vietnamese, such as processing compound sentences with the occurrence of two or more verbs, leading to confusion in determining the meaning of the sentence. Furthermore, when considering the similarity between pairs of sentences, it also needs to be adjusted to improve the accuracy and processing speed of the problem. At the same time, it is necessary to collect more sample data - synonyms, antonyms, and source data warehouses for complete testing.

Acknowledgement This study is funded in part by the Can Tho University, Code: TDH2022-04.

References

1. V. Thatcher, *The New Webster Encyclopedic Dictionary of the English Language* (Consolidated Book Pub, Chicago, 1974)
2. T. Foltýnek, D. Dlabolová, A. Anohina-Naumeca, S. Razı, J. Kravjar, L. Kamzola, J. Guerrero-Dib, Özgür Çelik, D. Weber-Wulff, Testing of support tools for plagiarism detection. Int. J. Educ. Technol. High. Educ. **17**(1) (2020). https://doi.org/10.1186/s41239-020-00192-4
3. P.H.L. Pham Thi Hai Van, A research on the text comparison method using cosine similarity. Version B Vietnam J. Sci. Technol. **59**(1) (2017). https://b.vjst.vn/index.php/ban_b/article/view/543
4. M. Potthast, M. Hagen, T. Gollub, M. Tippmann, J. Kiesel, P. Rosso, E. Stamatatos, B. Stein, Overview of the 5th international competition on plagiarism detection, in *Working Notes for CLEF 2013 Conference , Valencia, Spain, September 23–26, 2013, ser. CEUR Workshop Proceedings*, ed. by P. Forner, R. Navigli, D. Tufis, N. Ferro, vol. 1179. CEUR-WS.org, 2013. http://ceur-ws.org/Vol-1179/CLEF2013wn-PAN-PotthastEt2013.pdf
5. M.A.C. Jiffriya, M.A. C.A. Jahan, R.G. Ragel, Plagiarism detection on electronic text based assignments using vector space model, in *7th International Conference on Information and Automation for Sustainability* (2014), pp. 1–5. https://doi.org/10.1109/iciafs.2014.7069593
6. T.T. Dien, H.N. Han, N. Thai-Nghe, An approach for plagiarism detection in learning resources, in *Future Data and Security Engineering*. Lecture Notes in Computer Science. (Springer International Publishing, 2019), pp.722–730. https://doi.org/10.1007/978-3-030-35653-8_52
7. I. Indriyanto, I.D. Sumitra, Measuring the level of plagiarism of thesis using vector space model and cosine similarity methods, in *IOP Conference Series: Materials Science and Engineering*, vol. 662 (2019), p. 022111. https://doi.org/10.1088/1757-899x/662/2/022111
8. R. Saptono, H. Prasetyo, A. Irawan, Combination of cosine similarity method and conditional probability for plagiarism detection in the thesis documents vector space model. J. Telecommun. **10**(2–4), 139–143 (2018). https://jtec.utem.edu.my/jtec/article/view/4332
9. P. Donner, Identifying constitutive articles of cumulative dissertation theses by bilingual text similarity. Evaluation of similarity methods on a new short text task. Quant. Sci. Stud. **2**(3), 1071–1091 (2021). https://doi.org/10.1162/qss_a_00152
10. R. Singh, S. Singh, Text similarity measures in news articles by vector space model using NLP. J. Inst. Eng. (India) Ser. B **102**(2), 329–338 (2020). https://doi.org/10.1007/s40031-020-00501-5
11. M. Arora, U. Kanjilal, D. Varshney, Evaluation of information retrieval: precision and recall. Int. J. Indian Cult. Bus. Manage. **12**(2), 224 (2016). https://doi.org/10.1504/ijicbm.2016.074482
12. M. Sanderson, J. Zobel, Information retrieval system evaluation, in *Proceedings of the 28th annual international ACM SIGIR conference on Research and development in information retrieval - SIGIR 05* (ACM Press, 2005). https://doi.org/10.1145/1076034.1076064

A Study of Causal Modeling with Time Delay for Frost Forecast Using Machine Learning from Data

Shugo Yoshida, Yosuke Tamura, Kenta Owada, Liya Ding, Kosuke Noborio, and Kazuki Shibuya

Abstract Causal modeling with time delay has been proposed as a method for predicting frost occurrence in a short period of time. In this method, environment factors are considered as cause, and used as input variables for prediction of frost. For coping with the uncertainty of prediction rooted in randomness of environment, a granulation of environment factors offers potential. In this study, we show that the accuracy of predicting frost occurrence can be improved by appropriately granulating each of the input environment factors involved.

1 Introduction

Frost destroys plants or fruits and severely damages crops. Farmers may have had entire fields destroyed in just a few frosty nights [1]. According to the Ministry of Agriculture, Forestry and Fisheries (MAFF), Japan, the total amount of damage caused by low temperatures and frost since late March 2010 was around 10 billion yen (about 98 million dollars) in 21 prefectures across Japan [2].

Farmers are taking measures (fans, sprinkler, combustion methods) against frost [3], but the operation can be costly. At present, farmers implement these measures based on the frost warnings issued by the Japan Meteorological Agency. However, such warnings are simply made by forecasted temperature [4], and lack accuracy. Another factor is that frost occurrence varies with regional differences. Therefore, prediction of frost occurrence using regional environment factors is important for farmers to take countermeasures against frost at right time.

S. Yoshida (✉) · Y. Tamura · K. Owada · L. Ding
School of Science and Technology, Meiji University, 1-1-1, Higashimita, Tama-ku, Kawasaki, Kanagawa, Japan
e-mail: ce211070@meiji.ac.jp

K. Noborio · K. Shibuya
School of Agriculture, Meiji University, 1-1-1, Higashimita, Tama-ku, Kawasaki, Kanagawa, Japan

N. H. Phuong and V. Kreinovich (eds.), *Biomedical and Other Applications of Soft Computing*, Studies in Computational Intelligence 1045,
https://doi.org/10.1007/978-3-031-08580-2_24

In our previous studies [5–9], we have used machine learning to predict frost occurrence. Among them, causal modeling focusing on short time periods is highly effective. In this causal modeling, for input data granulation moving time windows have been applied to all input factors involved in same model (Air Temperature, Vapor Pressure) with same magnitude. In this paper, we study how the accuracy of the model changes by varying the size of the granulation for each of input factors individually, and what characteristics of the input factors are responsible for the change in accuracy.

The rest of the article is arranged as follows: In Sect. 2, we explain the concept of causal modeling with time delay for forecasting frost occurrences. Section 3 describes the method of granulation in the input variables of causal modeling, and Sect. 4 describes the implementation. Experiments are provided in Sect. 5, followed by comparison and discussion. Section 6 gives conclusions and future works.

2 Causal Modeling with Time Delay

In this section, we introduce the concept of modeling in forecasting frost occurrence. Frost forecast always requires a time interval between the observation of input environment factors and the prediction of output frost labels. This is because it takes time for farmer or IoT device to act against frost, or for the effect to appear after the protection action is taken. If the time delay is P and the current time is t, we can expect to predict the occurrence of frost at time $t + P$, denoted by $Frost_{t+P}$, using the environment factors observed in time interval $[t - T^*, t]$, where time duration T^* is selected according to different modeling methods and setting. There are two modeling methods of predicting the future event of frost by using environment factors from the past to the present.

The first method is to obtain the values of environment factors at time $t + P$ from the time series forecasting (such as ARIMA model) with the corresponding time series data observed in time period $[t - q, t]$, and then use the predicted environment factors at time $t + P$ to predict frost occurrence at the same time point. This is shown in Fig. 1a, where q is the period of data used in fitting of time series model, and typically from a few months to a year for capturing seasonality of time series. As mentioned above, this method involves two steps. Since both of these two steps have random elements, the output obtained includes uncertainties. In other words, the environmental variables at the time point $t + P$ obtained by the time series forecasting include uncertainties, so forecasting the presence or absence of frost using the predicted environment factors means producing an uncertain output using uncertain inputs.

The other method is to predict the occurrence of frost at time $t + P$ directly making use of the environment factors observed in time period $[t - k, t]$. This is shown in Fig. 1b. We call this method causal modeling [5, 7]. This method is effective for forecasting in a short term when the value is relatively small so the cause-effect can be well captured.

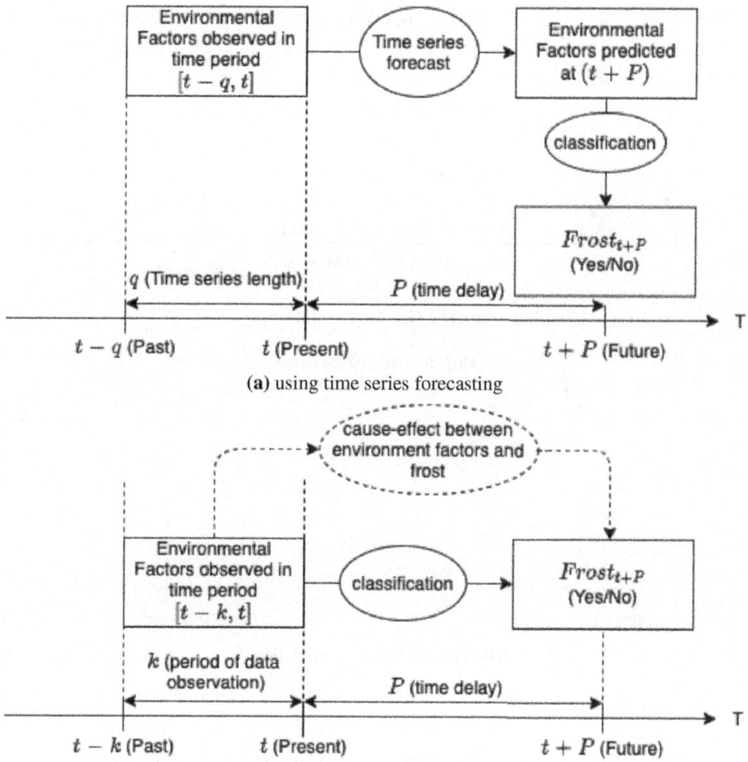

(a) using time series forecasting

(b) using causal modeling

Fig. 1 Two modeling methods of frost prediction

In the past, it required prediction of future frost occurrence over a long period of time because it was necessary for people to actually go to the farm to work on frost prevention. However, with the widespread use of IoT, it has become possible for machines to remotely implement frost prevention methods [10]. Therefore, it becomes more necessary to predict frost occurrence effectively in a short period of time.

For these reasons, this paper focuses on causal modeling with time delay.

3 Input Variable Granulation

One of the key issues in dealing with causal modeling with time delay is how to handle granulation of environment factors at time. The key idea of causal modeling is based on the understanding that the relation between environment factors and frost is of cause-effect with time delay. To cope with the randomness involved in

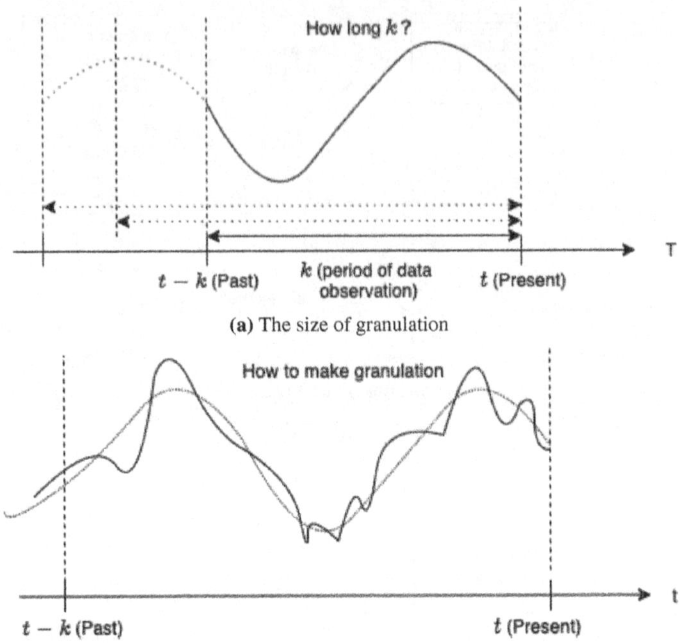

(a) The size of granulation

(b) The method of granulation

Fig. 2 Input granulation

such cause-effect, we take granulation on environment factors and consider that the movements of environment factors during time period $[t - k, t]$ make impact to future frost occurrence at time $t + P$. Since causal modeling uses the influence of environment factors from time period $[t - k, t]$ to predict frost at time $t + P$, the accuracy of prediction is expected to improve if the data observed in $[t - k, t]$ can be appropriately granularized to well capture their impact on frost at time $t + P$. There are two points of interest in this granulation: the size of granulation and the method of granulation. The idea is given in Fig. 2, where (a) is for the size of the granulation and (b) is for the method of granulation.

The size of the granulation is the length of moving time windows in time series data for the duration of environment factors taken into account when one prediction to be made. Environment factors are observed every 5 min by sensors installed in the farm, and the question is how to find an appropriate length of k for the granulation. A short length relies on more recent data and instantaneous features, but reflects less pattern of overall trend. On the other hand, a long length is advantageous in capturing trend and seasonality with the price of missing instantaneous movements.

After the size of granulation k is determined, we need select a method of calculation for granulation. One of the methods for granulation of time series data is smoothing, for finding important patterns in data while removing less important ones [11]. In this paper, we adopt on moving averages, one of the smoothing methods.

There are two types of moving averages that are often used: simple moving averages and exponential moving averages. A simple moving average is an unweighted simple average of the data over a period k. An exponential moving average is an exponentially weighted average of the data in order to emphasize the most recent data. The equations for simple moving average and exponential moving average are given respectively as follows

$$SMA_t = \frac{\sum_{i=t}^{t-k}(EF)_i}{k} \tag{1}$$

$$EMA_t = (1-a)EMA_{t-1} + a(EF)_t \tag{2}$$

$$EF = Environment\ factor$$

$$t = Present\ time, k = Moving\ time\ window\ length$$

$$a = \frac{2.0}{k+1.0}$$

These two moving averages are used to granularize the input data. We will compare their performance in experiments.

Having multiple environment factors involved in frost prediction, each of them may affect the occurrence of frost in different way, be more instantaneous, or more periodical. We shall carry out experiments to investigate the impact of granulation of environment factors individually.

4 Implementation and Experiments

4.1 Datasets

There are two datasets used in this experiment, both of which were observed at a vineyard in Ikeda-cho, Hokkaido, Japan in every five minutes by an actual instrument installed there. One is collected during the period from April 1, 2020, 00:00 to May 31, 2020, 23:55 (index: 17281, frost was observed 21 times) (hereafter referred to as data1), and the other is from April 1, 2021, 00:00 to May 31, 2021, 23:55 (index 17281, frost was observed 14 times) (hereafter referred to as data2). Figure 3 shows the actual frost status contained in data1 and data2.

4.2 Input and Output

From the data (environment factors) observed by the instruments, Air Temperature and Vapor Pressure are selected as input variables based on previous studies [9], such

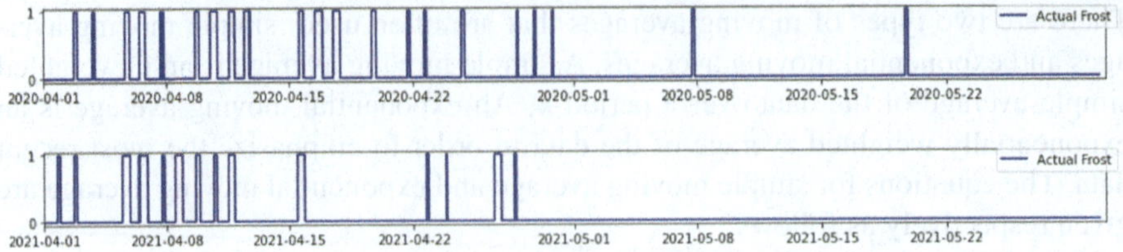

Fig. 3 Frost datasets

as their correlation with frost. Other environment factors such as solar radiation and wind speed will be investigated in the future.

The output is the presence or absence of frost (Yes = 1, No = 0), and the presence or absence of frost at time $t + P$ is denoted as $FrostAF_P$, because it is the frost at P hours after the present.

4.3 *Model Implementation*

Causal modeling with time delay is a supervised learning process that uses current environment factors (Air Temperature and Vapor Pressure) to determine if there will be a future frost event. It is a task of binary classification with the output being the presence or absence of frost.

The data is imbalanced because the total time period of frost occurrence observed is much shorter compared with that of no frost. With this in mind, we adopt support vector machine, which is a frontier that best segregates the two classes by maximizing the margin between them [12, 13].

We will use data1 as the training data and data2 as the test data.

5 Experiments

5.1 *Granulation for Air Temperature and Vapor Pressure*

5.1.1 Compare Granulation Size

We first carried out experiments with different sizes of granulation for Air Temperture and Vapor Pressure and then made comparision. For Air Temperture, we performed granulation at 1 point (no granulation), 30 min, 1, 3, and 6 h, and for Vapor Pressure, we performed granulation at 1 point (no granulation), 3, 6, 12 h, 1, and 2 days. The results are summarized in Table 1, using simple moving average with target $FrostAF_1$.

Table 1 Performance with different sizes of granulation for Air Temperture and Vapor Pressure

Air temperature	Vapor pressure	Recall	Precision
1 point	1 point	0.966	0.451
1 point	3 h	0.979	0.480
1 point	6 h	0.977	0.577
1 point	12 h	0.966	0.587
1 point	1 day	0.924	0.546
1 point	2 days	0.965	0.534
30 min	1 point	0.966	0.448
30 min	3 h	0.980	0.465
30 min	6 h	0.973	0.561
30 min	12 h	0.966	0.578
30 min	1 day	0.928	0.538
30 min	2 days	0.970	0.529
1 h	1 point	0.966	0.453
1 h	3 h	0.977	0.452
1 h	6 h	0.965	0.537
1 h	12 h	0.960	0.564
1 h	1 day	0.930	0.526
1 h	2 days	0.968	0.516
3 h	1 point	0.953	0.448
3 h	3 h	0.978	0.408
3 h	6 h	0.937	0.443
3 h	12 h	0.893	0.470
3 h	1 day	0.957	0.463
3 h	2 days	0.966	0.447
6 h	1 point	0.968	0.356
6 h	3 h	0.971	0.330
6 h	6 h	0.963	0.321
6 h	12 h	0.840	0.345
6 h	1 day	0.944	0.339
6 h	2 days	0.920	0.295
12 h	12 h	0.995	0.196

Table 1 shows that the results for recall are almost the same regardless of the size of the granulation. On the other hand, there is an obvious difference in precision with the size charge of granulation.

The relation between precision and granulation size for Air Temperature is shown in Fig. 4, and between precision and granulation size for Vapor Pressure in Fig. 5.

Figure 4 shows that in terms of granulation size of Air Temperature, precision reaches the highest at 1 point (i.e. without granulation), and drops as the size increas-

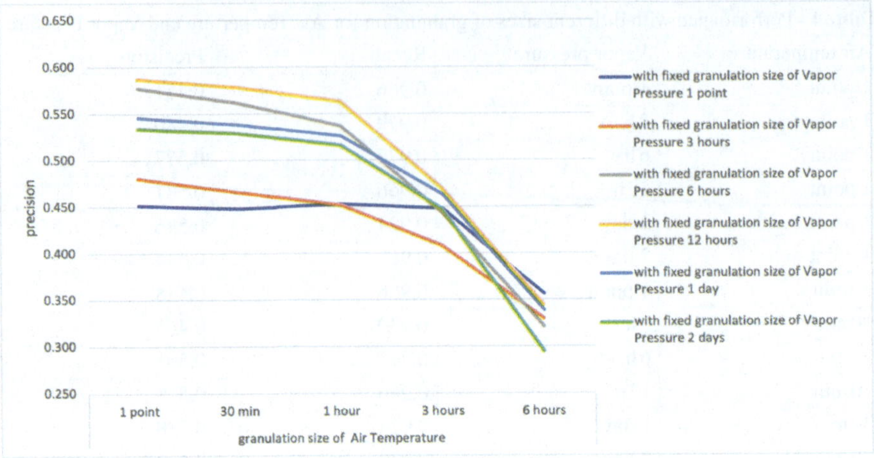

Fig. 4 The relation between precision and granulation size of Air Temperature

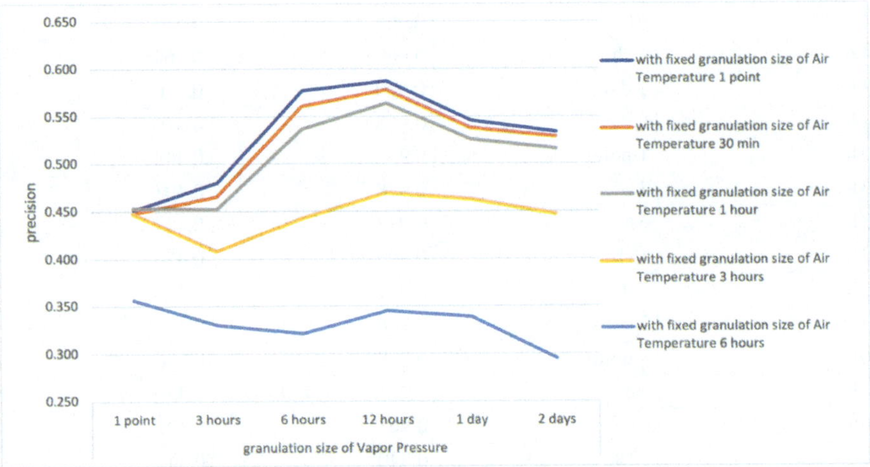

Fig. 5 The relation between precision and granulation size of Vapor Pressure

ing. Figure 5 shows that in terms of granulation size of Vapor Pressure precision increases from 1 point to 12 h and starts dropping when the size further increasing.

This suggests that Air Temperature has an instantaneous effect on frost occurrence as shown in the high precision at 1 point. On the other hand, the effect of Vapor Pressure on frost occurrence shows a different behavior with high precision of prediction with granulation size of 6 to 12 h, which suggests that the characteristics over a certain period of time have an effect on frost occurrence.

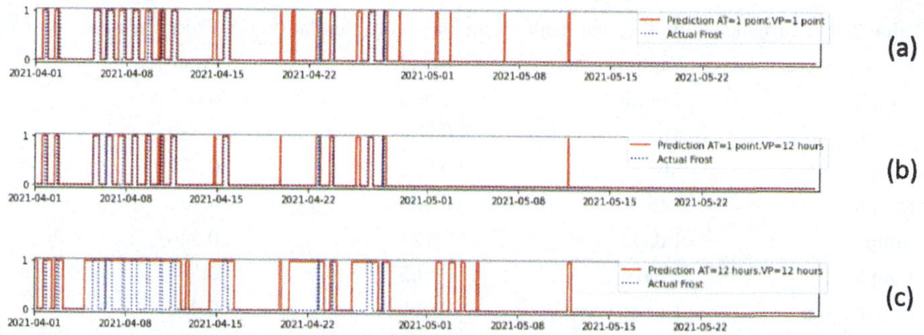

Fig. 6 Results of frost prediction with various size of granulation

Figure 6 gives the performance of three models with different settings of granulation size: (a) 1 point Air Temperture and 1 point Vapor Pressure; (b) 1 point Air Temperture and 12 h Vapor Pressure; (c) 12 h Air Temperture and 12 h Vapor Pressure.

Figure 6 shows that the prediction accuracy is best in (b) with 1 point in Air Temperature and 12 h in Vapor Pressure. In (a) there are many false alarms that predict the frost occurrence during the time of no frost. It confirmed that the false alarm of frost can be reduced by taking an appropriate size of granulation.

5.1.2 Compare Granulation Method

Now we test two methods of moving averages for granulation, simple moving average and exponential moving average, and compare their performance. For each of the methods, the size of granulation for Vapor Pressure is changed and compared as shown in Table 2, with the size of granulation for Air Temperature fixed at 1 point because the precision at one point is higher as confirmed in previous experiment.

The relation between precision and the size of granulation of Vapor Pressure for the two methods of moving average is provided in Fig. 7.

Figure 7 shows that the precision reaches the highest at 12 h of granulation size for both methods. In both methods, precision keeps increasing with granulation size until 12 h and then starts dropping. At 6 h, simple moving average outperforms exponential moving average, and at 12 h, the measures of precision are about the same. For longer than 1 day, exponential moving average outperforms simple moving average.

The result of two methods suggest consistently that the granulation size of the Vapor Pressure from 6 to 12 h is important for frost occurrence. Simple moving average is an unweighted average and performs better at granulation size of 6 h, while exponential moving average gives higher weights to more recent and performs better with granulation size longer than 12 h.

Table 2 Performance with different methods and sizes of granulation for Vapor Pressure

Moving average	Vapor pressure	Recall	Precision
Simple	1 point	0.966	0.451
Simple	3 h	0.979	0.480
Simple	6 h	0.977	0.577
Simple	12 h	0.966	0.587
Simple	1 day	0.924	0.546
Simple	2 days	0.965	0.534
Exponential	1 point	0.966	0.451
Exponential	3 h	0.975	0.486
Exponential	6 h	0.974	0.559
Exponential	12 h	0.955	0.591
Exponential	1 day	0.965	0.568
Exponential	2 days	0.972	0.547

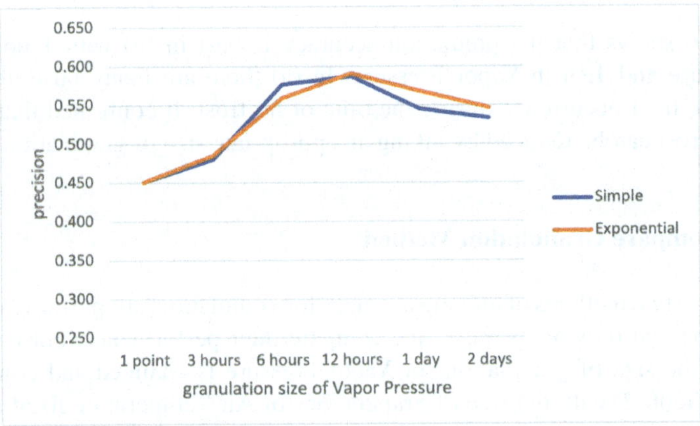

Fig. 7 Comparision of SMA and EMA

Therefore, it can be concluded that there is no significant difference in terms of highest precision between simple moving average and exponential moving average in the method of granulation of Vapor Pressure. The difference is in that simple moving average is advantageous for granulation of data in a short period of time while exponential moving average is for granulation of data in a long period of time when precision of prediction is our key concern.

6 Conclusion and Future Work

6.1 Conclusion

The findings from this study are as follows:

- The accuracy of frost prediction can be improved with the number of false alarms reduced by taking appropriate granulation of input variables in causal modeling with time delay.
- Air Temperature has more instantaneous effect on frost occurrence while Vapor Pressure has more periodic effect over a period of time.
- The difference between simple moving average and exponential moving average is that simple moving average is more advantageous for granulation of data in a short period of time while exponential moving average for granulation of data in a long period of time, when precision of prediction is concerned.

6.2 Future Work

In this study, we have focused on Air Temperature and Vapor Pressure as environment factors for frost occurrence. However, there are many other influences to frost including Solar Radiation and Wind Speed, so we will investigate these factors as well and report our new findings in the near future.

Acknowledgements The authors thank METER Group, Inc. USA for providing the equipment.

References

1. Nationnal Geographic Society, Frost. https://www.nationalgeographic.org/encyclopedia/frost/
2. Japan Meteorological Agency, Damage to agriculture forestry and fisheries related to low temperature and frosts since late March 2010 and responses (2013). https://www.maff.go.jp/j/saigai/teion/0621.html
3. FAO (Food and Agriculture Organization of the United Nations), RECOMMENDED METHODS OF FROST PROTECTION, https://www.fao.org/3/y7223e/y7223e08.htm
4. Japan Meteorological Agency, Requirements for issuing a frost warning in Ikeda-cho, Hokkaido, Japan. https://www.jma.go.jp/jma/kishou/know/kijun/kushiro/kijun_0164400.pdf
5. L. Ding, K. Noborio, K. Shibuya, Frost forecast using machine learning -from association to causality. Proc. Comput. Sci. **159**, 1001–1010 (2019)
6. L. Ding, Y. Tamura, K. Noborio, K. Shibuya, Frost forecast: a practice of machine learning from data. Int. J. Reas.-Based Intell. Syst. **13**(4), 191–203 (2021). https://doi.org/10.1504/IJRIS.2021.10038162
7. L. Ding, K. Noborio, K. Shibuya, Modelling and learning cause-effect - application in frost forecast. Proc. Comput. Sci. **176**, 2264–2273 (2020)

8. Y. Tamura, L. Ding, K. Noborio & K. Shibuya, Frost prediction for vineyard using machine learning, in *2020 Joint 11th International Conference on Soft Computing and Intelligent Systems and 21st International Symposium on Advanced Intelligent Systems (SCIS-ISIS)* (2020). https://doi.org/10.1109/SCISISIS50064.2020.9322770.
9. L. Ding, Y. Tamura, S. Yoshida, K. Owada, T. Toyoda, Y. Morishita, K. Noborio, K. Shibuya, Ensemble causal modelling for frost forecast in vineyard. Proc. Comput. Sci. **192**, 3194–3203 (2021)
10. LOXONE, Frost protection for crops. https://www.loxone.com/enen/kb/frost-protection-for-crops
11. A. Raudys, V. Lenčiauskas, E. Malčius, Moving averages for financial data smoothing. ICIST 2013: Information and Software Technologies (2013), pp. 34–45
12. S. Ray, Understanding Support Vector Machine(SVM) algorithm from examples (along with code) (2017). https://www.analyticsvidhya.com/blog/2017/09/understaing-support-vector-machine-example-code/#h2_4
13. N.I. Sapankevych, R. Sankar, Time series prediction using support vector machines: a survey. IEEE Comput. Intell. Mag. **4**(2), 24–38 (2009)

The manufacturer's authorised representative in the EU is Springer
Nature Customer Service Centre GmbH, Europaplatz 3, 69115 Heidelberg,
Germany. If you have any concerns regarding our products, please
contact ProductSafety@springernature.com

Printed and bound by CPI Group (UK) Ltd, Croydon, CR0 4YY
28/09/2026
02224073-0008